CAMBRIDGE TRACTS IN MATHEMATICS

General Editors

B. BOLLOBAS, F. KIRWAN, P. SARNAK, C.T.C. WALL

131 Bipartite Graphs and their Applications

T0269194

Armen S. Asratian   Tristan M.J. Denley   Roland Häggkvist
Luleå University   *University of Mississippi*   *Umeå University*

Bipartite Graphs and their Applications

CAMBRIDGE
UNIVERSITY PRESS

CAMBRIDGE UNIVERSITY PRESS
Cambridge, New York, Melbourne, Madrid, Cape Town, Singapore, São Paulo

Cambridge University Press
The Edinburgh Building, Cambridge CB2 8RU, UK

Published in the United States of America by Cambridge University Press, New York

www.cambridge.org
Information on this title: www.cambridge.org/9780521593458

First published 1998
This digitally printed version 2008

A catalogue record for this publication is available from the British Library

ISBN 978-0-521-59345-8 hardback
ISBN 978-0-521-06512-2 paperback

Contents

Preface

The graphs in figure 0.1.1 all have one property in common: their vertices can be divided into two parts such that no two vertices in the same part are joined by an edge. In the diagrams we have indicated one possible division by colouring the vertices black and white. Graphs which have this property are called bipartite, and their properties form the subject of this book.

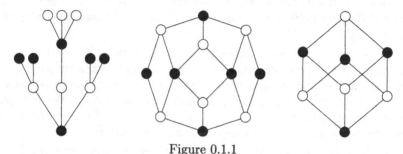

Figure 0.1.1

The first systematic investigation of the properties of bipartite graphs was begun by König (1914, 1915, 1916). His work was motivated by an attempt to give a new approach to the investigation of matrices, in particular to find a simple proof of a theorem of Frobenius (1912) on determinants of matrices. Much of the material from this early work is presented in König's famous book (1936). But this is not the beginning of the story. Trees, a large subclass of bipartite graphs, had already been defined and investigated by several authors much earlier. Kirchhoff (1847), Cayley (1857), Sylvester (1873) and Jordan (1869) each independently developed theories of trees in the mid nineteenth century.

From a practical point of view, bipartite graphs form a model of the interaction between two different types of objects, be they sets and their elements, jobs and workers, or telephone exchanges and cities. The desire to model such interaction is extremely common, and many recreational and much more serious problems can be phrased in terms of problems on bipartite graphs.

From a theoretical point of view, bipartite graphs at first glance seem to have a much simpler structure than graphs in general, but this is not altogether the case. They are certainly much easier to envisage, but actually almost all the difficulties which are inherent in general graphs are already present in bipartite graphs. For instance, wide classes of computational problems are already NP-hard even for bipartite graphs with small maximal degrees. Historically, many results on bipartite graphs have been the starting points for various generalisations to results on general graphs. For these reasons alone, bipartite graphs have been considered in every book about graph theory, but up until now only as a special class in some wider context.

However, over the past few years many new and interesting results about bipartite graphs have been obtained, which are difficult to include in a traditional book on graph theory. For this reason, among others, we decided to write this, the first book about bipartite graphs alone. Our aim was to describe properties of this class of graphs using only their own structure, together with occasionally some simple algebraic techniques. We have by and large avoided results which are proved using random methods and linear programming (the interested reader can find such material in the books of Alon and Spencer (1992), Bollobás (1985), Ford and Fulkerson (1962) and Lawler (1976)).

Together with the traditional topics which we have tried to present in a new light, the reader will also find many new and unusual results. In particular, we have concentrated much attention on edge colouring problems with a variety of restrictions, and their applications to timetabling. We also give applications of many other results in the areas of algebra, combinatorics, chemistry, communication networks and computer science.

Let us briefly survey the contents of the chapters, pointing out some of the highlights. We begin from the beginning, and introduce the basic tools and concepts to be used throughout the book. Next follow the basic characterisation theorems of both general bipartite graphs and some particular special classes. We move on to some results about global structure, first considering metric properties of bipartite graphs, trees and hypercubes in particular, and then results concerning connectivity. A method for efficiently routing information about a computer network, and a construction of linear superconcentrators which are sparse and yet highly connected networks, form the applications for these chapters.

Next is a chapter about various types of matching problems. We begin with maximum matchings, both with their properties and with the practical task of actually finding such an object. Together with the well-known algorithm of Hopcroft and Karp (1973) to find a maximum matching, we also give a modified algorithm due to Alt, Blum, Mehlhorn and Paul (1991). One section is devoted to stable matchings, and the last section to a polynomial algorithm for finding the k best perfect matchings in a weighted complete bipartite graph.

The chapter which follows begins from somewhat similar ground, but from a slightly different standpoint. It is concerned with what can be said about graphs which have a so-called 'expanding' property, with some additive or multiplicative factor. The chapter continues to the more contemporary subject of expanders, and we show how the structure of these graphs can be used for powerful sorting algorithms.

Chapter 7 is concerned with subgraphs which satisfy degree restrictions of one sort or another. We discuss existence theorems for (g, f)-factors, and f-factors, and go on to discuss the important special cases of 2-factors, and connected 2-factors (or, as they are better known, Hamilton cycles). Restricting only the parity of the vertex degrees, T-joins are also considered.

Of all bipartite graph topics, edge colourings are among the most widely studied problems in the field. In the chapter on edge colourings, there are a variety of results about the existence of edge colourings under many different restrictions and with many different properties. Edge colourings of bipartite graphs have a rather natural interpretation in terms of timetables. We explore this interplay quite fully throughout the chapter; indeed many of the styles of restrictions we consider are inspired by timetabling scenarios. Among the results is a proof of the now famous conjecture about the list colouring problem for bipartite graphs.

The connection between bipartite graphs and matrices with non-negative entries is fundamental both historically and in reality. In this chapter we give the theorem of Birkhoff and von Neumann on a decomposition of a doubly stochastic matrix, and a criterion for when this decomposition is unique. As an application, but this time in somewhat the opposite sense to elsewhere, we apply a proof of van der Waerden's conjecture, to give results about numbers of perfect matchings in bipartite graphs.

The chapter on coverings begins with a formulation of a general covering problem. Throughout this chapter we give a broad spectrum of different instances of this problem in the context of bipartite graphs.

The penultimate chapter employs techniques and results from throughout the book, to prove a collection of combinatorial results. Theorems of edge colouring, connectivity and Hamilton cycles are used to prove results about Gray codes, systems of distinct representatives, sums of integer divisors, and completing partial latin squares.

In the final chapter we turn to problems concerning bipartite subgraphs of arbitrary graphs. We give results on the maximum size of bipartite subgraphs and several forms of edge coverings and edge-decompositions by bipartite graphs. The book ends with an appendix of NP-complete problems concerning bipartite graphs.

This book is based on graduate courses taught by A.S. Asratian at Yerevan State University and lectures given by R. Häggkvist at Umeå University. The selection of the material was of course heavily influenced by our personal interests and also limitations of space. For the most part the material discussed is accessible to any reader with a graduate understanding of mathematics. However, the book also contains advanced sections requiring much more specialised knowledge, which will be of interest to specialists in combinatorics and graph theory.

Although we include exercises designed to clarify the material, many of the exercises given are present more to supplement the results in the text. We have made some attempt to indicate the difficulty of each exercise. The easiest exercises are denoted with a $\triangle$, exercises which are hard or require much work are denoted with a $\triangledown$, and a few very challenging exercises which require techniques beyond this book are denoted with a $\square$. The remaining exercises are deemed to be tractable. The book would serve amply for teaching graduate courses, as well as a reference text.

We send thanks to our colleagues at the universities in Cambridge, Moscow, Twente, Umeå and Yerevan, for their help and remarks. We are also thankful to C. Berge, B. Bollobás, J.A. Bondy, N.N. Kuzjurin and O.B. Lupanov for their encouragement at various stages of preparation. Dr. Asratian would also like to thank the Netherlands Organisation for Scientific Research and the Natural Sciences and Engineering Research Council of Canada for partial support. Finally, and most wholeheartedly, we would like to thank the University of Umeå for providing the opportunity for this venture.

Armen S. Asratian
Tristan M.J. Denley
Roland Häggkvist
Umeå, 26 June 1997

Notation

$V(G)$	the vertex set of a graph G
$E(G)$	the edge set of G
$d_G(v)$	the degree of a vertex v in G
$\delta(G)$	the minimum degree of G
$\Delta(G)$	the maximum degree of G
$G[W]$	the subgraph of G induced by the vertex or edge set W
$d_G(u,v)$	the distance between vertices u and v in G
$N_i(v)$	the set of vertices at distance i from v
$N(S)$	the set of vertices which are neighbours of some vertex in S
$G-S$	the subgraph of G obtained by deleting the set of vertices S
$G-v$	the subgraph of G obtained by deleting the vertex v
$G-e$	the subgraph of G obtained by deleting the edge e
$G+e$	the graph obtained by adding the edge e to G
$E_G(X,Y)$	the set of edges joining a vertex in X to a vertex in Y in G
$\mathbf{M}(G)$	the incidence matrix of G
$\mathbf{A}(G)$	the adjacency matrix of G
$\mathbf{B}(G)$	the biadjacency matrix of G
$\chi'(G)$	the chromatic index of G
$r(G)$	the radius of G
$d(G)$	the diameter of G
$\kappa(G)$	the vertex connectivity of G
$\lambda(G)$	the edge connectivity of G
$bip(G)$	the graph obtained by subdividing every edge of G
(V_1, V_2)	a bipartition of a bipartite graph
$H \subseteq G$	the graph H is a subgraph of the graph G
$H \subset G$	the graph H is a proper subgraph of the graph G
$G \times H$	the graph product of the graphs G and H
uv	an edge joining vertices u and v in a simple graph
$K_{r,s}$	complete bipartite graph with colour classes of r & s vertices
K_n	the complete graph on n vertices
Q_n	the n–dimensional hypercube
$M(f,r)$	the set of edges coloured r in the edge colouring f
$\rho(\boldsymbol{\alpha}, \boldsymbol{\beta})$	the Hamming distance between $\boldsymbol{\alpha}$ and $\boldsymbol{\beta}$
$\vec{G}$	an orientation of the graph G
$\overrightarrow{uv}$	a directed edge from u to v in a simple directed graph
$\lceil x \rceil$	the smallest integer at least as large as x
$\lfloor x \rfloor$	the greatest integer at most as large as x
S_n	the set of permutations of the set $\{1, \dots, n\}$
$A \triangle B$	the symmetric difference of the sets A and B
$A \subseteq B$	A is a subset of B
$A \subset B$	A is a proper subset of B

Chapter 1

Basic concepts

1.1 Graphs

In this opening chapter it is our intention to give the basic definitions and some of the notation which we shall use throughout this book. We begin with some basic concepts concerning graphs.

A *graph* is an ordered triple $(V(G), E(G), \psi_G)$ consisting of a finite, non-empty set of *vertices*, $V(G)$, a finite set $E(G)$ of *edges*, disjoint from $V(G)$, and an *incidence function* ψ_G that associates an unordered pair of distinct vertices with each edge. We say that e joins u and v if $\psi_G(e) = \{u, v\}$, written uv, and that e has *ends* u and v. An edge is *incident* with a vertex v if v is one of its ends, and two vertices joined by an edge are *adjacent*. Edges which have the same ends are called *multiple* or *parallel edges*: the case when an edge joins a vertex to itself will not be considered in this book. We shall call a graph with no multiple edges *simple*. If G is simple and $e \in E(G)$ an edge with $\psi_G(e) = uv$, we shall write $e = uv$. The *degree* $d_G(v)$ of a vertex v is the number of edges incident with v. We denote the minimum and maximum degrees of G by $\delta(G)$ and $\Delta(G)$, respectively. A vertex with degree zero is called an *isolated vertex*. A graph with only one vertex is called *trivial*, and all other graphs *non-trivial*. A graph is *regular* if every vertex has the same degree, and it is *k-regular* if that degree is k. A 3-regular graph is called a *cubic* graph. A graph G is *planar* if it can be drawn in the plane so that its edges intersect only at their ends.

A graph H is a *subgraph* of another graph G if $V(H) \subseteq V(G)$, $E(H) \subseteq E(G)$ and ψ_H is the restriction of ψ_G to $E(H)$. We denote this by $H \subseteq G$. When $H \subseteq G$ without equality we say that H is a *proper subgraph* of G and write $H \subset G$. H is a *spanning subgraph* of G when $V(G) = V(H)$. Suppose now that $V \subseteq V(G)$ is a subset of vertices. We may consider the subgraph of G with vertex set V, and with edge set consisting of all those edges which have

1

both ends in V. We call this the subgraph *induced* by V and write it as $G[V]$. Similarly we can define a subgraph induced by a subset of edges $E \subseteq E(G)$; this graph will have edge set E and a vertex set consisting of all the end vertices of edges in E. This we denote by $G[E]$. Two subgraphs H_1 and H_2 of G are said to be *vertex-disjoint* or simply *disjoint* if $V(H_1) \cap V(H_2) = \emptyset$, and similarly H_1 and H_2 are *edge-disjoint* subgraphs if $E(H_1) \cap E(H_2) = \emptyset$.

A *walk* in a graph G is a finite sequence of vertices and edges, $W = v_0 e_1 v_1 \ldots e_n v_n$ where $\psi_G(e_i) = v_{i-1} v_i$ for each $1 \leq i \leq n$. When it is clear which edges are involved we shall write the walk $W = v_0 e_1 v_1 \ldots e_n v_n$ as simply $W = v_0 v_1 \ldots v_n$. A walk from u to v we shall call a (u,v)-*walk* and by the *length* of a walk we mean the number of edges which are traversed. We say that a non-trivial (u,u)-walk is *closed*. Given a (u,v)-walk W, the (v,u)-walk obtained by traversing W in the opposite direction is denoted by W^{-1}. A *trail* is a walk in which the edges are distinct and a *path* is a walk in which the edges and vertices are distinct. A closed trail, in which the origin and internal vertices are distinct, is called a *cycle*. We shall also use the words 'path' and 'cycle' to denote the graph or subgraph whose vertices and edges are those of a path or a cycle, respectively. The cycle of length 3 is called a *triangle*. A graph without triangles is called *triangle-free*.

Two vertices u and v of G are *connected* if there is a (u,v)-path in G. 'Connection' is an equivalence relation on the set of vertices. Thus, we can form a partition of the vertex set $V(G) = V_1 \cup \ldots \cup V_n$ so that two vertices are connected if and only if they are from the same subset. Then the subgraphs $G[V_i]$ are called the *connected components* of G, and G is called *connected* if it has only one such component.

In a connected graph, we may consider the length of the shortest path between two vertices u and v. This length is called the *distance* between the u and v, $d_G(u,v)$ or simply $d(u,v)$. We write $N_{i,G}(v)$, or simply $N_i(v)$, for the set of vertices at distance i from v. $N_{1,G}(v)$ is called the *neighbourhood* of v; we shall write $N_G(v)$, $N_1(v)$, or even $N(v)$, for the neighbourhood of v, when the context is clear. Analogously for a subset of vertices $S \subset V(G)$ we denote the set of all vertices of G which are adjacent to at least one vertex in S by $N_G(S)$, or simply $N(S)$. Let $k = \max\{i : N_i(v) \neq \emptyset\}$, then we call the partition of the vertex set $V(G) = N_0(v) \cup N_1(v) \cup \ldots \cup N_k(v)$ the *level representation of G with respect to v*. Algorithmically this partition is easily constructed, by inductively defining $N_i(v)$ to be the set of neighbours of $N_{i-1}(v)$ which have not already been accounted for. If X and Y are disjoint subsets of vertices in a graph G then we write $E_G(X,Y)$, or simply $E(X,Y)$, for the set of edges with one end in X and the other in Y. If $X = \{x\}$ then we write $E(x,Y)$ for $E(\{x\},Y)$.

We also need some operations to combine and change graphs. Given two graphs G and H we denote by $G \cup H$ the graph with vertex set $V(G) \cup V(H)$, and edge set $E(G) \cup E(H)$. Correspondingly, if $X \subset V(G)$ and $Y \subset E(G)$, we write $G - Y$ for the graph obtained by removing Y from the edge set of G and $G - X$ for $G[V(G)\backslash X]$. In the cases when $X = \{v\}$ and $Y = \{e\}$ we write $G - v$ and $G - e$, respectively, for $G - \{v\}$ and $G - \{e\}$. We denote the graph obtained from G by joining two non-adjacent vertices u and v by $G + uv$. An edge e is said to be *subdivided* when it is deleted and replaced by a path of length 2 connecting its ends, the internal vertex of this path being a new vertex. The bipartite graph obtained from a graph G by subdividing every edge is denoted by $bip(G)$. The (*lexicographic*) *product* of the disjoint simple graphs G and H is the graph denoted by $G \times H$ with vertex set $V(G \times H)$ consisting of all ordered pairs (u, v) with $u \in V(G)$ and $v \in V(H)$, and with edge set $E(G \times H) = \{(u_1, v)(u_2, v) : u_1 u_2 \in E(G), v \in V(H)\} \cup \{(u, v_1)(u, v_2) : v_1 v_2 \in E(H), u \in V(G)\}$. Although there are many others (see Sabidussi (1960)), this is the only graph product we shall consider.

We say that two graphs G and H are *isomorphic* if there are two bijections $\theta : V(G) \longleftrightarrow V(H)$ and $\phi : E(G) \longleftrightarrow E(H)$ so that $\psi_G(e) = uv$ if and only if $\psi_H(\phi(e)) = \theta(u)\theta(v)$. In particular, when G and H are simple graphs, they are isomorphic if there is a bijection $\theta : V(G) \longleftrightarrow V(H)$ so that $uv \in E(G)$ if and only if $\theta(u)\theta(v) \in E(H)$.

A *directed graph* D is also an ordered triple $(V(D), E(D), \xi_D)$ where $V(D)$ is a non-empty set of vertices, $E(D)$ is a set of *arcs* or *directed edges* (disjoint from $V(D)$) and ξ_D is an incidence function which associates an ordered pair of distinct vertices with each edge in $E(D)$. Let $v_0 e_1 v_1 e_2 v_2 \ldots e_n v_n$ be a sequence where $v_0, \ldots, v_n$ are vertices, $e_1, \ldots, e_n$ are arcs, and $\xi_D(e_i) = (v_{i-1}, v_i)$ for each $i = 1, \ldots, n$. If $v_0, v_1, \ldots, v_n$ are distinct this sequence is called a *directed path*; if $v_0 = v_n$ and $v_1, \ldots, v_n$ are distinct the sequence is a *directed cycle*. From each graph G we can obtain a directed graph by specifying, for each edge, an order of its ends. We call such a directed graph an *orientation of* G and denote it by $\vec{G}$. On the other hand from each directed graph D we can obtain a graph by ignoring the directions of the edges; this graph is called the *underlying graph* of D. If the underlying graph of D is simple and e is an arc of D with $\xi_D(e) = (u, v)$ then we denote e by $\overrightarrow{uv}$. In a natural way all the concepts we have defined for graphs have their counterparts for directed graphs by considering their underlying graphs. If v is a vertex of D then we denote the set of neighbours of v adjacent to v via an arc starting at v by $N^+(v)$ and that via one ending at v by $N^-(v)$. The cardinalities of $N^+(v)$ and $N^-(v)$ are denoted by $d_G^+(v)$ and $d_G^-(v)$ respectively.

The terminology above is broadly consistent with that used in a variety of other books. In particular, that used in the book of Bondy and Murty (1976) is perhaps the closest.

1.2 Partially ordered sets

A finite *partially ordered set* (or *poset*) $\mathcal{P} = (P, \succeq)$ consists of a finite set P and a relation $\succeq$ which satisfies

(1) $x \succeq x, \forall x \in P$ (reflexivity),

(2) $x \succeq y$ and $y \succeq x \Rightarrow x = y, \forall x, y \in P$ (antisymmetry),

(3) $x \succeq y$ and $y \succeq z \Rightarrow x \succeq z, \forall x, y, z \in P$ (transitivity).

If $x \succeq y$ and $x \neq y$ then we write $x \succ y$. If $x \succ y$ and there is no element $w \in P$ with $x \succ w \succ y$ then we say that x *covers* y. An element x is called *maximal* in P if there exists no element y with $y \succ x$. Using the cover relation we can define a graphical representation of $\mathcal{P}$. The *Hasse diagram* of $\mathcal{P}$ is the directed graph $\vec{G}$ with vertex set P where vertices x and y are joined by an directed edge $\vec{xy}$ if x covers y in $\mathcal{P}$. The underlying simple graph G is called the *non-oriented Hasse diagram* of $\mathcal{P}$.

The *greatest lower bound* of a subset of elements $T \subseteq P$ is an element g such that $x \succeq g$ for every $x \in T$ and for any $h \in P$ with $x \succeq h$ for every $x \in T$, we have $g \succeq h$. If such a greatest lower bound exists, its uniqueness is guaranteed by the antisymmetry of $\succeq$. Similarly the *least upper bound* of a subset $T \subset P$, if it exists, is an element l such that $l \succeq x$ for every $x \in P$, and for any k with $k \succeq x$ for every $x \in P$ we have $k \succeq l$.

A finite *lattice* is a finite partially ordered set, in which every two elements have a greatest lower bound, and a least upper bound. A finite *modular lattice* is a lattice satisfying the additional conditions that for every $x, y \in P$, the least upper bound of $\{x, y\}$ covers x and y, and both x and y cover the greatest lower bound of $\{x, y\}$. Finally, the *boolean lattice* with 2^n elements is the partially ordered set consisting of all subsets of some n-element set ordered by inclusion.

1.3 Reducibility of problems and NP-completeness

The questions in discrete mathematics which we might wish answered are many and various, but most can be rephrased into questions which require a 'YES' or 'NO' answer. Such questions are called *decision problems*, and they normally concern some object or *input*. For instance, we might ask

Problem 1: *Can the vertices of a graph be each labelled with one of two colours, so that no edge has both its ends labelled with the same colour?*
or
Problem 2: *Does the graph G have a cycle which includes every vertex?*

In each of these questions the graph G is the input to the problem. There are various ways of representing this input graph to perhaps some computer, for instance as a collection of vertices and edges or as an adjacency matrix, but we shall envisage this representation as being coded as a binary string in some fashion. This done, we can identify the question with a collection of binary strings – the collection of binary strings which represent graphs for which the answer to the question is 'YES'. More formally, given a decision problem $\mathcal{D}$ we define the *language* $\mathcal{L}(\mathcal{D})$ to be the set of input strings to $\mathcal{D}$ for which the answer to $\mathcal{D}$ is 'YES'. In other words the decision problem $\mathcal{D}$ could be thought of as the problem of deciding whether a given input string is a member of the language $\mathcal{L}(\mathcal{D})$ or not.

The two problems we posed above are of somewhat different natures. As we shall see later in this volume, there is a rather efficient algorithm to answer the first, but no such algorithm is known for the second. We can formalise the distinction in terms of the recognition of binary strings.

We say that a decision problem belongs to the *P class* (is a P problem) if there are an algorithm $\mathcal{A}$ and a number α so that for every binary input string of length β, the algorithm $\mathcal{A}$ will have decided whether the input string is a member of $\mathcal{L}(\mathcal{D})$ after β^α operations. Such an algorithm, which always stops after a number of operations which is a polynomial function of the length of the input, is said to operate in *polynomial time* and is called a *polynomial algorithm*.

We say (see also Wilf (1986)) that a decision problem $\mathcal{D}$ belongs to the *NP class* (is an NP problem) if there is a polynomial algorithm $\mathcal{A}$ which carries out the following:

(1) For every input string I which is in the language $\mathcal{L}(\mathcal{D})$ there is a *certificate* string $C(I)$ so that when I and $C(I)$ are input to $\mathcal{A}$, the algorithm recognises that I is in $\mathcal{L}(\mathcal{D})$.

(2) If some input string I does not belong to the language $\mathcal{L}(\mathcal{D})$ then there is no certificate string $C(I)$ which will cause $\mathcal{A}$ to conclude that I is a member of $\mathcal{L}(\mathcal{D})$.

We shall see later that Problem 1 is a P problem, but that both Problems 1 and 2 can be easily shown to be members of the NP class. For Problem 1 the certificate might be a labelling of the vertices with colours: the algorithm $\mathcal{A}$ then only has to check that each vertex of the graph is labelled with one of only two colours and that no edge has both its ends labelled with the same colour – a task which can easily be carried out in a polynomial number of operations. In a similar way a sequence of vertices can be a certificate for Problem 2. For to check that this sequence actually describes a cycle in the graph which passes through each vertex is an easy task. It is perhaps worth pointing out that to find a certificate which verifies the 'YES' answer to some NP problem may be very difficult, but to establish that the problem is an NP problem requires only that the certificate can be shown to be valid in polynomial time. More generally, it is clear that P⊆NP, but the question as to whether this containment is strict remains as yet unanswered. A problem $\mathcal{D}$ is a member of the *co-NP class* (is a co-NP problem) if the problem of deciding whether the answer to $\mathcal{D}$ is 'NO' is an NP problem.

It is often the case that a solution to one problem also provides solutions to many others, by employing some judicious rephrasing or reformulation. Consequently, it is natural to define some formal sense of reformulation for formal decision problems.

We say, of two decision problems $\mathcal{D}$ and $\mathcal{E}$, that $\mathcal{D}$ is *polynomially reducible* to $\mathcal{E}$, if there is a polynomial algorithm $\mathcal{A}$ which, given an input string to $\mathcal{D}$, constructs an input string for $\mathcal{E}$ which has the same answer ('YES' or 'NO'). We say that a problem is *NP-complete* if it itself is an NP problem, and every other NP problem can be polynomially reduced to it. Thus in some sense NP-complete problems are the hardest NP problems of all. Indeed polynomial reducibility can be defined not only for decision problems, but also for wider classes of problems. We shall say that a problem is *NP-hard* if any NP problem can be polynomially reduced to it. For instance, the problem of deciding whether, in a graph G, there is a subset C of at most k vertices such that any edge of G is incident with some vertex in C is NP-complete (Karp (1972)), but the problem of finding the smallest such C is NP-hard.

Clearly to show that a particular problem is NP-complete it is enough to show that some other NP-complete problem is polynomially reducible to it. Several hundreds of NP-complete problems from all over mathematics have been identified (see Garey and Johnson (1979)). As far as bipartite graphs go, we give a list of NP-complete problems in the Appendix at the back of this book.

Chapter 2

Introduction to bipartite graphs

2.1 Recognising bipartite graphs

A graph G is *bipartite* if the vertex set $V(G)$ can be partitioned into two sets V_1 and V_2 in such a way that no two vertices from the same set are adjacent. The sets V_1 and V_2 are called the *colour classes* of G and (V_1, V_2) is a *bipartition* of G. In fact a graph being bipartite means that the vertices of G can be coloured with at most two colours, so that no two adjacent vertices have the same colour. Throughout this book we will depict bipartite graphs with their vertices coloured black and white to show one possible bipartition. We shall call a graph *m by n bipartite*, if $|V_1| = m$ and $|V_2| = n$, and a graph a *balanced bipartite graph* when $|V_1| = |V_2|$; for example the graph shown in figure 2.1.1 is a 4 by 4 bipartite graph, since figure 2.1.2 shows that it has a bipartition in which each of the colour classes has four vertices. Given these basic definitions, let us begin by making some simple observations about the structure of bipartite graphs.

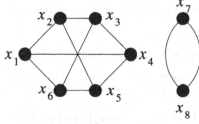

Figure 2.1.1

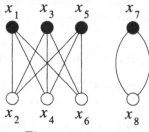

Figure 2.1.2

7

Property 2.1.1 *A connected bipartite graph has a unique bipartition.*

Property 2.1.2 *A bipartite graph, without isolated vertices, which has t connected components has 2^{t-1} bipartitions.*

For example, the bipartite graph in figure 2.1.2 has two bipartitions. One is, of course, shown in the figure and the other has $V_1 = \{x_1, x_3, x_5, x_8\}$ and $V_2 = \{x_2, x_4, x_6, x_7\}$.

During the course of this book we shall meet several characterisations of bipartite graphs, but let us begin with one of the most widely used, which was obtained by König (1916).

Theorem 2.1.3 *A graph G is bipartite if and only if G has no cycle of odd length.*

Proof. Suppose that G is a bipartite graph with bipartition (V_1, V_2) and $C = v_0 v_1 v_2 \ldots v_k v_0$ is a cycle of G. Without loss of generality, we may assume that $v_0 \in V_1$. Then, since G is bipartite, v_1 must be a vertex of V_2. Indeed we must have that $v_{2i} \in V_1$ and $v_{2i+1} \in V_2$. Hence k must be odd, and C is an even cycle.

Clearly it suffices to prove the converse when G is connected. Let G contain only even cycles, and let v be an arbitrary vertex of G. We can define a partition of $V(G)$ by setting

$$V_1 = \{u \in V(G) : d_G(u, v) \text{ is even}\},$$
$$V_2 = \{u \in V(G) : d_G(u, v) \text{ is odd}\}.$$

It remains to show that (V_1, V_2) is indeed a bipartition of G. Suppose that x and y are two vertices of V_1, and that $xy \in E(G)$. Let P be a shortest (x, v)-path, Q a shortest (y, v)-path, and v_1 be the first common vertex of P and Q. Clearly, since P and Q were shortest paths, their (v_1, v) sections must also be shortest (v_1, v)-paths. In particular they are of the same length. Let P_1 and Q_1 be the (x, v_1) and (y, v_1) sections of P and Q, respectively. Then, since P and Q are both of even length, it follows that P_1 and Q_1 have the same parity. However, this gives rise to the odd cycle $C = P_1 Q_1^{-1} x$, and thus the required contradiction. Hence no two vertices of V_1 are adjacent. Similarly, no two vertices of V_2 are adjacent, and (V_1, V_2) is indeed a bipartition of G. □

A similar argument gives the following corollary.

Corollary 2.1.4 *A connected graph G is bipartite if and only if for every vertex v there is no edge xy with $d(v, x) = d(v, y)$.*

Proof. Suppose that G is bipartite with bipartition (V_1, V_2), and let v be a fixed vertex of G. Then if $d(v, y) = d(v, y)$ x and y are members of the same colour class and $xy \notin E(G)$.

On the other hand, given a vertex $v \in V$ we define the partition $V_1 = \bigcup_{i \geq 1} N_{2i-1}(v)$ and $V_2 = V(G) \backslash V_1$. Then it follows from our assumptions that G only has edges joining vertices from different V_i's, and therefore that G is bipartite. $\square$

There are many characterisations of bipartite graphs, and therefore many algorithmic ways to recognise them. Corollary 2.1.4 gives rise to one such algorithm: choose a vertex $v \in V(G)$ and consider the level representation of G with respect to v. If each $N_i(v)$ spans no edges then G is bipartite, otherwise G is not bipartite. The following variation of Theorem 2.1.3 will be useful later.

Corollary 2.1.5   *A graph G is bipartite if and only if it contains no closed walk of odd length.*

Proof. Since an odd cycle is also an odd walk the condition is certainly sufficient. Thus it suffices to show that a bipartite graph contains no closed walk of odd length. Let G be bipartite and $W = v_0 v_1 v_2 \dots v_k v_0$ be a closed walk in G. Consider the level representation of G with respect to v_0. We define the sequence $\alpha_1, \alpha_2, \dots \alpha_{k+1}$ by

$$\alpha_i = \begin{cases} 1 & \text{if } 1 \leq i \leq k \text{ and the level of } v_{i-1} \text{ is less than the level of } v_i, \\ 0 & \text{otherwise.} \end{cases}$$

Then, since W is closed, the sequence must contain equal numbers of 1's and 0's, and hence must be of even length. Therefore, W is also of even length. $\square$

Exercises

2.1.1   Let $V_1 = \{v_1, v_2, \dots, v_m\}$ and $V_2 = \{u_1, \dots, u_n\}$. Show that

    (a) $\triangle$ there are 2^{mn} simple bipartite graphs with bipartition (V_1, V_2),

    (b) $\triangledown$ the number of simple bipartite graphs with even vertex degrees and bipartition (V_1, V_2) is $2^{(m-1)(n-1)}$.

2.1.2 $\triangle$ Show that a graph G is bipartite if and only if every subgraph H has a set of at least $|V(H)|/2$ mutually non-adjacent vertices.

2.1.3 $\triangledown$ Prove that a triangle-free simple graph in which no three vertices have equal degree is bipartite. (Erdős, Fajtlowits, Staton (1991))

2.1.4 $\triangledown$ Show that a triangle-free graph G on p vertices with $\delta(G) > 2p/5$ is bipartite. (Andrásfai, Erdős, Sós (1974))

2.1.5 $\triangle$ Let G be a connected, simple graph with $V(G) = \{v_1, \dots, v_p\}$ and H be a bipartite graph with bipartition (V_1, V_2) where the colour classes are $V_1=\{x_1,\ldots,x_p\}$, $V_2=\{y_1,\ldots,y_p\}$ and $v_iv_j \in E(G)$ if and only if x_iy_j and x_jy_i are in $E(H)$. Show that G is bipartite if and only if H is not connected.

2.1.6 △ Show that any arbitrary pair of simple graphs G and H are isomorphic if and only if the graphs $bip(G)$ and $bip(H)$ are isomorphic.

2.2 Bipartite graphs of certain types

Several special classes of bipartite graphs are of particular interest, and in this section we shall briefly introduce some of these natural special cases.

The first of these is a *tree* – a connected, simple graph which contains no cycles; amusingly a union of disjoint trees is called a *forest*. Trees and forests are certainly bipartite, since they contain no cycles of either parity, and they have an abundance of different characterisations; in the following theorem we give only two, but several more can be found throughout this book.

Proposition 2.2.1 *The following statements are equivalent for a graph G:*

(1) G is a tree,

(2) each pair of vertices is joined by a unique path,

(3) G is connected and $|V(G)| = |E(G)| + 1$.

Proof. *(1)* implies *(2)*

Since G is connected every two vertices are connected by some path. Let u and v be two vertices of G which are connected by two distinct (u,v)-paths P_1 and P_2. We shall show that $P_1 \cup P_2$ contains a cycle. Let w be the first vertex of P_1 which is also on P_2, but whose successor in P_1 is not in P_2, and let w' be the next vertex of P_1 which also lies on P_2. Then the segments of P_1 and P_2 which lie between w and w' together form a cycle.

(2) implies *(3)*

Clearly G is connected. Let $p = |V(G)|$ and $q = |E(G)|$. We shall prove the relationship between the numbers of edges and vertices by induction on p.

The assertion is clear for connected graphs with one or two vertices. Thus let $p \geq 3$ be a value for which the implication holds for graphs with fewer than p vertices. Delete an edge, e, from G. Then by assumption $G - e$ will consist of two connected components to which we may apply our induction hypothesis. Each component will have one more vertex than edge and so we must also have $p - 1$ edges in G.

(3) implies *(1)*

Once again G is connected, and we need only show that G has no cycle. We shall again prove this by induction on the number of vertices. The implication is clear when there are only one or two vertices. Thus suppose that the implication holds for any graph with fewer vertices than G. Since $|E(G)| = |V(G)| - 1$ there must be some vertex u with $d_G(u) = 1$. Then certainly $G - u$ is connected and by induction is a tree. Since the addition of u can create no cycles G must also be a tree.   □

As we saw in the proof of the last result, every non-trivial tree has at least one vertex of degree 1; these are called *pendant vertices.*

Corollary 2.2.2   *Let T be a tree on p vertices and with $\Delta(T) = \Delta$. Then the number of pendant vertices, n_T, satisfies the inequality*

$$2 \leq n_T \leq \frac{p(\Delta - 2) + 2}{\Delta - 1}.$$

Proof. Each of the inequalities can be obtained by counting the number of edges in the tree. To see the first inequality from the above theorem we have $2(p-1) = 2|E(T)| = \sum_{v \in V(T)} d_T(v) \geq n_T \cdot 1 + (p - n_T) \cdot 2$ and rearranging gives $n_T \geq 2$.

The other inequality follows immediately from

$$2(p-1) = 2|E(T)| = \sum_{v \in V(T)} d_T(v) \leq n_T \cdot 1 + \Delta(p - n_T). \qquad □$$

We now turn to our next type of bipartite graph. The *complete m by n bipartite graph*, $K_{m,n}$, is the simple bipartite graph with bipartition (V_1, V_2) in which $|V_1| = m$, $|V_2| = n$ and each vertex of V_1 is adjacent to each vertex of V_2. In particular, $K_{1,n}$ is called an *n-star* or simply a *star*. Clearly the graph $K_{m,n}$ has mn edges, and it is a simple task to check that amongst all bipartite graphs on p vertices $K_{\lceil p/2 \rceil, \lfloor p/2 \rfloor}$ has the most edges. Rather more surprising, though, is that this graph is also extremal for a rather larger class of graphs.

Proposition 2.2.3 (Mantel (1907)) *Among simple graphs on p vertices which are triangle-free $K_{\lceil p/2 \rceil, \lfloor p/2 \rfloor}$ has the maximum number of edges.*

Proof. It is convenient to break the proof into two similar arguments depending on the parity of p. Here we give only the proof for p even.

The statement is obvious for $p = 2$. We shall proceed by induction on p. Suppose that the statement is true for all $p \leq 2n$ and let G be a graph on $2n + 2$ vertices which contains no triangles. Choose an edge $uv \in E(G)$ and consider $G' = G - \{u, v\}$. Then G' has $2n$ vertices and hence, by induction, has at most $4n^2/4$ edges. We need now only consider the degrees of u and v.

If u and v have some common neighbour G would contain a triangle, thus if $d_G(u) = k \ d_G(v) \leq 2n - k$, and the number of edges of G is at most

$$n^2 + k + 2n - k + 1 = n^2 + 2n + 1 = \lceil p^2/4 \rceil. \qquad \square$$

A simple bipartite graph is *chordal* if in every cycle of length strictly greater than 4 there are two non-consecutive vertices which are adjacent. For example, the graphs $K_{m,n}$ and $P_n \times P_2$ are both chordal, bipartite graphs. We say that a pair of edges, e and f, is *separable* if there exists a set of vertices, S, whose removal from the graph leaves the edges e and f in different connected components of the remaining subgraph. The set S is called an *edge separator* for e and f and we say that S is *minimal* if no proper subset of S is an edge separator for e and f. The following theorem gives a characterisation of chordal graphs in terms of its edge separators.

Proposition 2.2.4 (Golumbic, Goss (1978)) *A simple bipartite graph G is chordal if and only if every minimal edge separator induces a complete bipartite subgraph.*

Proof. Let $C = v_1v_2 \ldots v_kv_1$ be a cycle of G with $k > 4$. Then the set $S = (N(v_2) \cup N(v_3)) \backslash \{v_2, v_3\}$ separates v_2v_3 from v_5v_6, and $S \cap V(C) = \{v_1, v_4\}$. Let $S' \subseteq S$ be a minimal such separator which clearly must contain v_1 and v_4. Then if $G[S']$ is a complete bipartite subgraph v_1v_4 is an edge, since v_1 and v_4 cannot be in the same colour class, and C has a chord.

Conversely, let S be a minimal edge separator, and G_1 and G_2 be the connected components of $G - S$. Let x and y be arbitrary vertices of S from different colour classes of G. Then since G_1 and G_2 are connected there exist paths $P_1 = xu_1u_2 \ldots u_ky$ and $P_2 = xv_1v_2 \ldots v_ly$ with $u_i \in V(G_1)$ and $v_i \in V(G_2)$. Certainly these paths are of length at least 3, and thus $C = P_1P_2^{-1}$ is a cycle of length at least 6. If this cycle has a chord it can only be the edge xy and thus $xy \in E(G)$ and $G[S]$ is a complete bipartite subgraph. $\qquad \square$

Another type of graphs which we shall mention are graphs which are products of other graphs. These need not be bipartite, but the following theorem gives a criterion on which to decide.

Proposition 2.2.5 *The product $G_1 \times G_2$ of two disjoint simple graphs is bipartite if and only if both G_1 and G_2 are bipartite.*

Proof. Suppose that $G = G_1 \times G_2$ is bipartite. Then, by Theorem 2.1.3, every cycle of G is of even length. By definition G contains induced subgraphs isomorphic to G_1 and G_2, and so every cycle of these subgraphs must be of even length. Thus, by Theorem 2.1.3, G_1 and G_2 are also bipartite.

Conversely, suppose that G_1 and G_2 are bipartite. Let $V(G_1) = \{x_1, \ldots, x_n\}$ and $V(G_2) = \{y_1, \ldots, y_m\}$, and consider a cycle C in $G_1 \times G_2$. Let $C = z_1 z_2 \ldots z_k z_1$ where $z_l = (x_{i_l}, y_{j_l})$ for $l = 1, \ldots, k$. Clearly the sequences $x_{i_1} x_{i_2} \ldots x_{i_k} x_{i_1}$ and $y_{j_1} y_{j_2} \ldots y_{j_k} y_{j_1}$ induce closed walks W_1 and W_2 in G_1 and G_2, respectively. Moreover, the length of C equals the sum of the lengths of W_1 and W_2. Since G_1 and G_2 are bipartite, by Corollary 2.1.5 W_1 and W_2 are of even length, and so also is C. Thus by Theorem 2.1.3 $G_1 \times G_2$ is bipartite. □

Given this concept of product it is natural to consider a concept of *prime factorisation*. A simple graph G is *prime* if there exist no two graphs G_1 and G_2 such that $|V(G_1)| \geq 2$, $|V(G_2)| \geq 2$ and $G_1 \times G_2$ is isomorphic to G. A *prime factorisation* of G is a representation $G_1 \times G_2 \times \ldots \times G_k$ such that G_i is a prime graph with at least two vertices, $i = 1, \ldots, k$, and G is isomorphic to $G_1 \times G_2 \times \ldots \times G_k$. Sabidussi (1960), and later Vizing (1963), showed that the graph product enjoys unique factorisation for simple connected graphs, up to ordering of the factors and isomorphisms of each factor. Combining the above theorem with the result of Sabidussi and Vizing we obtain the following.

Corollary 2.2.6 *Every simple bipartite graph has a unique prime factorisation in which all the factors are bipartite.* □

A polynomial algorithm for finding a prime factorisation of a simple graph has been given by Winkler (1986) and Feigenbaum, Hershberger and Schäffer (1985).

Regular bipartite graphs are amongst the most important particular classes of bipartite graphs. Later we shall see that regular bipartite graphs have some very nice properties. The final result of this section shows that we can always embed a bipartite graph into a regular bipartite graph. By using this method of regularisation we may be able to use some of the structure of regular graphs to our advantage. Some applications of this type of approach are in the chapter on edge colourings.

Proposition 2.2.7 *Given a bipartite graph G and any integer $k \geq \Delta(G)$ there exists a k-regular bipartite graph H such that G is an induced subgraph of H.*

Proof. The result is trivial if G is k-regular. Otherwise, we define a new graph G_1 consisting of two disjoint copies of G with an edge joining any vertex $v \in V(G)$ for which $d_G(v) < k$ to its copy. Notice that if G has bipartition (V_1, V_2) then the bipartition $(V_1 \cup V_2', V_2 \cup V_1')$ (where V_i' are the colour classes of the copy of G) shows that G_1 is still a bipartite graph. If G_1 is k-regular then set $H = G_1$. Otherwise we continue inductively, defining G_{i+1} to be the graph obtained by taking two disjoint copies of G_i and joining any vertex not of degree k to its copy. We then set $H = G_t$ where $t = \Delta(G) - \delta(G)$. □

In fact when beginning with a simple graph the construction in the above proof gives a simple regular graph. Thus for every simple bipartite graph G there is a simple k-regular bipartite graph, which contains G as an induced subgraph, for any $k \geq \Delta(G)$.

Exercises

2.2.1 △ Show that a connected graph G is a tree if and only if for each edge $e \in E(G)$ the graph $G - e$ is not connected.

2.2.2 Let T be a tree with vertex set $\{v_1, \ldots, v_p\}$, $p \geq 2$. Prove that the number of pendant vertices of T is given by $1 + \sum_{i=1}^{p} \dfrac{|d_G(v_i) - 2|}{2}$.

2.2.3 Let G be a simple connected bipartite graph with $|V(G)| \geq 4$. Show that G is a complete balanced bipartite graph if and only if $d(u) + d(v) \geq |N(u) \cup N(v) \cup N(w)|$ for any path uwv of G. (Asratian, Sarkisian (1991))

2.2.4 A graph is called *magic* if it is possible to label the edges by non-negative numbers in such a way that every edge has a unique label and the sum of the labels on edges incident with any vertex is always the same. Prove that $K_{m,n}$ is magic if and only if $m = n \neq 2$. (Doob (1978))

2.2.5 A simple graph G is said to have a *graceful labelling* if there is a labelling of the vertices $f : V(G) \longrightarrow \{0, 1, 2, \ldots\}$ so that $\{|f(x) - f(y)| : xy \in E(G)\} = \{1, 2, \ldots, |E(G)|\}$. Prove that $K_{m,n}$ has a graceful labelling. (Rosa (1967), Golomb (1972))

2.2.6 ▽ Let G be a bipartite graph with bipartition (V_1, V_2). G is called *convex* on V_2 if there is an ordering of V_2 so that for any $x \in V_1$ the set $N(x)$ forms an interval in the ordering. Furthermore, three vertices x, y, z are an *asteroidal triple* of G if there exists a path between any two with no vertex of that path adjacent to the third. Prove that G is convex on V_2 if and only if V_1 contains no asteroidal triple. (Tucker (1972))

2.2.7 Prove that every simple bipartite graph which is convex on V_2 is chordal.

2.2.8 A simple graph is called a *permutation graph* if there exist two permutations of $V(G)$, π_1 and π_2, for which $xy \in E(G)$ if and only if x precedes y in π_1 and y precedes x in π_2. Prove that a simple bipartite graph with bipartition (V_1, V_2) is a permutation graph if and only if there are orderings on V_1 and V_2 so that whenever

$xy, x'y' \in E(G)$ and $x < x'$, $y < y'$ in the respective orders then $xy', x'y \in E(G)$. (Brandstädt, Spinrag, Stewart (1987))

2.2.9 △ A simple bipartite graph with bipartition (V_1, V_2) is doubly convex if it is convex on both V_1 and V_2. Show that a bipartite permutation graph is doubly convex.

2.2.10   Let G be a bipartite graph on p vertices and q edges. Prove that there is a $\Delta(G)$-regular bipartite graph H, on $2p - 2\lfloor q/\Delta(G)\rfloor$ vertices, which contains G as an induced subgraph, but there is no such graph on fewer vertices. (Lovász (1979))

2.3 Matrix characterisations of bipartite graphs

There is a close connection between matrices and bipartite graphs. In this section we shall explore this link and introduce some characterisations of bipartite graphs using matrix representations.

Let G be a graph with vertex set $V(G) = \{v_1, v_2, \ldots, v_p\}$ and edge set $E(G) = \{e_1, e_2, \ldots, e_q\}$. The *incidence matrix* of G is the $p \times q$ matrix $\mathbf{M}(G) = [m_{ij}]$, where m_{ij} is 1 if the edge e_j is incident with vertex v_i and 0 otherwise. A graph G and its incidence matrix $\mathbf{M}(G)$ are shown in figure 2.3.1.

The matrix $\mathbf{M}(G)$ is called *totally unimodular* if every square submatrix of $\mathbf{M}(G)$ has determinant 0, 1 or -1. The matrix shown in figure 2.3.1 is not totally unimodular, since the top left 3×3 submatrix has determinant 2. Also, the graph G shown in the figure is not bipartite. This is no coincidence, since total unimodularity provides our first matrix characterisation of bipartite graphs.

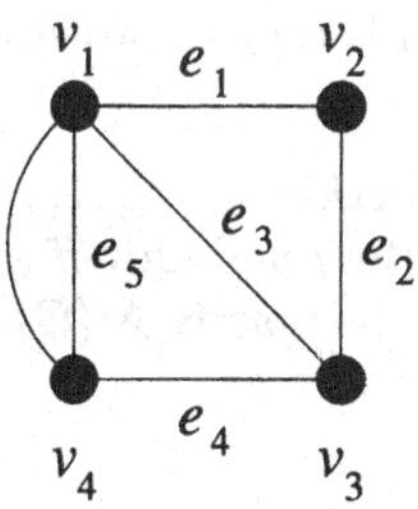

$$\mathbf{M}(G) = \begin{array}{c} \\ v_1 \\ v_2 \\ v_3 \\ v_4 \end{array} \begin{array}{cccccc} e_1 & e_2 & e_3 & e_4 & e_5 & e_6 \\ \begin{pmatrix} 1 & 0 & 1 & 0 & 1 & 1 \\ 1 & 1 & 0 & 0 & 0 & 0 \\ 0 & 1 & 1 & 1 & 0 & 0 \\ 0 & 0 & 0 & 1 & 1 & 1 \end{pmatrix} \end{array}$$

Figure 2.3.1

Theorem 2.3.1 *A graph G is bipartite if and only if its incidence matrix is totally unimodular.*

Proof. Suppose that $\mathbf{M}(G)$ is totally unimodular, but that G is not bipartite. Then, by Theorem 2.1.3, G possesses an odd cycle of length $2l + 1$ say. Let $\mathbf{M}_1$ be the $(2l + 1) \times (2l + 1)$ submatrix of $\mathbf{M}(G)$ corresponding to the edges and vertices of this cycle. Then $\mathbf{M}_1$, up to permutation of the rows and columns, has the form

$$\mathbf{M}_1 = \begin{pmatrix} 1 & 0 & 0 & \dots & 0 & 1 \\ 1 & 1 & 0 & \dots & 0 & 0 \\ 0 & 1 & 1 & \dots & 0 & 0 \\ \vdots & \vdots & \vdots & \ddots & \vdots & \vdots \\ 0 & 0 & 0 & \dots & 1 & 1 \end{pmatrix}.$$

It is simple to check that $\det(\mathbf{M}_1) = 1 + (-1)^{2l} = 2$, which contradicts the total unimodularity of $\mathbf{M}(G)$.

Conversely, let G be a bipartite graph, and consider any arbitrary $k \times k$ submatrix of $\mathbf{M}(G)$, $\mathbf{Q}$. We must show that $\det(\mathbf{Q}) = 0, 1$ or -1. We do so by induction on k.

The case $k = 1$ is clear. If $\mathbf{Q}$ has a column consisting only of 0's, then clearly $\det(\mathbf{Q}) = 0$, and if $\mathbf{Q}$ has a column which contains a single 1 then we may expand the determinant about this, and the result follows by induction. Thus we may assume that every column contains exactly two 1's. The colour classes of G partition the rows of $\mathbf{M}(G)$ into two sets. We may assume that the first r rows correspond to one colour, and the last $k - r$ to the other. Then, since G is bipartite, every column of $\mathbf{Q}$ must contain precisely one 1 in the first r rows, and precisely one 1 in the last $k - r$ rows. Thus the sum of the first r row vectors is equal to the sum of the last $k - r$, the rows of $\mathbf{Q}$ are linearly dependent and $\det(\mathbf{Q}) = 0$. $\square$

Another matrix associated with a graph G is the *adjacency matrix*. If, once again, G has vertex set $V(G) = \{v_1, v_2, \dots, v_p\}$ and edge set $E(G) = \{e_1, e_2, \dots, e_q\}$ the adjacency matrix of G is the $p \times p$ matrix $\mathbf{A}(G) = [a_{ij}]$ where a_{ij} is the number of edges joining v_i and v_j. $\mathbf{A}(G)$ is a symmetric matrix with non-negative integer entries and it is easy to prove the following result.

Proposition 2.3.2 *Let G be a graph with vertices $v_1, v_2, \dots, v_p$ and adjacency matrix $\mathbf{A}(G) = [a_{ij}]$. Then G is bipartite if and only if there exists a permutation π of the set $\{1, 2, \dots, n\}$ so that the matrix $\mathbf{A}'(G) = [a_{\pi(i)\pi(j)}]$ has the following form:*

$$\begin{pmatrix} 0 & \mathbf{B} \\ \mathbf{B}^{\mathrm{T}} & 0 \end{pmatrix}$$

where $\mathbf{B}^{\mathrm{T}}$ is the transpose of $\mathbf{B}$. $\square$

$$G = $$

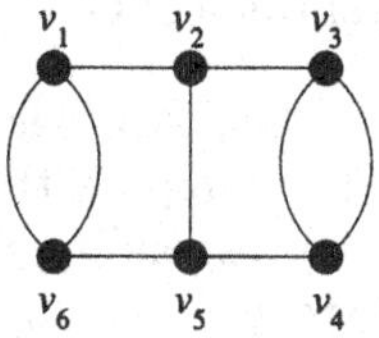

$$
A(G) = \begin{array}{c} \\ v_1 \\ v_2 \\ v_3 \\ v_4 \\ v_5 \\ v_6 \end{array}
\begin{array}{c}
\begin{array}{cccccc} v_1 & v_2 & v_3 & v_4 & v_5 & v_6 \end{array} \\
\begin{pmatrix}
0 & 1 & 0 & 0 & 0 & 2 \\
1 & 0 & 1 & 0 & 1 & 0 \\
0 & 1 & 0 & 2 & 0 & 0 \\
0 & 0 & 2 & 0 & 1 & 0 \\
0 & 1 & 0 & 1 & 0 & 1 \\
2 & 0 & 0 & 0 & 1 & 0
\end{pmatrix}
\end{array}
\qquad
A'(G) = \begin{array}{c} \\ v_1 \\ v_3 \\ v_5 \\ v_2 \\ v_4 \\ v_6 \end{array}
\begin{array}{c}
\begin{array}{cccccc} v_1 & v_3 & v_5 & v_2 & v_4 & v_6 \end{array} \\
\begin{pmatrix}
0 & 0 & 0 & 1 & 0 & 2 \\
0 & 0 & 0 & 1 & 2 & 0 \\
0 & 0 & 0 & 1 & 1 & 1 \\
1 & 1 & 1 & 0 & 0 & 0 \\
0 & 2 & 1 & 0 & 0 & 0 \\
2 & 0 & 1 & 0 & 0 & 0
\end{pmatrix}
\end{array}
$$

Figure 2.3.2

A graph G, its adjacency matrix $\mathbf{A}(G)$ and transformed matrix $\mathbf{A}'(G)$ are shown in figure 2.3.2. This transformation suggests a simpler way to represent bipartite graphs as matrices. Let $V_1 = \{v_1, v_2, \ldots, v_m\}$ and $V_2 = \{v_{m+1}, \ldots, v_{m+n}\}$ and denote by $\mathcal{B}(m, n)$ the set of all bipartite graphs with bipartition (V_1, V_2). Then the *biadjacency matrix* of a graph $G \in \mathcal{B}(m, n)$ is the $m \times n$ matrix $\mathbf{B}(G) = [b_{ij}]$, where b_{ij} is the number of edges joining v_i to v_{j+m}, $1 \le i \le m$, $1 \le j \le n$. Clearly, there is a one-to-one correspondence between the set $\mathcal{B}(m, n)$ and the set of all non-negative integral $m \times n$ matrices. Also there is a simple relationship between $\mathbf{A}(G)$ and $\mathbf{B}(G)$ given by

$$
\mathbf{A}(G) = \begin{pmatrix} 0 & \mathbf{B}(G) \\ \mathbf{B}(G)^{\mathrm{T}} & 0 \end{pmatrix}. \tag{2.3.1}
$$

Some algebraic properties of the adjacency matrix have implications for the underlying graph. We shall now investigate some results of this type, but first let us recall some standard definitions and facts from linear algebra.

A number λ is called an *eigenvalue* of an $m \times n$ matrix $\mathbf{A}$ if there is a non-zero vector $\mathbf{x}$ satisfying $\mathbf{Ax} = \lambda\mathbf{x}$. Each such vector is called an *eigenvector* of $\mathbf{A}$ with eigenvalue λ. It is well-known that if $\mathbf{A}$ is a real symmetric matrix then all its eigenvalues are also real. Thus for each graph G the adjacency matrix $\mathbf{A}(G)$ has only real eigenvalues. The eigenvalues of $\mathbf{A}(G)$ are often referred to as simply eigenvalues of G and the collection of eigenvalues as the *spectrum* of G. The following result was first recorded in chemistry literature, by Coulson and Roushbrooke (1940). Chemists often refer to it as the 'Pairing Theorem'.

Proposition 2.3.3 *If G is a bipartite graph with at least one edge, then its spectrum is symmetrical with respect to 0, i.e. if a number λ is an eigenvalue of G then $-\lambda$ is also an eigenvalue of G.*

Proof. Let $G \in \mathcal{B}(m,n)$. Then $\mathbf{A}(G)$ is an $(m+n) \times (m+n)$ matrix of the form shown in (2.3.1). Suppose that λ is an eigenvalue of G and $\mathbf{x} = (x_1, x_2, \ldots, x_{m+n})$ is a corresponding eigenvector. Consider the vector $\mathbf{y} = (y_1, y_2, \ldots, y_{m+n})$ where $y_j = x_j$ if $1 \leq j \leq m$ and $y_j = -x_j$ if $m+1 \leq j \leq m+n$. Then

$$\sum_{j=1}^{m+n} a_{ij}y_j = \sum_{j=m+1}^{m+n} a_{ij}y_j = -\sum_{j=m+1}^{m+n} a_{ij}x_j = -\lambda x_i = -\lambda y_i, \quad \text{if } 1 \leq i \leq m,$$

and

$$\sum_{j=1}^{m+n} a_{ij}y_j = \sum_{j=1}^{m} a_{ij}y_j = \sum_{j=1}^{m} a_{ij}x_j = \lambda x_i = -\lambda y_i, \quad \text{if } m+1 \leq i \leq m+n.$$

So $\mathbf{y}$ satisfies $\mathbf{A}(G)\mathbf{y} = -\lambda \mathbf{y}$, and $-\lambda$ is an eigenvalue of G. □

H. Sachs (1966) proved that symmetry of the spectrum of G is also sufficient for G to be bipartite. Thus in fact, we have the following theorem.

Theorem 2.3.4 *A graph G with at least one edge is bipartite if and only if its spectrum is symmetrical with respect to 0.* □

Indeed there is a much stronger characterisation of regular bipartite graphs.

Theorem 2.3.5 (Hoffman (1963)) *A connected k-regular graph G is bipartite if and only if $-k$ is an eigenvalue of G.*

Proof. First suppose that G is a bipartite k-regular graph. Then in each row and each column of $\mathbf{A}(G)$ there are precisely k 1's. Thus, the vector $\mathbf{x} = (1, 1, \ldots, 1)^{\mathrm{T}}$ satisfies $\mathbf{A}\mathbf{x} = k\mathbf{x}$, and k is an eigenvalue of G and so, by Proposition 2.3.3, $-k$ is also an eigenvalue of G.

Let G be a k-regular graph with $V(G) = \{v_1, v_2, \ldots, v_p\}$. To prove the converse we must construct a bipartition of G. Let $\mathbf{x} = (x_1, x_2, \ldots, x_p)$ be an eigenvector of G, with eigenvalue $-k$. We may assume, without loss of generality, that $\max_{1 \leq j \leq p} x_j = 1$ and that $|x_j| \leq 1$ for every j.

Suppose that $x_{i_0} = 1$. Then since each row or column of $\mathbf{A}(G)$ contains precisely k 1's and $\sum_{j=1}^{p} a_{i_0 j}x_j = -kx_{i_0} = -k$, x_j must equal -1 for each j such that $v_j \in N(v_{i_0})$. Similarly, we have that if $x_{i_0} = -1$ then $x_j = 1$ for each j with $v_j \in N(v_{i_0})$. Since G is connected all of the coordinates of x must thus be 1 or -1. Let $V_1 = \{v_j : x_j = 1, 1 \leq j \leq p\}$ and

$V_2 = \{v_j : x_j = -1,\ 1 \le j \le p\}$. Then (V_1, V_2) is a bipartition of G, as required.   □

More results about the spectral properties of bipartite graphs can be found in the book of Cvetković, Doob and Sachs (1979).

Exercises

2.3.1 △ Let G be a graph with vertex set $V(G) = \{v_1, v_2, \ldots, v_p\}$. Prove that

  (a)   the ijth entry of $\mathbf{A}(G)^n$ is the number of walks of length n from v_i to v_j.

  (b) G is bipartite if and only if the diagonal entries of $\mathbf{A}(G)^{2n-1}$ are zero, for $1 \le n \le \lceil p/2 \rceil$.

2.3.2   Show that a connected graph on p vertices is bipartite if and only if the rank of its incidence matrix is $p-1$. (Sachs (1967), Van Nuffelen (1973))

2.3.3 ▽ Construct two non-isomorphic bipartite graphs with the same spectrum.

2.3.4   Show that the spectrum of $K_{1,t}$ is $-\sqrt{t}, 0, \ldots, 0, \sqrt{t}$ where 0 has multiplicity $t - 1$.

2.3.5 △ Prove that the maximum eigenvalue of a connected k-regular graph G is k.

2.3.6 □ Prove that a connected graph G with maximum eigenvalue λ is bipartite if and only if $-\lambda$ is also an eigenvalue of G. (Lovász (1979))

2.3.7   Prove that the spectrum of a non-connected graph is the union of the spectra of its connected components, in the sense that if an eigenvalue λ occurs in the spectra of the components with multiplicities $m_1, m_2, \ldots, m_s$ then it has multiplicity $m_1 + m_2 + \ldots + m_s$ in the spectrum of G.

2.3.8   Using the two previous exercises prove Theorem 2.3.4.

Application

2.4 Gaussian elimination

Let $\mathbf{Q} = [q_{ij}]$ be an $n \times n$ matrix over some field, then we may reduce $\mathbf{Q}$ to the identity matrix $\mathbf{I}$ by a series of basic matrix operations: choose a non-zero entry q_{ij} to act as a *pivot* and then use elementary row operations to make $q_{ij} = 1$ and all other entries in the ith row and jth column equal to 0. This familiar process is, of course, Gaussian elimination and the reduction may be carried out in a variety of ways, some of which are clearly more efficient than others. If, for instance, the matrix $\mathbf{Q}$ is sparse then reducing in an arbitrary way may well result in some zero entries becoming non-zero during the calculation, whereas some more careful reduction may not; for an example of this see the two reductions in figure 2.4.1. In the first of these reductions, various non-zero entries are changed, whilst in the second there are no such changes. Since the most efficient way to store a sparse matrix in a computer is by storing only the non-zero elements, the ability to reduce a matrix by referring only to the non-zero elements is clearly a great advantage.

$$\mathbf{Q}=\begin{pmatrix} 4 & 1 & 1 & 1 \\ 1 & 1 & 0 & 0 \\ 1 & 0 & 1 & 0 \\ 1 & 0 & 0 & 1 \end{pmatrix} \rightarrow \begin{pmatrix} 1 & 0 & 0 & 0 \\ 0 & 3 & -1 & -1 \\ 0 & -1 & 3 & -1 \\ 0 & -1 & -1 & 3 \end{pmatrix} \rightarrow \begin{pmatrix} 1 & 0 & 0 & 0 \\ 0 & 1 & 0 & 0 \\ 0 & 0 & 8 & -4 \\ 0 & 0 & -4 & 8 \end{pmatrix} \rightarrow \begin{pmatrix} 1 & 0 & 0 & 0 \\ 0 & 1 & 0 & 0 \\ 0 & 0 & 1 & 0 \\ 0 & 0 & 0 & 12 \end{pmatrix}$$

$$\downarrow$$

$$\mathbf{Q}=\begin{pmatrix} 4 & 1 & 1 & 1 \\ 1 & 1 & 0 & 0 \\ 1 & 0 & 1 & 0 \\ 1 & 0 & 0 & 1 \end{pmatrix} \rightarrow \begin{pmatrix} 3 & 1 & 1 & 0 \\ 1 & 1 & 0 & 0 \\ 1 & 0 & 1 & 0 \\ 0 & 0 & 0 & 1 \end{pmatrix} \rightarrow \begin{pmatrix} 2 & 1 & 0 & 0 \\ 1 & 1 & 0 & 0 \\ 0 & 0 & 1 & 0 \\ 0 & 0 & 0 & 1 \end{pmatrix} \rightarrow \begin{pmatrix} 1 & 0 & 0 & 0 \\ 0 & 1 & 0 & 0 \\ 0 & 0 & 1 & 0 \\ 0 & 0 & 0 & 1 \end{pmatrix}$$

Figure 2.4.1

In this section we shall investigate this problem of reducing a matrix $\mathbf{Q}$ to the identity matrix $\mathbf{I}$ without ever changing a zero entry. Let us reformulate this problem in terms of a bipartite graph.

Let $\mathbf{B}$ be the matrix obtained from $\mathbf{Q}$ by replacing every non-zero entry by a 1, and $G = G(\mathbf{Q})$ be the graph with biadjacency matrix $\mathbf{B}$. An edge $xy \in E(G)$ is called *bisimplicial* if $G[N(x) \cup N(y)]$ is a complete bipartite graph. It is not difficult to check that a non-singular matrix $\mathbf{Q}$ has a sequence of pivots which preserve the zeros in $\mathbf{Q}$ if and only if the graph $G(\mathbf{Q})$ has a sequence of pairwise non-adjacent edges $\sigma = [x_1y_1, x_2y_2, \ldots, x_ky_k]$ such that

(1) x_1y_1 is a bisimplicial edge of G and, for each i, $2 \leq i \leq k$, x_iy_i is a bisimplicial edge of $G - \{x_1, \ldots, x_{i-1}, y_1 \ldots, y_{i-1}\}$,

(2) $G - \{x_1, \ldots, x_k, y_1, \ldots, y_k\}$ is empty.

G is called a **perfect elimination bipartite graph** if such a sequence σ exists, and the sequence is called a **scheme** (a sequence satisfying only condition *(1)* is called a **partial scheme**). The problem of recognising when $\mathbf{Q}$ may be reduced to $\mathbf{I}$ whilst preserving the zero entries now corresponds to recognising if $G(\mathbf{Q})$ is a perfect elimination bipartite graph.

Theorem 2.4.1 (Golumbic, Goss (1978)) *If xy is a bisimplicial edge of a perfect elimination bipartite graph G then $G - \{x, y\}$ is also a perfect elimination bipartite graph.*

Proof. Clearly we need only show that if there is a scheme for the graph G, $\sigma = [x_1y_1, \ldots, x_ky_k]$, then there is a scheme beginning with xy. The proof breaks into three cases.

Case A. $x = x_i$ and $y = y_i$ for some i, $1 \leq i \leq k$

We need only notice that if $e \in E(G)$ is a bisimplicial edge then e is also a bisimplicial edge of any induced subgraph of G in which it is an edge. Thus, there is a scheme $\sigma' = [xy, x_1y_1, \ldots, x_{i-1}y_{i-1}, x_{i+1}y_{i+1}, \ldots, x_ky_k]$.

Case B. $x = x_i$ and $y = y_j$ for some i, j, $1 \leq i, j \leq k$, $i \neq j$

For simplicity we use the notation $G_i = G - \{x_1, \ldots, x_{i-1}, y_1, \ldots, y_{i-1}\}$. Possibly by interchanging the colour classes, we may assume that $i < j$ and hence $[x_iy_j, x_1y_1, \ldots, x_{i-1}y_{i-1}]$ is a partial scheme.

Let us show that if $j - i \geq 2$ and $i < h < j$ then x_hy_h is a bisimplicial edge in $G'_h = G - \{x_1, \ldots, x_{h-1}, y_j, y_1, \ldots, y_{i-1}, y_{i+1}, \ldots, y_{h-1}\}$. Since x_hy_h is a bisimplicial edge of G_h and $V(G'_h)\backslash V(G_h) = \{y_i\}$ it is sufficient to prove that whenever $x_hy_i \in E(G)$, and there exists an $m > h$ with $x_my_h \in E(G)$, then x_my_i is also an edge of G. To see this consider the following implications:

$$x_iy_i \text{ bisimplicial in } G_i \text{ implies that } x_hy_j \in E(G),$$

$$x_hy_h \text{ bisimplicial in } G_h \text{ implies that } x_my_j \in E(G),$$

$$x_iy_j \text{ bisimplicial in } G \text{ implies that } x_my_i \in E(G).$$

Thus x_hy_h is a bisimplicial edge in G'_h for each h, $i < h < j$, and

$$\sigma' = [x_iy_j, x_1y_1, \ldots, x_{i-1}y_{i-1}, x_{i+1}y_{i+1}, \ldots, x_{j-1}y_{j-1}]$$

is also a partial scheme.

Since x_iy_j is bisimplicial in G and $x_iy_i, x_jy_j \in E(G)$ certainly x_jy_i is an edge of G. We shall see that to finish this case it is enough to show that x_jy_i is a bisimplicial edge in G'_j. This is certainly the case, unless for some $s, t > j$ we have x_iy_t and x_sy_i as edges of G. However, then we have that

$x_i y_i$ bisimplicial in G_i implies that $x_s y_j \in E(G)$,

$x_j y_j$ bisimplicial in G_j implies that $x_s y_t \in E(G)$,

and $x_j y_i$ is a bisimplicial edge of G'_j. Thus

$$\sigma' = [x_i y_j, x_1 y_2, \ldots, x_{i-1} y_{i-1}, x_{i+1} y_{i+1}, \ldots,$$
$$x_{j-1} y_{j-1}, x_j y_i, x_{j+1} y_{j+1}, \ldots, x_n y_n]$$

is a scheme as required.

Case C. One of x and y is not among the x_i and y_j

We may assume, without loss of generality, that $x = x_i$ and $y \neq y_j$ for any j. Then clearly $\sigma' = [x_i y, x_1 y_1, \ldots, x_{i-1} y_{i-1}]$ is a partial scheme. We need only show that $x_h y_h$ is a bisimplicial edge of G'_h, for each $h > i$. This will certainly be true, unless there is some $h > m$ with $x_m y_h$ and $x_h y_i$ edges of G. However, then we may proceed as before:

$x_i y_i$ bisimplicial in G_i implies that $x_h y_i \in E(G)$,

$x_h y_h$ bisimplicial in G_h implies that $x_m y_i \in E(G)$.

Thus $\sigma' = [x_i y, x_1 y_1, \ldots, x_{i-1} y_{i-1}, \ldots, x_{i+1} y_{i+1}, \ldots, x_n y_n]$ is a scheme. $\square$

This theorem immediately provides an effective algorithm for recognising a perfect elimination bipartite graph, and producing a corresponding scheme if one exists. We simply have to repeatedly find a bisimplicial edge and delete its vertices from the graph. If, after some number of steps in this process, we succeed in eliminating all the vertices from the graph, then the sequence of edges gives us a scheme. Otherwise no such scheme exists.

Exercises

2.4.1   Prove that every chordal graph is a perfect elimination bipartite graph. (Golumbic, Goss (1978))

2.4.2 ▽ Construct a perfect elimination bipartite graph that is not chordal.

2.4.3   Show that a graph is chordal bipartite if and only if every induced subgraph is a perfect elimination bipartite graph.

Chapter 3

Metric properties

3.1 Radius and diameter

The notion of distance is an invaluable conceptual and practical tool to examine the structure of graphs. This chapter is devoted to some of its implications.

If G is a connected graph then the distance function $d_G(u, v)$ forms a classical *metric*, that is for $u, v, w \in V$

> (1) $d(u, v) \geq 0$ with $d(u, v) = 0$ if and only if $u = v$,
>
> (2) $d(u, v) = d(v, u)$,
>
> (3) $d(u, v) + d(v, w) \geq d(v, w)$.

We shall introduce some new simple parameters, which give a guide as to how concentrated the graph might be. We define the *eccentricity*, $e(v)$, of a vertex $v \in V(G)$ to be $e(v) = \max_{u \in V(G)} d(u, v)$. The maximum eccentricity of a vertex of G is called the *diameter*, $d(G)$ of the graph G, and the minimum eccentricity is called the *radius*, $r(G)$, i.e.

$$d(G) = \max_{v \in V(G)} e(v) = \max_{v \in V(G)} \max_{u \in V(G)} d(u, v),$$

$$r(G) = \min_{v \in V(G)} e(v) = \min_{v \in V(G)} \max_{u \in V(G)} d(u, v).$$

Alternatively, the diameter can be thought of as the maximum distance between any two vertices of G. We say that two vertices, u and v, for which $d(u, v) = d(G)$ are *diametrical*. In similar vein, a vertex $v \in V(G)$ is *central* if $e(v) = r(G)$, and the *centre* of G consists of its central vertices.

Proposition 3.1.1 *For any connected graph G $r(G) \leq d(G) \leq 2r(G)$.*

Proof. The first inequality, of course, follows directly from the definitions. The second is a consequence of the triangle inequality. Choose diametrical vertices u and v, and let w be a central vertex. Then we have

$$d_G(u,v) \leq d_G(u,w) + d_G(w,v) \leq 2r(G). \qquad \square$$

The actual values of $r(G)$ and $d(G)$ can be calculated fairly easily for a particular graph by considering the level representation of G with respect to each vertex in turn. More theoretically, we should like to connect $d(G)$ and $r(G)$ with more familiar graph parameters. To this end we begin with the following simple result.

Theorem 3.1.2 *Let G be a connected, simple, bipartite graph on p vertices, $p \geq 3$, and maximum degree $\Delta \geq 3$. Then*

$$p \leq 2\frac{(\Delta - 1)^{d(G)} - 1}{\Delta - 2}.$$

Moreover, if equality is attained then G is Δ-regular and the minimum length of a cycle is $2d(G)$.

Proof. Choose a vertex x of minimum degree and an edge $xy \in E(G)$ and consider the sets S_i where

$$S_0 = \{x, y\}, S_1 = N(S_0),$$
$$S_{i+1} = N(S_i)\backslash(S_{i-1} \cup S_i) \quad \text{if } 1 \leq i \leq d(G) - 2.$$

Clearly, then, by the definition of the diameter, $V(G) = \bigcup_i S_i$ and, since Δ is the maximum degree,

$$p \leq 2(1 + (\Delta - 1) + (\Delta - 1)^2 + \ldots + (\Delta - 1)^{d(G)-1}) = 2\frac{(\Delta - 1)^{d(G)} - 1}{\Delta - 2}$$

as required. It is also clear that equality can hold if and only if $d_G(x) = \Delta$ and thus G is Δ-regular.

Finally, letting xy be an edge on a cycle of minimum length and repeating the argument above, we see that equality holds if and only if the length of a minimum cycle is $2d(G)$. $\qquad \square$

The bipartite graphs which attain equality in Theorem 3.1.2 are called bipartite Moore graphs. They can exist only when $d = 2$ (the graph $K_{\Delta,\Delta}$) and $d = 3, 4, 6$ (see Biggs (1974) Chapter 5). For $d = 3, 4, 6$ bipartite Moore graphs actually exist when $\Delta - 1$ is a prime power, and are based on the incidence graphs of projective geometries. The Moore graphs for $\Delta = 3$ and $d = 3,4$ are respectively the Heawood graph and the Tutte-Coxeter graph and are shown in figure 3.1.1.

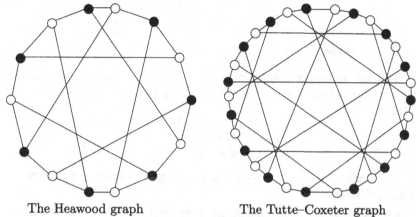

The Heawood graph   The Tutte–Coxeter graph

Figure 3.1.1

The next two theorems, proved by Erdős, Pach, Pollack and Tuza (1989), investigate the relationship between the diameter, radius and minimum degree. The results and proofs are identical for bipartite and triangle-free graphs, and so we shall formulate the theorems in their more general form.

Theorem 3.1.3 *Let G be a connected, triangle-free simple graph on $p \geq 3$ vertices, and minimum degree $\delta \geq 2$. Then $d(G) \leq 4 \left\lceil \dfrac{p - \delta - 1}{2\delta} \right\rceil$.*

Proof. Let x and y be diametrical vertices of G, and consider the sets $N_i = N_i(x)$, for $0 \leq i \leq d$. For convenience we also define $N_{-1} = N_{d+1} = \emptyset$. Then for every i exactly one of two possibilities occurs: either N_i spans no edge of G in which case

$$|N_{i-1}| + |N_{i+1}| \geq \delta, \tag{3.1.1}$$

or we have some pair of vertices v, $v' \in N_i$ with $vv' \in E(G)$ and so, since G is triangle-free, the neighbourhoods of v and v' are disjoint. In this latter case we have

$$|N_{i-1}| + |N_i| + |N_{i+1}| \geq 2\delta. \tag{3.1.2}$$

Notice that equations (3.1.1) and (3.1.2) immediately imply that

$$|N_{i-1}| + |N_i| + |N_{i+1}| + |N_{i+2}| \geq 2\delta \quad \text{for every } 0 \leq i \leq d-1 \tag{3.1.3}$$

simply by considering whether or not one of N_i and N_{i+1} spans an edge.

Now, by summing, we have that $p \geq 2\delta(\lfloor d/4 \rfloor + 1) + \phi(d)$ where

$$\phi(d) = \begin{cases} -\delta + 1 & \text{if } d \equiv 0 \pmod 4, \\ 0 & \text{if } d \equiv 1 \pmod 4, \\ 1 & \text{if } d \equiv 2 \pmod 4, \\ 2 & \text{if } d \equiv 3 \pmod 4, \end{cases}$$

and hence obtain our bound for the diameter. $\square$

In fact the following construction shows that this bound may hold with equality for infinitely many values of p. Let $V(G) = W_0 \cup W_1 \cup \ldots \cup W_{4k}$ with

$$|W_i| = \begin{cases} 1 & \text{if } i \equiv 0 \text{ or } 1 \pmod 4 \text{ and } i \neq 1, \\ \delta & \text{if } i = 1 \text{ or } 4k - 1, \\ \delta - 1 & \text{otherwise,} \end{cases}$$

and let each of W_i and W_{i+1} induce a complete bipartite subgraph of G. Then $d(G) = 4k$.

Before formulating the second theorem, we shall prove a very useful auxiliary lemma.

Lemma 3.1.4 *Let G be a connected graph and x any vertex of G. Then there is a tree $T(x)$ so that $d_G(x, y) = d_{T(x)}(x, y)$ for every vertex $y \in V(G)$.*

Proof. Consider the level representation of G with respect to the vertex x and for each vertex $v \in N_i(x)$ choose a vertex $v' \in N_{i-1}(x)$ with $vv' \in E(G)$, for $i = 1, \ldots, e(x)$. Then the collection of all these edges vv' defines a spanning tree $T(x) \subseteq G$ with the property that $d_{T(x)}(x, y) = d_G(x, y)$ for all $y \in V(G)$. $\square$

Theorem 3.1.5 *Let G be a connected triangle-free simple graph on $p \geq 3$ vertices, and minimum degree $\delta \geq 2$. Then $r(G) \leq \dfrac{p-2}{\delta} + 12$.*

Proof. Let x be a central vertex of G, let $r = r(G)$ and let $T = T(x)$ be the tree constructed in Lemma 3.1.4. Let $T(x, y)$ denote the (x, y)-path in T and put

$$N_{\leq j} = \bigcup_{0 \leq i \leq j} N_j(x) \quad \text{and} \quad N_{\geq j} = \bigcup_{j \leq i \leq r} N_j(x).$$

Now fix a vertex $y' \in N_r(x)$. We say that a vertex $y'' \in V(G)$ is *related* to the vertex y' if there exist vertices $\bar{y}' \in T(x, y') \cap N_{\geq 9}$ and $\bar{y}'' \in T(x, y'') \cap N_{\geq 9}$ so that

$$d_G(\bar{y}', \bar{y}'') \leq 4. \tag{3.1.4}$$

The argument now breaks into two cases.

Case A. There exists a vertex $\bar{y}'' \in N_{\geq r-9}$ which is not related to y'

For any $0 \leq i \leq r$ we denote by S_i' and S_i'' the sets of all elements of $N_i(v)$ whose distance from at least one vertex of $T(x,y') \cap N_{\geq 9}$, or $T(x,y'') \cap N_{\geq 9}$ respectively, is at most 2. Then since y' and y'' are not related we have

$$\left(\bigcup_{i=7}^{r} S_i'\right) \cap \left(\bigcup_{i=7}^{r} S_i''\right) = \emptyset,$$

and by a similar argument to that used to obtain equation (3.1.3) we have

$$|S_{i-1}'| + |S_i'| + |S_{i+1}'| + |S_{i+2}'| \geq 2\delta \quad \text{for } 8 \leq i \leq r-1,$$
$$|S_{i-1}''| + |S_i''| + |S_{i+1}''| + |S_{i+2}''| \geq 2\delta \quad \text{for } 8 \leq i \leq s-1,$$

where $s = d_G(x,y'') \geq r-9$. This yields that

$$p \geq |N_{\leq 6}| + \sum_{i=7}^{r} |S_i'| + \sum_{i=7}^{s+1} |S_i''| \geq \delta(r-12) + 2,$$

and the required inequality for the radius follows immediately.

Case B. Every vertex of $N_{\geq r-9}$ is related to y'

Then let x' denote the vertex in $T(x,y') \cap N_9(x)$. Then, for any $y \in N_{\leq r-10}$

$$d_G(x',y) \leq d_G(x',x) + d_g(x,y) \leq 9 + r - 10 = r - 1.$$

On the other hand, every $y'' \in N_{\geq r-9}$ is related to y', and so by equation (3.1.4) we have

$$\begin{aligned}
d_G(x',y'') &\leq d_G(x',\bar{y}') + d_G(\bar{y}',\bar{y}'') + d_G(\bar{y}'',y'') \\
&\leq (d_G(x,\bar{y}') - 9) + 4 + (r - d_G(x,\bar{y}'')) \\
&\leq r - 5 + d_G(\bar{y}',\bar{y}'') \leq r - 1.
\end{aligned}$$

Thus, $d_G(x',y) \leq r-1$ for every $y \in V(G)$, which contradicts the definition of the radius. □

Exercises

3.1.1 △ Given $n \geq 1$ construct a connected bipartite graph G such that $d(G-e) = d(G) + n$ for some edge $e \in E(G)$.

3.1.2 △ Show that for each connected graph G there is a spanning tree with the same diameter. Is there a spanning tree with the same radius?

3.1.3   A connected graph G is called a *median graph* if G is connected and
          if for any three vertices u, v, w of G there exists a unique vertex
          x, called the median of u, v, w, such that

$$d_G(u, x) + d_G(x, v) = d(u, v),$$
$$d_G(v, x) + d_G(x, w) = d(v, w),$$
$$d_G(u, x) + d_G(x, w) = d(u, w).$$

Prove that a median graph is bipartite, and give some examples of
median graphs.

3.2 Metric properties of trees

Many metric properties for trees have a much more explicit form than
is the case for general graphs. This is in the main due to the role of pendant vertices. For instance the diameter of a tree is the maximum distance
amongst the list of the distances between pendant vertices. Indeed, this
list of distances actually contains all the information about a tree. We formalise this in the following theorem. The result is a combination of results
of Smolenskiĭ (1962) and Zaretskiĭ (1965).

Theorem 3.2.1 *A tree can be uniquely reconstructed (up to isomorphism)
from the list of pairwise distances between its pendant vertices.*

Proof. Let T be a tree with $n \geq 2$ pendant vertices $x_1, \ldots, x_n$, let P_{ij}
be the unique (x_i, x_j)-path between x_i and x_j, and $d_{ij} = d_T(x_i, x_j)$, for
$1 \leq i < j \leq n$.

Property 1 $d_{ij} + d_{jk} - d_{ik} \geq 0$ and $d_{ij} + d_{jk} - d_{ik} \equiv 0 \pmod 2$, *for any
three indices i, j and k.*

Proof of Property 1. Let v be the nearest vertex to x_k which is also a
vertex of the path P_{ij}. Clearly, since T is a tree, this vertex is uniquely
defined. Then $d_{ij} + d_{jk} - d_{ik} = 2d_T(x_j, v)$, which proves the property.   □

Property 2 *For any four indices i, j, k, l two of the numbers $d_{ij} + d_{ik}$,
$d_{ik} + d_{jl}$ and $d_{il} + d_{jk}$ are equal and the third is not greater than these two.*

Proof of Property 2. Consider the two paths P_{ij} and P_{kl}. They may
have no, one or at least two vertices in common. The proof now breaks into
three cases.

Case A. The paths do not intersect

Let $u \in P_{ij}$ and $v \in P_{kl}$ be the pair of vertices whose (u, v)-path is of
minimum length. Let Q be this shortest path, R_i be the (x_i, u)-path, and

R_j, R_k and R_l be defined similarly. Then we have the following relationships between these paths.

$$P_{ik} = R_i \cup Q \cup R_k, \quad P_{jl} = R_j \cup Q \cup R_l,$$
$$P_{jk} = R_j \cup Q \cup R_k, \quad P_{il} = R_i \cup Q \cup R_l.$$

So $d_{ik} + d_{jl} = d_{il} + d_{jk} = d_{ij} + d_{kl} + 2d_T(u,v)$ which proves the property for this case.

Case B. The paths have one vertex in common

Let v be this vertex. Let R_i be the (v, x_i)-path, and let R_j, R_k and R_l be defined similarly. Then by the same method as we used in case A we have that $d_{ik} + d_{jl} = d_{ij} + d_{kl} = d_{il} + d_{jk}$.

Case C. The paths have two or more vertices in common

Then either P_{ik} and P_{jl} or P_{il} and P_{jk} do not meet and we have case A by exchanging indices.   □

Property 3 *Every tree S with pendant vertices $x_1, \ldots, x_n$ and pairwise distances $d_S(x_i, x_j) = d_{ij}$, for $1 \le i < j \le n$, is isomorphic to T.*

Proof of Property 3. We shall proceed by induction on n. The assertion is clear when $n = 2$. Let $n \ge 2$ be a value for which the assertion holds and consider trees S and T with $n + 1$ pendant vertices.

Let v_S be the closest vertex in S to x_{n+1} which has degree at least 3, and define v_T analogously. Then let S' and T' be the trees obtained by deleting the (v_S, x_{n+1})-path and (v_T, x_{n+1})-path except the vertices v_S and v_T from S and T respectively. Clearly now S' and T' are isomorphic, by induction. Let $\phi : S' \longrightarrow T'$ be an isomorphism.

We shall show that $\phi(v_S) = v_T$. Consider the unique $(\phi(v_S), v_T)$-path in T and extend it to a path joining x_i to x_j in T, for some $1 \le i < j \le n$. Then

$$d_T(x_i, x_j) + d_T(x_i, x_{n+1}) - d_T(x_j, x_{n+1}) = 2d_T(x_i, v_T),$$

while

$$d_S(x_i, x_j) + d_S(x_i, x_{n+1}) - d_S(x_j, x_{n+1}) = 2d_S(x_i, v_S).$$

So by assumption $d_T(x_i, v_T) = d_S(x_i, v_S)$ and $v_T = \phi(v_S)$, as required. Similarly since

$$d_T(x_{n+1}, x_i) + d_T(x_{n+1}, x_j) - d_T(x_i, x_j) = 2d_T(x_{n+1}, v_T),$$

while

$$d_S(x_{n+1}, x_i) + d_S(x_{n+1}, x_j) - d_S(x_i, x_j) = 2d_S(x_{n+1}, v_S),$$

we have $d_T(x_{n+1}, \phi(v_S)) = d_S(x_{n+1}, v_S)$ and so we may easily extend ϕ into an isomorphism of S and T.   □

To complete the proof of Theorem 3.2.1 we must construct a tree S with pendant vertices $x_1, \ldots, x_n$ and distances $d_S(x_i, x_j) = d_{ij}$, for $1 \le i < j \le n$. We shall construct S inductively.

Suppose that a tree T' with pendant vertices $x_1, \ldots, x_{n-1}$ has been constructed so that $d_{T'}(x_i, x_j) = d_{ij}$ for each $1 \le i < j \le n - 1$. Let x_i and x_j be chosen so that $d_{ni} + d_{nj} - d_{ij}$ is minimal. Let v be the vertex in the (x_i, x_j)-path in T' for which $d_{T'}(v, x_i) = (d_{ni} + d_{nj} - d_{ij})/2$ – this is certainly possible by Property 1 – and let S be the tree T' with a path of length $d_{T'}(v, x_i)$ joining x_n to v. We claim that S is the tree we require. We need only show that $d_S(x_k, x_n) = d_{kn}$ for every $1 \le k \le n - 1$. Indeed, from the above construction, we already know this to be true for $k = i, j$.

To see the identity for the remaining values consider Property 2 which implies that two of $d_{ij} + d_{nk}$, $d_{ik} + d_{jn}$ and $d_{in} + d_{jk}$ are equal and not less than the third. We claim that $d_{ij} + d_{nk}$ is not less than the other two. For suppose that $d_{ij} + d_{nk} < d_{in} + d_{jk}$; then $d_{nk} + d_{nj} - d_{jk} < d_{ni} + d_{nj} - d_{ij}$ which contradicts the choice of i and j. So, for suitable choice of the indices i and j, we have

$$d_{ij} + d_{nk} = d_{in} + d_{jk} \ge d_{jn} + d_{ik}. \tag{3.2.1}$$

Now S itself satisfies Property 2 and in fact the (x_n, x_k)-path and (x_i, x_j)-path intersect in S. Hence by the proof of Property 2 we must have

$$d_S(x_i, x_j) + d_S(x_k, x_n) = d_S(x_{i_0}, x_n) + d_S(x_{j_0}, x_k)$$
$$\ge d_S(x_{j_0}, x_n) + d_S(x_{i_0}, x_k) \tag{3.2.2}$$

where $\{i_0, j_0\} = \{i, j\}$. Now since we have by (3.2.1) and (3.2.2) that $d_{ij} = d_S(x_i, x_j)$, $d_{in} = d_S(x_i, x_n)$ and $d_{jn} = d_S(x_j, x_n)$, we must also have $i = i_0$, $j = j_0$ and $d_{kn} = d_S(x_k, x_n)$. □

In fact Yushmanov (1984) proved that to reconstruct a tree with n pendant vertices we do not even need to know all the pairwise distances between the pendant vertices. It is sufficient to know $2n - 3$ specially chosen distances from this list.

A characterisation of the centre of a tree also follows from considering pendant vertices. The following theorem was proved originally by Jordan (1869) and independently by Sylvester (1873).

Proposition 3.2.2 *Every tree has a centre consisting of either one vertex, or two adjacent vertices.*

Proof. Certainly the result is obvious for the trees K_1 and K_2. We shall prove the theorem by demonstrating that a tree T has the same centre as the tree T' obtained from T by deleting its pendant vertices.

For any vertex $v \in V(T)$ the eccentricity $e(v)$ is attained by some pendant vertex of T. Thus the eccentricity of each vertex of T' will be one less than that of the same vertex in T. In particular, those vertices with minimum eccentricity in T will still be the vertices with minimum eccentricity in T'. Hence T and T' have the same centre. The result follows by induction. $\square$

As a corollary of this theorem we may also establish a connection between the radius and diameter of a tree.

Corollary 3.2.3   *For any tree T we have $r(T) = \left\lceil \dfrac{d(T)}{2} \right\rceil$.*

Proof. The proof breaks into two cases.

Case A. The centre of T consists of one vertex, v_0

Consider the level representation of T with respect to v_0. For convenience we set $r = r(T)$. Then from the definition of the radius it follows that there must be two vertices $a, b \in N_r(v_0)$ so that the (v_0, a)-path and the (v_0, b)-path intersect only at v_0. Then $d_T(a, b) = 2r(G)$. Recall that by Proposition 3.1.1 $d(T) \le 2r(T)$ and hence we must have equality.

Case B. The centre of T consists of two adjacent vertices, v_1, v_2

Consider the new tree T' obtained by adding a new vertex v_0 to T subdividing the edge $v_1 v_2$. Let $V(T') = V(T) \cup \{v_0\}$ and $E(T') = E(T) \cup \{v_0 v_1, v_0 v_2\}$. Clearly then the centre of T' is the single vertex v_0 and $r(T') = 2d(T')$ by the previous case. Furthermore, $r(T') = r(T)$ and $d(T') = d(T) + 1$. Thus $d(T)$ is odd, and $2r(T) = d(T) + 1$. $\square$

Indeed we have proved the following corollary.

Corollary 3.2.4   *The diameter of a tree T is even if the centre consists of one vertex, and odd if it consists of two.* $\square$

Exercises

3.2.1   Show that a graph G is a tree if and only if it is connected, contains no triangles, and satisfies the condition that for any four vertices x, y, z, t,

$$d(x, y) + d(z, t) \le \max\{d(x, z) + d(y, t), d(x, t) + d(y, z)\}.$$

(Buneman (1974))

3.2.2   Prove that a tree on p vertices with diameter at least $2k - 3$ has at least $p - k$ paths of length k.

3.2.3 △ Let $t(p, d)$ be the number of trees having vertices $v_1, \ldots, v_p$ and diameter d. Show that $t(p, p - 1) = p!/2$, $t(p, p - 2) = (p - 4)p!/2$, and $t(p, 2) = p$.

3.2.4   Show that the following inequalities hold for a tree T on p vertices:

(a) $p - 2 \le f_2(T) \le \binom{p-1}{2}$,

(b) $p - 3 \le f_3(T) \le \lceil \frac{p-2}{2} \rceil \lfloor \frac{p-2}{2} \rfloor$,

where $f_i(G)$ is the number of paths of length i in G.

3.2.5   Let the trees T_1 and T_2 have the same vertex set V, and the graphs $T_1 - v$ and $T_2 - v$ be isomorphic for every $v \in V$. Show that T_1 and T_2 have the same diameter.

3.2.6   Let T be a tree with radius r, and let $M_r(x) = \{y \in V(T) : d_T(x, y) \le r\}$ for every vertex x. Prove that the set $\bigcap\limits_{x \in V(T)} M_r(x)$ is the centre of T.

3.2.7   Let T be a tree and $V(T) = X_1 \cup X_2$ where $X_1 \cap X_2 = \emptyset$ and $||X_1| - |X_2|| \le 1$. Prove the following inequality:

$$\sum_{i=1}^{2} \sum_{u, v \in X_i} d_T(u, v) \le \sum_{u \in X_1, v \in X_2} d_T(u, v).$$

3.2.8   For every tree T let $\phi(T) = \sum_{u, v \in V(T)} d_T(u, v)$. Show that the function $\phi(T)$ achieves its minimum value among all trees on p vertices if and only if $T = K_{1, p-1}$ and achieves its maximum value if and only if T is a path.

3.2.9   Show that in a tree T the function $\phi(x) = \sum\limits_{y \in V(T)} d_T(x, y)$ achieves its minimum value at one vertex or at two adjacent vertices of T.

3.3 Metric properties of the n-cube

One of the most important classes of bipartite graphs is the class of *hypercubes*. Let B^n denote the set of all ordered n-tuples of 0's and 1's. For any pair $\boldsymbol{\alpha} = (\alpha_1, \ldots, \alpha_n)$ and $\boldsymbol{\beta} = (\beta_1, \ldots, \beta_n)$ from B^n the number
$$\rho(\boldsymbol{\alpha}, \boldsymbol{\beta}) = \sum_{i=1}^{n} |\alpha_i - \beta_i|$$
is called the *Hamming distance* between $\boldsymbol{\alpha}$ and $\boldsymbol{\beta}$. The hypercube of dimension n (or the n-cube for short), denoted by Q_n, is the graph with vertex set B^n where vertices $\boldsymbol{\alpha}$ and $\boldsymbol{\beta}$ are adjacent if and only if $\rho(\boldsymbol{\alpha}, \boldsymbol{\beta}) = 1$. The graphs Q_1, Q_2 and Q_3 are shown in figure 3.3.1. Usually the n-tuples $(0, \ldots, 0)$ and $(1, \ldots, 1)$ are respectively denoted by $\mathbf{0}$ and $\mathbf{1}$. If

we denote by V_1 the set of all n-tuples with an even number of 1's, and by V_2 the set with an odd number, it is clear that Q_n is bipartite with bipartition (V_1, V_2). Notice that $Q_{n+m} = Q_n \times Q_m$ for any positive integers n and m; in particular this means that $Q_{n+1} = Q_n \times K_2$. We begin by considering the simplest properties of Q_n.

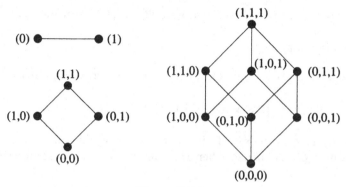

Figure 3.3.1

Property 3.3.1 *Q_n is an n-regular simple graph with 2^n vertices and $n2^n$ edges.* □

Property 3.3.2 *Every pair of vertices at distance 2 have precisely two common neighbours.* □

Property 3.3.3 *For any pair of vertices $\alpha, \beta \in V(Q_n)$ the length of the shortest path connecting α and β is the Hamming distance $\rho(\alpha, \beta)$ and the number of shortest (α, β)-paths is $\rho(\alpha, \beta)!$.*

Proof. Let $\alpha_1 \alpha_2 \ldots \alpha_l$ be a shortest path connecting α and β. Since, for each $1 \leq i \leq l - 1$, α_i and α_{i+1} differ in exactly one coordinate, certainly $l \geq \rho(\alpha, \beta)$. On the other hand, it is easy to construct a path between α and β of length $\rho(\alpha, \beta)$ simply by consecutively changing those coordinates in which α and β differ. Since these changes can be carried out in $\rho(\alpha, \beta)!$ different ways, the number of shortest (α, β)-paths is as required. □

Property 3.3.4 *$d(Q_n) = n$ and $r(Q_n) = n$.*

Proof. Clearly $\rho(\alpha, \beta) \leq n$ for every pair of vertices $\alpha, \beta \in B^n$. On the other hand, for each $\alpha \in B^n$ there is a vertex $\bar{\alpha} = (1 - \alpha_1, \ldots, 1 - \alpha_n)$ with $\rho(\bar{\alpha}, \alpha) = n$. Now the property follows from Property 3.3.3 . □

The n-cube has an abundance of symmetry, as can be demonstrated by the following property.

Property 3.3.5 *For any vertex* $\boldsymbol{\alpha} \in V(Q_n)$ *there is an isomorphism* φ : $B^n \longrightarrow B^n$ *such that* $\varphi(\boldsymbol{\alpha}) = \mathbf{0}$. *More plainly, we may consider any vertex of* Q_n *as the* $\mathbf{0}$.

Proof. Let $\boldsymbol{\alpha} = (\alpha_1, \ldots, \alpha_n)$. Then it is easy to check that

$$\varphi\left((\beta_1, \ldots, \beta_n)\right) = ((\beta_1 + \alpha_1)(\text{mod}2), \ldots, (\beta_n + \alpha_n)(\text{mod}2))$$

is the required mapping. $\square$

Let G be a connected graph and for each pair of vertices $u, v \in V(G)$ let

$$I_G(u, v) = \{x \in V(G) : d_G(u, x) + d_G(x, v) = d_G(u, v)\}.$$

Actually $I_G(u, v)$ is the set of all vertices each of which lies on a shortest (u, v)-path in G. As usual, when the context is clear we shall write only $I(u, v)$.

Property 3.3.6 *For any vertex* $\boldsymbol{\alpha}$ *of* Q_n

$$\left|I(\mathbf{0}, \boldsymbol{\alpha}) \cap N(\boldsymbol{\alpha})\right| = d_{Q_n}(\mathbf{0}, \boldsymbol{\alpha}) \quad \text{and} \quad \left|I(\mathbf{1}, \boldsymbol{\alpha}) \cap N(\boldsymbol{\alpha})\right| = d_{Q_n}(\mathbf{1}, \boldsymbol{\alpha}).$$

Proof. By Property 3.3.3 we have that $d_{Q_n}(\mathbf{0}, \boldsymbol{\alpha}) = \rho(\mathbf{0}, \boldsymbol{\alpha})$ and $d_{Q_n}(\mathbf{1}, \boldsymbol{\alpha}) = \rho(\mathbf{1}, \boldsymbol{\alpha})$. Thus $d_{Q_n}(\mathbf{0}, \boldsymbol{\alpha})$ actually equals the number of 1's in $\boldsymbol{\alpha}$ and $d_{Q_n}(\mathbf{0}, \boldsymbol{\alpha})$ is the number of 0's. By definition, we can construct a shortest path from $\boldsymbol{\alpha}$ to $\mathbf{0}$ passing through any member of $I(\mathbf{0}, \boldsymbol{\alpha}) \cap N(\boldsymbol{\alpha})$, and since by Property 3.3.3 we must move from $\boldsymbol{\alpha}$ to a vertex with one fewer 1's $|I(\mathbf{0}, \boldsymbol{\alpha}) \cap N(\boldsymbol{\alpha}))| = d_{Q_n}(\mathbf{0}, \boldsymbol{\alpha})$ as required. The other identity follows in an entirely analogous fashion. $\square$

The simple metric properties of the n-cube are now evident. What perhaps is less evident is that these basic properties are enough to characterise the n-cube. In preparation to see this we shall prove the following two results.

Proposition 3.3.7 *A graph* G *is regular if for each pair of vertices* $u, v \in V(G)$ *with* $d_G(u, v) = 2$ *we have* $|N(u) \cap N(v)| = 2$.

Proof. Let uv be an edge of G and let us colour $N(u) \backslash \{v\}$ red and colour $N(v) \backslash \{u\}$ blue, colouring common neighbours with both colours. Now choose an arbitrary blue vertex w. Then w and v have u as a common neighbour, so w is adjacent to precisely one red vertex. Similarly every red vertex is adjacent to precisely one blue vertex, and so there are as many red vertices as blue vertices. Thus $d_G(u) = d_G(v)$ and since G is connected G is regular. $\square$

Proposition 3.3.8 *Let G be a graph where $|N(x) \cap N(y)| = 2$ for every pair of vertices with $d_G(x, y) = 2$. Then for each pair of vertices $u, v \in V(G)$ with $d_G(u, v) \geq 2$*

$$\big|I(u, v) \cap N(u)\big| \geq d_G(u, v).$$

Proof. The proof is by induction on $d_G(u, v)$. Firstly, if $d_G(u, v) = 2$ then $I(u, v) \cap N(u) = N(v) \cap N(u)$ so $|I(u, v) \cap N(u)| = 2$ by assumption. Now let $d_G(u, v) \geq 3$ and let w be a neighbour of u in $I(u, v)$. Then $d_G(w, v) = d_G(u, v) - 1$ and so by induction, $\big|I(w, v) \cap N(w)\big| \geq d_G(u, v) - 1$. To complete the proof we now use a colouring argument similar to that used in the previous proof. Colour the vertices of $I(w, v) \cap N(w)$ red and colour the vertices of $(I(u, v) \cap N(u)) \backslash \{w\}$ blue. Any red vertex and u have w as a common neighbour, and so every red vertex is adjacent to exactly one blue vertex. Any blue vertex and w have u as a common neighbour, so blue vertices are joined to at most one red vertex. Hence there are at least as many blue vertices as red vertices and

$$\big|I(u, v) \cap N(u)\big| \geq 1 + \big|I(w, v) \cap N(w)\big| \geq d_G(u, v). \qquad \square$$

Now we are in a position to establish our first metric characterisation of the n-cube. This is a result of Mulder (1979) which is a slight strengthening of a result of Foldes (1977).

Proposition 3.3.9 *A connected, bipartite graph G is isomorphic to Q_n if and only if it satisfies the following conditions:*

(1) $d(G) = n$,

(2) $|N(x) \cap N(y)| = 2$ for each pair of vertices $x, y \in V(G)$ with $d_G(x, y) = 2$,

(3) if x_0 and x_1 are diametrical vertices of G then for every vertex $u \in V(G)$

$$|I(x_0, u) \cap N(u)| = d_G(x_0, u) \text{ and } |I(x_1, u) \cap N(u)| = d_G(x_1, u).$$

Proof. The necessity of the conditions follows immediately from Properties 3.3.4, 3.3.2 and 3.3.6. We shall instead turn to the sufficiency.

Suppose that G satisfies conditions *(1)-(3)*. Then it follows from Proposition 3.3.7 that G is regular. To find the degree we need only observe that since G is bipartite $N(x_0) \subset I(x_0, x_1)$. Thus we have

$$|N(x_0)| = |I(x_0, x_1) \cap N(x_0)| = d(G) = n$$

and G is n-regular. Furthermore, for any vertex $u \in V(G)$

$$n \geq |I(x_0, u) \cap N(u)| + |I(u, x_1) \cap N(u)|$$
$$= d_G(x_0, u) + d_G(x_1, u) \geq d(G) = n.$$

Thus $V(G) = I(x_0, x_1)$. Indeed, we can use property *(3)* again to let us count the number of vertices. By observing that $N_i(x_0) = N_{n-i}(x_1)$ and counting the edges between $N_i(x_0)$ and $N_{i-1}(x_0)$ we have that

$$i|N_i(x_0)| = (n - i + 1)|N_{i-1}(x_0)| \qquad \text{for } 1 \leq i \leq n. \qquad (3.3.1)$$

Now since $|N_0(x_0)| = 1$ we have $|N_i(x_0)| = \binom{n}{i}$, by using (3.3.1) and induction. To complete the proof we now construct an isomorphism of G and the n-cube level by level. By induction we shall show that the map

$$\phi : u \mapsto I(x_0, u) \cap N(x_0) \qquad \text{for } u \in V(G)$$

is the map we need.

Let G_j be the subgraph of G induced by the vertices $\bigcup_{i=0}^{j} N_i(x_0)$. It is clear that G_1 is isomorphic to $\phi(G_1)$, so let $j \geq 1$ be a value of j for which the assertion holds. Then by induction the vertices of $N_j(x_0)$ are represented by subsets of j elements of $N(x_0)$ in such a way that if A and B represent two vertices of $N_j(x_0)$ they have a common neighbour in $N_{j-1}(x_0)$ if and only if $|A \cap B| = j - 1$. If it exists, their common neighbour must be represented by $A \cap B$, and their other common neighbour must lie in $N_{j+1}(x_0)$.

By counting, it is easy to see that the number of pairs of vertices in $N_j(x_0)$ with a common neighbour in $N_{j-1}(x_0)$ is equal to the number of pairs of vertices in $N_j(x_0)$ with a common neighbour in $N_{j+1}(x_0)$. Thus, since the common neighbour in $N_{j-1}(x_0)$ is unique, any pair of vertices in $N_j(x_0)$ with a common neighbour in $N_{j+1}(x_0)$ must have only one such neighbour. Let w be a vertex in $N_{j+1}(x_0)$ and let $A_1, \ldots, A_{j+1}$ be the neighbours of w in $N_j(x_0)$. Then, for every $1 \leq s < t \leq j + 1$, A_s and A_t must also have a common neighbour in $N_{j-1}(x_0)$ and so $|A_s \cap A_t| = j - 1$. Thus

$$A = N(x_0) \cap I(x_0, w) = \bigcup_{i=1}^{j+1} A_i$$

is indeed a set of $j + 1$ elements which behaves in the required manner.

To finish the proof we need only notice that distinct vertices in $N_{j+1}(x_0)$ receive distinct sets of $j + 1$ elements under this representation and thus every set of $j + 1$ elements is assigned to some vertex. The result follows by induction. □

Proposition 3.3.9 provides a tool to prove two more pleasing characterisations of the n-cube.

Theorem 3.3.10 (Mulder (1980)) *A connected, n-regular graph G is isomorphic to the n-cube if and only if it has 2^n vertices and every pair of vertices at distance 2 have precisely two common neighbours.*

Proof. Once again the conditions of the theorem are necessary, and so suppose that G is a graph satisfying these conditions. Take a vertex $u \in$

$V(G)$. Clearly, each vertex in $N_2(u)$ must have precisely two neighbours in $N_1(u)$ (its common neighbours with u), and each vertex in $N_1(u)$ has at most $n-1$ neighbours in $N_2(u)$. Thus by counting the edges between $N_1(u)$ and $N_2(u)$ we have

$$2|N_2(u)| \leq |N_1(u)|(n-1),$$

that is

$$|N_2(u)| \leq \frac{n(n-1)}{2}. \tag{3.3.2}$$

By applying Proposition 3.3.8 we may count the edges between $N_i(u)$ and $N_{i-1}(u)$ and have

$$n|N_{i-1}(u)| - (i-1)|N_{i-1}(u)| \geq i|N_i(u)|,$$
$$(n-i+1))|N_{i-1}(u)| \geq i|N_i(u)|. \tag{3.3.3}$$

Thus, just as we did in the proof of Proposition 3.3.9, we may apply induction using (3.3.2) and (3.3.3) to get $|N_i(u)| \leq \binom{n}{i}$. However, recall that $|V(G)| = 2^n$. Thus we must have that $|N_i(u)| = \binom{n}{i}$ and equality in (3.3.3). Thus each vertex in $N_i(u)$ must have precisely i neighbours in $N_{i-1}(u)$ and $n-i$ neighbours in $N_{i+1}(u)$, for $1 \leq i \leq n$. In particular, there is a vertex v so that $N_n(u) = \{v\}$, and furthermore $N_i(u) = N_{n-i}(v)$. So G satisfies the conditions of Proposition 3.3.9 and G is isomorphic to the n-cube.   □

Theorem 3.3.11 (Foldes (1977)) *A connected graph G is isomorphic to the n-cube if and only if it is bipartite, has diameter n, and for any two vertices $u, v \in V(G)$ the number of shortest (u,v)-paths is $d_G(u,v)!$.*

Proof. If G is isomorphic to the n-cube then by Property 3.3.3 the condition of the theorem is satisfied. Conversely let G be a graph satisfying the conditions of the theorem. Then $d_G(x,y) = 2$ implies $|N(x) \cap N(y)| = 2$, since the number of shortest (x,y)-paths must be 2 by assumption. Now consider two diametrical vertices x_0 and x_1 and let u be a vertex of G. On each shortest (x_0,u)-path there is exactly one vertex of $N(u)$. On the other hand there are only $(d_G(x_0,u) - 1)!$ shortest paths from each neighbour of u which lies on a shortest (x_0,u)-path to x_0. Thus $|I(x_0,u) \cap N(u)| = d_G(x_0,u)$. Similarly $|I(x_1,u) \cap N(u)| = d_G(x_1,u)$. So we have satisfied the conditions of Proposition 3.3.9 and so G is isomorphic to the n-cube.   □

Many other metric characterisations of the n-cube can be found in the monograph of Mulder (1980).

To conclude this section we shall say a little about another interpretation of an n-cube. We can define Q_n to be the graph with the collection of subsets of $\{1, 2, \ldots, n\}$ as its vertex set, in which two sets, A and B, are adjacent if and only if $|A \triangle B| = 1$ (where as usual $\triangle$ is the symmetric difference). In other words Q_n is a graphic representation of a boolean lattice.

Now, using the above results, we can recognise a non-oriented Hasse diagram as being that of a boolean lattice. Actually, this recognition is possible not only for boolean lattices, but also for finite modular lattices.

Theorem 3.3.12 (Alvarez (1965)) *A connected simple graph G is a non-oriented Hasse diagram of a finite modular lattice if and only if G is bipartite and satisfies the following conditions:*

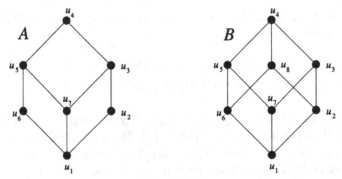

Figure 3.3.2

(1) G contains two diametrical vertices x_0 and x_1 so that whenever $u, v \in N_i(x_j)$ have a common neighbour in $N_{i-1}(x_j)$ u and v also have a unique common neighbour in $N_{i+1}(x_j)$ for $1 \leq i \leq d(G)$ and $j \in \{0, 1\}$.

(2) If the graph A in figure 3.3.2 is a subgraph of G, then there also exists a vertex u_8 such that B is also a subgraph. □

Exercises

3.3.1 △ Find the centre of the n-cube.

3.3.2 △ Show that Q_4 is the product of two cycles. Is it true that Q_{2n} is a product of cycles, for any $n \geq 2$?

3.3.3   Let $\boldsymbol{\alpha} = (\alpha_1, \ldots, \alpha_n)$ and $\boldsymbol{\beta} = (\beta_1, \ldots, \beta_n)$ be vertices of Q_n with $\rho(\boldsymbol{\alpha}, \boldsymbol{\beta}) = k \leq n$, and let $\alpha_i \geq \beta_i$ for $i = 1, \ldots, n$. Show that the subgraph induced by the $I(\boldsymbol{\alpha}, \boldsymbol{\beta})$ is isomorphic to Q_k.

3.3.4 ▽ Prove that a graph G is isomorphic to the n-cube if and only if G is an n-regular median graph. (Mulder (1980))

3.3.5   Prove the following:

(a)   The minimum number of edges whose removal increases the diameter of the n-cube is $n - 1$.

(b)   The minimum number of new edges whose addition to the n-cube decreases the diameter is 2.

(c) ▽ The maximum number of edges that may be added to Q_n without decreasing the diameter is

$$\binom{2n}{n-1} + \frac{1}{2}\binom{2n}{n} - (n+1)2^{n-1}. \text{ (N. Graham, Harary (1992))}$$

3.3.6 ☐ Prove that the n-cube Q_n contains a path $v_1 v_2 \ldots v_m$ such that $d_{Q_n}(v_i, v_j) > 1$ whenever $|i - j| > 1$ and $m > c \cdot 2^n$ where c is a constant. (Evdokimov (1969))

3.3.7 ☐ Prove that the n-cube has a spanning subgraph H with $|E(H)| < 7 \cdot 2^n$ such that $d_H(u, v) \leq 3 \cdot d_{Q_n}(u, v)$ for each pair of vertices $u, v \in V(Q_n)$. (Peleg, Ullman (1989))

3.3.8 △ Show that every bipartite graph is a non-oriented Hasse diagram of a partially ordered set.

Application

3.4 Addressing schemes for computer networks

In any modern day computer, data is perpetually moving throughout the system. To route this data efficiently is an important task.

A natural course to pursue is to represent a computer network as a graph and provide each vertex of this graph with an address. Perhaps less obvious is that these addresses could be made to reflect the distances between the vertices in the graph. If this were the case then a parcel of data could be optimally routed around a network simply by examining local information at each step. When a parcel reaches a vertex, it need only compare the addresses of its neighbours with the address of its destination and move to the neighbour closest to its target. We have already met a graph which naturally has this addressing property. In the n-cube the distances between any two vertices is simply the Hamming distance between the corresponding binary vectors (see Property 3.3.3). Our most obvious course, then, is to generalise this – to label the vertices of a graph with binary strings whose Hamming distances agree with the distances in the graph. Below we give a theorem of Djoković (1973) which exactly says when this is possible. Firstly, however, we need to introduce some terminology.

Let G be a simple graph. We call a subset $U \subseteq V(G)$ *closed* if whenever $a, b \in U$, $x \in V(G)$ and $d_G(a, x) + d_G(b, x) = d_G(a, b)$ then x is also a member of U. For $ab \in E(G)$ we also define

$$G(a, b) = \{x \in V(G) : d_G(x, a) = d_G(x, b) - 1\}.$$

Then $G(a, b)$ and $G(b, a)$ are disjoint and non-empty subsets of $V(G)$. Moreover, if G is a bipartite graph then $G(a, b)$ and $G(b, a)$ form a partition of the vertex set $V(G)$. We are now in a position to give the promised theorem.

Theorem 3.4.1 (Djoković (1973)) *There is an addressing scheme consisting of binary strings for the vertices of a simple graph G so that Hamming distances between the strings agree with the distances in the graph if and only if*

(1) G is a connected bipartite graph, and

(2) $G(a, b)$ is a closed set for every edge $ab \in E(G)$.

Proof. If such a labelling exists then G is a subgraph of an n-cube, and furthermore can be embedded into this cube in such a way that the distances between vertices are preserved when G is embedded into the cube. It is then easy to see that G satisfies properties *(1)* and *(2)*.

Conversely, let G be a graph satisfying conditions *(1)* and *(2)*, and let $v_0 \in V(G)$ be a fixed vertex. We define a binary relation θ on the set of edges $E(G)$ as follows. Let $e_1 = ab$ and $e_2 = uv$ be edges of G. We say $e_1 \theta e_2$ if and only if e_2 joins a vertex of $G(a, b)$ to a vertex of $G(b, a)$. We shall show that θ is actually an equivalence relation.

It is clear that θ is reflexive. Suppose now that $e_1 \theta e_2$ with $u \in G(a, b)$ and $v \in G(b, a)$. We claim that $G(u, v) = G(a, b)$. By symmetry it suffices to show that $G(a, b) \subset G(u, v)$. Let $x \in G(a, b)$ and suppose that $x \in G(v, u)$ i.e. $d_G(x, v) = d_G(x, u) - 1$. Then since u and x are in $G(a, b)$ and $d_G(x, v) + d_G(u, v) = d_G(x, u) - 1 + 1 = d_G(x, u)$, v must be in $G(a, b)$ since $G(a, b)$ is closed, which contradicts property *(1)*. Hence $G(a, b) = G(u, v)$, and it follows easily that θ is both symmetric and transitive.

We can now use the set of equivalence classes of θ to define an addressing scheme for G. Let $S = \{s_1, s_2, \ldots, s_n\}$ be this set of equivalence classes and for each edge e let $\bar{e}$ be the equivalence class containing e. For each $x \in V(G)$ we define a subset of $F(x) \subset S$ as follows: if $e = ab$ then $\bar{e} \in F(x)$ if and only if x and the fixed vertex v_0 do not both belong to $G(a, b)$ or both to $G(b, a)$. From $F(x)$ we define an address $\mathbf{x}$ for the vertex x:

$$\mathbf{x} = (a_1, a_2, \ldots, a_n) \quad \text{where} \quad a_i = \begin{cases} 1 & \text{if } s_i \in F(x), \\ 0 & \text{if } s_i \notin F(x). \end{cases}$$

The Hamming distance between addresses $\mathbf{x}$ and $\mathbf{y}$ is then the cardinality of the symmetric difference between the two subsets $F(x)$ and $F(y)$.

Let e be an edge joining a and b, then we claim that $F(a) \triangle F(b) = \{\bar{e}\}$. Certainly $\bar{e} \in F(a) \triangle F(b)$, thus suppose that some $\bar{e}_1$ is also a member. If $e_1 = uv$ then this means that a and b belong to different sets $G(u,v)$ and $G(v,u)$; thus $e_1 \theta e$ and $\bar{e}_1 = \bar{e}$. It is now only a small task to show that the addressing scheme associated with these $F(x)$'s agrees with the distances in G. Let a and b be vertices of G with $d_G(a,b) = m$, and let $ae_1x_2 \ldots x_m e_m b$ be a shortest path joining them. Then

$$F(a) \triangle F(b) = (F(a) \triangle F(x_2)) \triangle (F(x_2) \triangle F(x_3)) \triangle \ldots \triangle (F(x_m) \triangle F(b))$$
$$= \{\bar{e}_1, \bar{e}_2, \ldots, \bar{e}_m\}.$$

We need only show that the edges $e_1, \ldots e_m$ are pairwise non-equivalent. Indeed, suppose that $e_i \theta e_j$ for some $i < j$. Then by definition $d_G(x_{j+1}, x_i) = d_G(x_{j+1}, x_{i+1}) - 1$ which contradicts the fact that $d_G(a,b) = m$. Thus $\rho(\mathbf{a}, \mathbf{b}) = |F(a) \triangle F(b)| = m = d_G(a,b)$ as required. $\square$

The proof of this theorem actually contains a polynomial algorithm to construct binary addressing schemes, when this is possible. These schemes consist of binary strings with length equal to the number of equivalence classes. It is not difficult to check that every tree T satisfies the conditions of the theorem and the number of equivalence classes under θ is the number of edges in T, $|V(T)| - 1$.

So binary addressing schemes exist only for certain classes of bipartite graphs. However, we would like to have some form of addressing scheme for any graph. R.L. Graham and Pollack (1972) showed that this can be done by using strings of three symbols rather than two. Let $\mathbf{a} = (a_1, a_2, \ldots, a_m)$ and $\mathbf{b} = (b_1, b_2, \ldots, b_m)$ be m-tuples of 0's, 1's and $*$'s, then we define the Hamming distance between $\mathbf{a}$ and $\mathbf{b}$, $\rho(\mathbf{a}, \mathbf{b})$ to be

$$\rho(\mathbf{a}, \mathbf{b}) = \left| \{k : \{a_k, b_k\} = \{0,1\}\} \right|;$$

in other words the Hamming distance is as before, but we ignore any coordinates that contain a $*$. Graham and Pollack showed that the minimum length of strings of three symbols in such a addressing scheme for a tree T is $|V(T)| - 1$. The next theorem, due to Winkler (1983), shows that a tree is extremal in this sense and that any graph G can be addressed with strings of 0's, 1's and $*$'s of length at most $|V(G)| - 1$.

Theorem 3.4.2 (Winkler (1983)) *For any graph G on p vertices there is a labelling $L : V(G) \longrightarrow \{0, 1, *\}^{p-1}$ so that for each pair of vertices $x, y \in V(G)$*

$$d_G(x,y) = H(L(x), L(y)).$$

Proof. Let G be a connected graph on p vertices and let v_0 be an arbitrary distinguished vertex. As we saw in Lemma 3.1.4 there is a spanning tree T

with the property that $d_T(v_0, u) = d_G(v_0, u)$ for every vertex $u \in V(G)$. This tree will be the basis for numbering the vertices. We impose a numbering on the vertices of G so that this numbering corresponds to a so-called depth-first search of T: in other words we choose a shortest path joining v_0 to a pendant vertex and number the vertices along this path $v_1, v_2, \ldots$. Once the pendant vertex has a number we choose a shortest path to another pendant vertex, but this time one which begins from the numbered vertex with the smallest index. Continuing in this way we eventually account for every vertex. This numbering is in no way unique, but it has the structure we require, namely that whenever $k \in P(i) \backslash P(j)$ and $l \in P(j) \backslash P(i)$ we have $i < j$ if and only if $k < l$, where

$$P(i) = \{j : v_j \text{ lies on the unique path in } T \text{ from } v_0 \text{ to } v_i\}, \quad 0 \le i < p.$$

For any $0 < i < p$ let $i' = \max\{j : j \in P(i) \backslash \{i\}\}$, and for any $i, j < p$ we define $i \bigtriangledown j = \max\{j : j \in P(i) \cap P(j)\}$. We write $i \sim j$ if $P(i) \subset P(j)$ or $P(j) \subset P(i)$. Finally, let $\mathbf{C} = [c(i, j)]$ be the matrix of the difference between the distances in T and in G, i.e. $c(i, j) = d_T(v_i, v_j) - d_G(v_i, v_j)$.

We can now define an addressing scheme in terms of this matrix. Label each vertex v_i with the label $L(v_i) = a_i(1)a_i(2)\ldots a_i(p-1)$ of 0's, 1's and *'s defined by

$$a_i(k) = \begin{cases} 1 & \text{if } k \in P(i), \\ * & \text{if } (a) \ c(i, k) - c(i, k') = 2, \text{ or} \\ & \quad (b) \ c(i, k) - c(i, k') = 1 \text{ and} \\ & \qquad i < k \text{ and } c(i, k) \text{ is even, or} \\ & \quad (c) \ c(i, k) - c(i, k') = 1 \text{ and} \\ & \qquad k < i \text{ and } c(i, k) \text{ is odd,} \\ 0 & \text{otherwise.} \end{cases}$$

To better appreciate this construction see the example in Exercise 3.4.5.

It remains only now to prove that the construction actually works. To do this we need to make some observations about the discrepancy matrix, $\mathbf{C}$.

Lemma 3.4.3 *Let $\mathbf{C}$ be a matrix of distance discrepancies. Then*

(1) for all $0 \le i, j < p$, $c(i, j) \ge 0$,

(2) if $i \sim j$ then $c(i, j) = 0$,

(3) if $i \not\sim j$ then $c(i, j') \le c(i, j) \le c(i, j') + 2$.

Proof. The first property is clear, since T is a subgraph of G. To see the second property suppose $i < j$. Then $i \in P(j)$ and, by the triangle inequality,

$$d_G(v_i, v_j) \ge d_G(v_j, v_0) - d_G(v_i, v_0) = d_T(v_j, v_0) - d_T(v_i, v_0) = d_T(v_j, v_i).$$

So $c(i,j) = 0$ as required. Finally the third property follows by observing that $d_T(v_i, v_j) = d_T(v_i, v_{j'}) + 1$ and $|d_G(v_i, v_j) - d_G(v_i, v_{j'})| \leq 1$. □

Armed with this lemma we can prove that this construction really does mimic the distance function on G. Let $x = v_i$ and $y = v_j$. The result is immediate when $i = j$, and by symmetry we need only consider when $i < j$. It is also worth observing that to calculate the Hamming distance between two addresses we need only consider those coordinates in each sequence which could possibly be 1. The proof now breaks into two cases.

Case A. $i \sim j$

Then $P(i) \subset P(j)$ and $d_G(v_i, v_j) = d_T(v_i, v_j) = |P(j) \backslash P(i)|$. Now $a_j(k) = 1$ for each $k \in P(j) \backslash P(i)$ and $a_i(k) = 0$, since then $c(i,k) = 0$. Similarly for $k \in P(i)$ $a_i(k) = 1 = a_j(k)$. Since this accounts for all the possible occurrences of a 1 in either sequence we have that $\rho(L(v_i), L(v_j)) = |P(j) \backslash P(i)|$ as required.

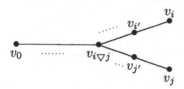

Figure 3.4.1

Case B. $i \not\sim j$

Consider the two sequences $c(i,j) \geq c(i,j') \geq c(i,j'') \geq \ldots \geq c(i, i \triangledown j) = 0$ and $c(i,j) \geq c(i',j) \geq c(i'',j) \geq \ldots \geq c(i \triangledown j, j) = 0$ (see figure 3.4.1). For each even m, $0 < m \leq c(i,j)$, let j_m be the unique index in the first sequence for which $c(i, j_m) \geq m > c(i, j_m')$. Then either $c(i, j_m) - c(i, j_m') = 2$, or $c(i, j_m) - c(i, j_m') = 1$ and $c(i, j_m)$ is even. In either case, the initial numbering makes $i < j_m$; we have that $a_i(j_m) = *$. Conversely, if $a_i(k) = *$ for some $k \in P(j) \backslash P(i)$, then $k = j_m$ for some even m.

Similarly, for each odd n, $0 < n \leq c(i,j)$, let i_n be the unique index in the second sequence for which $c(i_n, j) \geq n > c(i_n', j)$. This time $j > i_k$ for any $k \in P(i) \backslash P(j)$, so $a_j(k) = *$ just when $k = i_n$ for some odd n.

Now observe that the only values of k for which one of $a_i(k)$ and $a_j(k)$ can be 1 and the other not are those which lie in $P(i) \triangle P(j)$. Of these values, the number in which the other is $*$ is equal to the number of even numbers between 1 and $c(i,j)$ plus the number of odd numbers between 1 and $c(i,j)$; in other words there are $c(i,j)$ such places. Thus the Hamming distance between $L(v_i)$ and $L(v_j)$ is

$$\rho(L(v_i), L(v_j)) = |P(j) \backslash P(i)| + |P(i) \backslash P(j)| - c(i,j)$$
$$= d_T(v_i, v_j) - c(i,j) = d_G(v_i, v_j).$$ □

Exercises

3.4.1 ▽ Show that there is a simple graph G so that $bip(G)$ is not an induced subgraph of Q_n for any n. (Firsov (1965))

3.4.2 Is it true that if G is an induced subgraph of Q_n then $bip(G)$ is also an induced subgraph of $Q_{n'}$ for some n'?

3.4.3 △ Show that any tree T satisfies the conditions of Theorem 3.4.1, and has $|V(T)| - 1$ equivalence classes under the relation θ defined in the proof of the theorem.

3.4.4 ▽ Show that a tree T cannot be addressed with strings of 0, 1 and $*$ of length less than $|V(T)| - 1$. (R.L. Graham, Pollack (1972))

3.4.5 △ Verify that the addresses for the vertices of the graph in figure 3.4.2 mimic the graph distances, and also verify that, using the 'solid' edges as the spanning tree and the vertex numbering given, Winkler's construction gives these addresses.

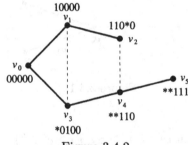

Figure 3.4.2

Chapter 4
Connectivity

4.1 k-connected graphs

$\mathbf{W}$e have already introduced the concept of a graph being connected, now we shall introduce some measure of this connectedness. The *vertex connectivity*, or simply *connectivity* $\kappa(G)$, of a graph is defined to be the minimum number of vertices whose removal disconnects the graph, or reduces it to a single vertex; for example $\kappa(K_p) = p - 1$, $\kappa(K_{n,n}) = n$ and $\kappa(T) = 1$ for any non-trivial tree T. A set of vertices which disconnects the graph in this way is called a *vertex cut*. If a vertex cut consists of only one vertex then that vertex is called a *cut vertex*. It is clear that there is always a vertex cut of cardinality at most $\delta(G)$ for any graph, giving an upper bound for the connectivity. If $\kappa(G) \geq k$ then we say that G is *k-connected*. To illustrate these concepts, we give the following simple result.

Proposition 4.1.1 *The n-cube Q_n has connectivity n, and furthermore, any minimum vertex cut consists of a vertex neighbourhood.*

Proof. The proof is by induction on n. The cases $n = 1$ and 2 are trivial. Let $n \geq 3$. Since $Q_{n+1} = Q_n \times K_2$ it may be partitioned into two copies of Q_n, G_1 and G_2, so that each vertex of G_1 has precisely one neighbour in G_2. By the induction hypothesis, to disconnect G_1 requires a vertex cut of n vertices which must be the neighbourhood of some vertex v. With one more vertex added to the cut, the neighbour of v in G_2, we may totally disconnect Q_{n+1}. Furthermore the removal of any other single vertex from G_2 instead, as part of the vertex cut, certainly leaves G_2 connected, and so the edges between G_1 and G_2 keep the remaining subgraph of Q_{n+1} connected. $\square$

The principal result about connectivity was proved by Menger (1927). This result gives a beautiful duality between vertex cuts and internally disjoint paths. Let u and v be two distinct vertices of a connected graph, then a collection of (u, v)-paths are *internally disjoint* if they have no vertex in common excluding u and v.

Theorem 4.1.2 (Menger's Theorem) *If s and t are two non-adjacent vertices of a graph G then the maximum number of internally disjoint paths between s and t is equal to the minimum number of vertices in a vertex cut which leaves s and t disconnected.*

Proof. We follow the proof of McCuaig (1984). The proof is by induction on the cardinality n of a minimum vertex cut disconnecting s from t. The theorem is trivial for $n = 1$. Suppose then that s and t are disconnected by the deletion of no set of less than $n + 1$ vertices. Certainly, by induction, there are n internally disjoint (s, t)-paths $P_1, P_2, \ldots, P_n$, and also, since the deletion of the second vertex in each of these paths leaves s and t connected, there is another (s, t)-path P whose first edge is not in any of the paths $P_1, \ldots, P_n$. Let x be the first vertex of P which is also a vertex of one of the P_i, and let P_{n+1} be the (s, x)-section of P.

Let us assume that the paths $P_1, \ldots, P_n, P_{n+1}$ have been chosen so that the distance from x to t in $G - s$ is minimal. If $x = t$ then we are done, so assume not.

In $G - x$ there are n internally disjoint (s, t)-paths $P_1', P_2', \ldots, P_n'$. Let us choose these so that $\left| \bigcup_{i=1}^{n} E(P_i') \cap \bigcup_{i=1}^{n+1} E(P_i) \right|$ is as large as possible, and let

$$H = G\left[\bigcup_{i=1}^{n} E(P_i') \cup \{x\} \right].$$ Choose some P_k, $1 \le k \le n + 1$, whose initial edge is not in $E(H)$ and let y be the first vertex on P_k, after s, which is in $V(H)$. If $y = t$ then we have found the extra path and we are done, so assume not. If $y = x$ then let R be the shortest (x, t)-path in $G - s$ and let z be the first vertex of R on some P_j', $1 \le j \le n$. Then the distance in $G - s$ from z to t is less than the distance from x to t, which contradicts the choice of the P_i's.

Finally, if y is some vertex on some P_i', $1 \le i \le n$, then the (s, y)-section of that path must be internally disjoint from the paths $P_1, \ldots, P_{n+1}$, for otherwise some P_i and P_j, $1 \le i < j \le n + 1$, would intersect in some vertex other than s, t and x. If we replace the (s, y)-section of P_i' with the (s, y)-section of P_k we have n internally disjoint (s, t)-paths which have a greater edge intersection with the P_i's, giving the final contradiction. $\square$

In fact the statement of Menger's Theorem is also true if we substitute 'a directed graph' for 'a graph'. Indeed Menger's original proof was for directed graphs. His theorem is of course a result about general graphs. What more might be said if our attention were restricted to bipartite graphs? The following result is a variation on a result of Whitney (1932).

Theorem 4.1.3 *Let G be a simple bipartite graph with bipartition (V_1, V_2) in which V_1 and V_2 both have at least $k+1$ vertices. Then G is k-connected if and only if there are k internally disjoint paths between any pair of vertices $s \in V_1$ and $t \in V_2$.*

Proof. The proof is by induction on k. The assertion is obvious when $k = 1$. Suppose first that G is k-connected, $k \geq 2$. If s and t are not adjacent then the assertion follows immediately from Menger's Theorem. Thus it is enough to consider when $s \in V_1$ and $t \in V_2$ are adjacent.

Suppose that there are at most $k - 1$ internally disjoint paths joining s and t. Let e be the edge st and $G' = G - e$. Then there are at most $k - 2$ internally disjoint (s, t)-paths in G'. Thus there exists a set $A \subseteq V(G)\backslash\{s, t\}$ of vertices, with $|A| \leq k - 2$, whose deletion disconnects s and t in G'. By assumption both V_1 and V_2 have at least $k + 1$ vertices, and so

$$|V_i \backslash A| \geq |V_i| - |A| \geq k + 1 - (k - 2) = 3,$$

showing the existence of a vertex u_i other than s and t in each $V_i \backslash A$, $i = 1, 2$. We shall show that there is an (s, u_2)-path P in G' which is internally disjoint from A.

Clearly this is true if s and u_2 are adjacent. On the other hand, if they are not adjacent Menger's Theorem gives k internally disjoint (s, u_2)-paths in G, and hence $k - 1$ internally disjoint (s, u_2)-paths in G'. Since $|A| \leq k - 2$, at least one of these paths will contain no vertex of A as required. Similarly we can show the existence of a (u_1, t)-path P' which does not contain any vertex of A.

Now if the vertices u_1 and u_2 are adjacent then the path $Q = PP'$ is an (s, t)-path avoiding A. If they are not adjacent, by applying Menger's Theorem just as above, we can find a (u_2, u_1)-path which avoids A, and so we again have an (s, t)-path which contains no vertex of A. The necessity is proved.

For sufficiency we assume that there are $k \geq 2$ internally disjoint paths between any two vertices $s \in V_1$ and $t \in V_2$. Suppose that A is a vertex cut with $|A| < k$. Then we consider the graph $G - A$. Certainly this graph contains at least two components, and furthermore, since each of V_1 and V_2 consist of at least $k + 1$ vertices, there must be components G_1 and G_2 of $G - A$ with $V(G_1) \cap V_1 \neq \emptyset \neq V(G_2) \cap V_2$. Select $s \in V(G_1) \cap V_1$ and $t \in V(G_2) \cap V_2$, then there are at most $|A| < k$ internally disjoint (s, t)-paths in G; a contradiction. $\square$

Exercises

4.1.1   Prove that if a graph is 2-connected then

(a)   every two vertices lie on a common cycle,

(b)   every two edges lie on a common cycle,

(c)   every vertex and edge lie on a common cycle.

4.1.2   Prove that in a k-connected graph every k vertices lie in a cycle.

4.1.3   Let G be a bipartite graph with bipartition (V_1, V_2) where $|V_1|$ and $|V_2| \geq k$. Prove that G is k-connected if and only if any two subsets $X \subseteq V_1$ and $Y \subseteq V_2$ of cardinality k are joined by k disjoint paths.

4.1.4 ▽ Let G be a bipartite graph with bipartition (V_1, V_2). Is it true that G is k-connected if and only if for each pair of vertices u, v from the same colour class there are k internally disjoint (u, v)-paths?

4.2 k-edge-connected graphs

As well as seeing how the removal of vertices from a graph affects its connectedness, we might consider the analogous problem with edges. In natural fashion, we define the *edge connectivity* $\lambda(G)$ of a graph G to be the minimum number of edges whose removal disconnects G. A set of edges whose removal disconnects G we call an *edge cut*. Furthermore, we say that G is *k-edge connected* if $\lambda(G) \geq k$. Then a non-trivial graph is 1-edge-connected if and only if it is connected. With these definitions we have a result corresponding to Theorem 4.1.3.

Theorem 4.2.1 *Let G be a bipartite graph with bipartition (V_1, V_2). Then G is k-edge-connected if and only if there are k edge-disjoint paths joining any two vertices $s \in V_1$ and $t \in V_2$.*

Proof. Let $V(G) = \{v_1, \ldots, v_p\}$, and $E(v_i)$ be the set of edges incident with v_i. For each $i = 1, \ldots, p$ we define the following two disjoint sets:

$$A_i = \left\{ a_i^e : e \in E(v_i) \right\} \quad \text{and} \quad B_i = \left\{ b_i^k : k = 1, \ldots, d_G(v_i) \right\}.$$

Consider a bipartite graph G_0 with vertex set $V(G_0) = \bigcup_{i=1}^{p} (A_i \cup B_i)$ in which every vertex from A_i is adjacent to every vertex of B_i, for $i = 1, \ldots, p$, and a_i^e is adjacent to a_j^e if and only of v_i and v_j are the ends of $e \in E(G)$. Then it is not difficult to check that G is k-edge-connected if and only if G_0 is k-vertex connected and that the theorem follows from applying Theorem 4.1.3 to G_0. □

For a subset $A \subseteq V(G)$ we shall write $E(A, \bar{A})$ for the set $E_G(A, V(G)\backslash A)$. Then the above theorem gives this rather useful corollary.

Corollary 4.2.2   *Let G be a bipartite graph with bipartition (V_1, V_2). Then, given vertices $x \in V_1$ and $y \in V_2$, there are k edge-disjoint (x, y)-paths if and only if for every $A \subseteq V(G)$ with $x \in A$ and $y \notin A$, we have $|E(A, \bar{A})| \geq k$.*

Proof. First suppose that for every $x \in V_1$ and $y \in V_2$ there are k edge-disjoint (x, y)-paths. Then by Theorem 4.2.1 G is k-edge-connected. Suppose that some $A \subset V(G)$ containing x but not y has $|E(A, \bar{A})| < k$. Then $E(A, \bar{A})$ forms an edge cut of less than k edges, contradicting that $\lambda(G) \geq k$.

Conversely, suppose that there is some pair of vertices $x \in V_1$ and $y \in V_2$ which are joined by less than k edge-disjoint paths. Then by Theorem 4.2.1 there is some edge cut of at most $k - 1$ edges which disconnects x and y. Then letting A be the set of vertices in the connected component containing x gives a subset with $|E(A, \bar{A})| < k$ as required. $\square$

Just as for vertex connectivity the minimum degree $\delta(G)$ gives a trivial upper bound for $\lambda(G)$. It is also easy to prove that $\lambda(G)$ is itself an upper bound for $\kappa(G)$. So we have the following theorem.

Proposition 4.2.3 *Let G be a graph. Then $\kappa(G) \leq \lambda(G) \leq \delta(G)$.* $\square$

There is no tendency, in general, to have equality in either of these inequalities. Indeed, it is not even true that a general k-regular bipartite graph is k-edge-connected. The following weaker property, however, does hold.

Theorem 4.2.4 (Hamidoune, Las Vergnas (1988)) *Let G be a k-regular bipartite graph with bipartition (V_1, V_2). Then for each vertex $x \in V_1$ there is a vertex $y \in V_2$ joined to x by k edge-disjoint paths.*

Proof. The proof is by induction on $|V(G)|$. The case for two vertices is trivial, for then the graph consists of two vertices joined by k parallel edges. Suppose then that $|V(G)| \geq 3$ and let $x \in V_1$.

If $|E(A, \bar{A})| \geq k$, for every $A \subset V(G)$ which contains x, the theorem follows immediately from Corollary 4.2.2. Let $A \subset V(G)$ be such that $x \in A$ and $|E(A, \bar{A})| < k$, and furthermore suppose that A is chosen with these properties so that $|E(A, \bar{A})|$ is minimal.

Let $V(G) = S \cup T$ be a partition of the vertex set, with $x \in S$, and let $e_1, \ldots, e_s$ and $f_1, \ldots, f_t$ be the edges in $E(A, \bar{A})$ with end vertices in $A \cap S$ and $A \cap T$, respectively. Then, by counting the edges of G with both end vertices in A we have

$$k|A \cap S| - s = k|A \cap T| - t.$$

Since $s + t = |E(A, \bar{A})| < k$ it follows that $s = t$.

Let G_1 be the graph induced by the vertex subset A together with an extra edge joining the end vertex in A of e_i to the end vertex in A of f_i, for $i = 1, \ldots, s$. Clearly G_1 is also a k-regular bipartite graph, but with fewer vertices than G. Thus by the induction hypothesis there are k edge-disjoint paths $R_1, \ldots, R_k$ joining x to some $y \in A \cap T$. We complete the proof by constructing k edge-disjoint paths from x to y in G.

Let G_2 be the graph obtained from G by identifying x with all vertices of A and deleting all edges with both ends in A. By the minimality of $|E(A, \bar{A})| = 2s$, the graph G_2 is $2s$-edge connected. Hence by Menger's Theorem there are $2s$ edge-disjoint paths joining x to any $z \in V(G) \backslash A$. Let these paths be $P_1, P_2, \ldots, P_s$ and $P'_1, \ldots, P'_s$ labelled so that $P_i \cap E(A, \bar{A}) = e_i$ and $P'_i \cap E(A, \bar{A}) = f_i$, for $i = 1, \ldots, s$. Then let R'_j be the path in G obtained from R_j by replacing each occurrence of an edge e_i in R_j by $P_i \cup P'_i$, for $j = 1, \ldots, k$. Clearly then $R'_1, \ldots, R'_k$ are k edge-disjoint paths joining x to y as required. $\qquad \square$

On the other hand there certainly are conditions to guarantee equality between edge connectivity and minimal degree.

Theorem 4.2.5 (Plesník, Znám (1989)) *Let G be a bipartite graph with $d(G) \leq 3$. Then $\lambda(G) = \delta(G)$.*

Proof. Let G have bipartition (V_1, V_2) and suppose that there is an edge cut E_0 of cardinality less than $\delta(G)$. Then $G - E_0$ consists of two components S and T. Let A_1 and A'_1 be the vertices in $S \cap V_1$ and $T \cap V_1$, and B_1 and B'_1 be the vertices in $S \cap V_2$ and $T \cap V_2$ which are ends of edges in E_0, respectively. Similarly let A_0, A'_0, B_0 and B'_0 be the remaining vertices of $V(G)$. Let the number of edges between A_1 and B'_1 be λ_1 and the number between A'_1 and B_1 be λ_2. Then $\lambda(G) = \lambda_1 + \lambda_2$.

Since $d(G) \leq 3$ and $d_G(x, y) = 4$ for any two vertices $x \in A_0$ and $y \in A'_0$, we have that either A_0 or A'_0 is empty. By symmetry we assume that $A_0 = \emptyset$. The proof now breaks into two cases.

Case A. $B_0 = \emptyset$

Either A_1 or B_1 is non-empty, and by symmetry we may assume that there is some vertex $x \in A_1$. Then we have $d_G(x) \leq \lambda_1 + |B_1|$. On the other hand $d_G(x) \geq \delta(G) \geq \lambda(G) + 1 = \lambda_1 + \lambda_2 + 1 \geq \lambda_1 + |B_1| + 1$; a contradiction.

Case B. $B_0 \neq \emptyset$

For every $x \in B_0$ we have $d_G(x) \leq |A_1| \leq \lambda_1 < \delta(G)$ which is impossible.

The result follows. $\qquad \square$

We close this section by giving a characterisation of bipartite graphs in terms of edge cuts. This result of McKee (1984) was originally proved using matroids; we shall give a direct proof.

Theorem 4.2.6 *A connected graph is bipartite if and only if every edge is in an odd number of minimal edge cuts.*

Proof. We begin with sufficiency. Suppose that G is a connected graph in which every edge is in an odd number of minimal edge cuts. Clearly when we remove a minimal edge cut we divide the graph into two connected components. Using these components, we can produce a labelling of the ends of the edges in the cut, by labelling all the vertices in one component with a 1 and all the vertices in the other with a 0. Given this, we form a new labelling of the vertices of G: at each vertex, we take the binary sum of its labels in the 0-1 labellings generated by all the minimum edge cuts – one labelling for each cut. It is now clear that since each edge lies in an odd number of minimal edge cuts, at each edge one end will receive the label 1 and the other 0 in this final labelling. Thus the final labelling defines a bipartition of G.

Let G be a connected bipartite graph. To see the necessity of the condition we proceed by induction on the number of edges q; the hypothesis is clear for $q = 1$. Suppose then that $q > 1$.

Let $e_0 \in E(G)$ be an edge, and x and y be its ends. Let K_G be the set of all minimal edge cuts of G, and let $K_G(e)$ be the set of minimal edge cuts of G which contain the edge e, for each $e \in E(G)$. If e_0 is a cut edge then $|K_G(e_0)| = 1$. Otherwise, e_0 must lie in a cycle in G, C say. Consider the graph $G' = G - e_0$. It is clear that if F is a minimal edge cut G which contains e_0 then $F \backslash \{e_0\}$ must be a minimal edge cut of G', but which minimal edge cuts of G' occur in this way? We shall show that $\{F \backslash \{e_0\} : F \in K_G(e_0)\}$ is precisely the set of $F' \in K_{G'}$ such that $|E(C) \cap F'|$ is odd. To see this, let us consider the (x, y)-path $P = C - e_0$, and a minimal edge cut $F \in K_G(e_0)$. Since F is minimal, $G - F$ must have two connected components, and no edge of F can have both its ends in the same component. Thus as we follow P from x to y, each time we traverse an edge of F we 'swap' from one component of $G - F$ to the other. It is now easy to see that x and y themselves lie in different components if and only if we swap components an odd number of times. Thus $F' = F \backslash \{e_0\}$ is a minimal edge cut in G' if and only if $|F' \cap E(C)|$ is odd.

To complete the proof, recall that each edge of $C - e_0$ lies in an odd number of minimal edge cuts in G', by induction. Thus, since G being bipartite implies that C is an even cycle, we have

$$1 \equiv \sum_{e \in E(C) \backslash \{e_0\}} |K_{G'}(e)| \, (\text{mod } 2)$$

$$\equiv |\{F \in K_{G'} : |F \cap E(C)| \equiv 1 \, (\text{mod } 2)\}| = |K_G(e_0)| \, (\text{mod } 2). \qquad \square$$

Exercises

4.2.1 △ Show that for any connected k-regular bipartite graph G ($k \geq 2$) and any edge e of G the graph $G - e$ is connected.

4.2.2 △ Show that $\kappa(G) = \lambda(G)$ if G is a cubic bipartite graph. Is this also true for any regular bipartite graph?

4.2.3 △ Show that for every $k \geq 3$ there exists a k-regular bipartite graph G with $\lambda(G) = 2$.

4.2.4 ▽ Prove that a regular bipartite graph G is magic if $\lambda(G) \neq 2$. (Doob (1978))

4.2.5 Prove that $\lambda(G) \leq \left\lceil \dfrac{2|E(G)|}{|V(G)|} \right\rceil$ for any graph G.

4.2.6 Prove that every edge cut is the union of disjoint minimal edge cuts.

4.2.7 Show that if G is a bipartite graph on p vertices with $\delta(G) \geq (p+1)/4$ then $\lambda(G) = \delta(G)$. (Volkmann (1988))

Application

4.3 The construction of linear super-concentrators

It is not difficult to imagine that having a graph with very good connective properties could be useful in many communication problems, for instance the design of telephone networks, parallel computers, and neural networks. In this section we will construct a special type of highly connected and yet very sparse graphs using bipartite graphs as building blocks.

Let G be a graph and let $X = \{x_1, \ldots, x_n\}$ and $Z = \{z_1, \ldots, z_m\}$ be disjoint subsets of vertices – we shall call X the *input vertices* and Z the *output vertices*. Such a graph is an *(n, m)-concentrator* ($n \geq m$) if each subset of m input vertices is connected to the output vertices by m mutually vertex-disjoint paths. The graph is an *n-superconcentrator* if $n = m$ and for each $r = 1, \ldots, n$ each subset of r input vertices is connected to each subset of r output vertices by r mutually vertex-disjoint paths.

Valiant (1976), using results based on previous work by Pinsker (1973), first showed the existence of n-superconcentrators with a linear number of edges relative to n (linear superconcentrators). Since then there have been various constructions of such superconcentrators. In this section we shall give a construction of Pippenger (1977). His construction is based on the existence of 'expansive' bipartite graphs which have very few edges. We begin with these building blocks.

Lemma 4.3.1 *For every m, there exists a bipartite graph with $6m$ input vertices (one colour class), and $4m$ output vertices (the other colour class) in which every input has degree at most 6, every output has degree at most 9, and for every $k \leq 3m$ and every set of k inputs, there exist k edges joining the given inputs to some k outputs.*

Proof. Let π be a permutation of $\mathcal{A} = \{0, 1, \ldots, 36m - 1\}$. From π we may construct a bipartite graph $G(\pi)$ by taking $V_1 = \{0, 1, \ldots, 6m - 1\}$ as the input vertices, $V_2 = \{0, 1, \ldots, 4m - 1\}$ as the output vertices, and for each $x \in \mathcal{A}$ joining $x \pmod{6m} \in V_1$ to $\pi(x) \pmod{4m} \in V_2$.

It is clear that $G(\pi)$ must satisfy the degree constraints required by the lemma, since there are only six elements of $\mathcal{A}$ in each residue class $\mod 6m$, and only nine elements of $\mathcal{A}$ in each residue class $\mod 4m$. It only remains to choose π in such a way that it ensures the last property in the statement of the lemma.

We shall say that a graph $G(\pi)$ is *good* if there do not exist $k \leq 3m$, a set I of k inputs, and a set O of k outputs, such that every edge from I has an end in O; otherwise $G(\pi)$ is *bad*. We shall show the existence of a good $G(\pi)$ by showing that the proportion of bad $G(\pi)$ is strictly less than 1.

Any set I of k inputs corresponds to a set of $6k$ elements $\mathcal{I} \subset \mathcal{A}$, and any set O of k outputs corresponds to a set of $9k$ elements $\mathcal{O} \subset \mathcal{A}$. For $G(\pi)$ to be bad every edge from I must end in O, in other words every element of $\mathcal{I}$ must have its image under π contained in $\mathcal{O}$. Of the $(36m)!$ permutations of $\mathcal{A}$ there are $\binom{9k}{6k}(6k)!(36m - 6k)!$ permutations which satisfy this condition. Thus the proportion of bad $G(\pi)$ is

$$I_m = \sum_{k=1}^{3m} \binom{6m}{k} \binom{4m}{k} \frac{\binom{9k}{6k}(6k)!(36m - 6k)!}{(36m)!} = \sum_{k=1}^{3m} \frac{\binom{6m}{k}\binom{4m}{k}\binom{9k}{6k}}{\binom{36m}{6k}}.$$

We now need only verify that this proportion is indeed always strictly less than 1. To do this we observe that since

$$\binom{36m}{6k} \geq \binom{6m}{k}\binom{4m}{k}\binom{26m}{4k}$$

we must have

$$I_m \leq J_m = \sum_{k=1}^{3m} \frac{\binom{9k}{6k}}{\binom{26m}{4k}}.$$

In J_m it is an easy check that the largest term is either the first term or the last. Then the verification that $I_m < 1$ reduces to an application of Stirling's formula.

So there exists a good graph $G = G(\pi_0)$. We shall show that for any set I of k inputs, $k \leq 3m$, there exist k edges joining I to some set of k outputs. Let $H_0 = G[I \cup N_G(I)]$ and let H be the graph obtained from H_0 by adding a new vertex v adjacent to every vertex of I and a new vertex w adjacent to every vertex in $N_G(I)$. Then the k required edges exist if and only of there are k internally vertex-disjoint (v, w)-paths in H. By Menger's Theorem it is enough to show that every vertex cut S which leaves v and w disconnected contains at least k vertices. Let $S = A \cup B$ where $A \subseteq I$ and $B \subseteq N_G(I)$. Then clearly $N_H(I \backslash A) \subseteq B$. On the other hand $|N_H(I \backslash A)| \geq |I \backslash A|$ since G is a good graph. Therefore, $|B| \geq |I \backslash A|$ and $|S| = |A| + |B| \geq |A| + |I \backslash A| = k$. $\square$

Now armed with the graphs of Lemma 4.3.1 we will give the promised construction.

Theorem 4.3.2 *There is an integer k such that for any n there is an n-superconcentrator with kn edges.*

Proof. We shall prove the result by induction on n. In fact Pippenger has shown that we may take $k = 40$ and other authors have reduced the constant still further (see Bassalygo (1981)), but to avoid detailed calculations we shall take $k = 60$. Then it is clear that for $n \leq 60$ $K_{n,n}$ is an n-superconcentrator with at most $60n$ edges. Now we turn to the induction step.

Suppose that the inductive hypothesis holds for all $n' < n$. Let $m = \lceil n/6 \rceil$ and let S_1 be a $4m$-superconcentrator with q edges, which comes equipped with its $4m$ input and $4m$ output vertices. By induction we have that $q \leq 4km = 240m$. We will show that we can 'boost' S_1 into an n-superconcentrator S with at most $312m + n$ edges. It is then easy to check that $312m + n \leq 60n$ when $n > 60$.

Let G_1 and G_2 be bipartite graphs which satisfy Lemma 4.3.1. As in the lemma we shall regard the $6m$ vertices as inputs of G_1 and the other colour class as outputs, but in G_2 we shall regard the roles as being reversed. Let S_1 be the graph obtained by deleting $6m - n$ inputs, together with any incident edges, from G_1 and identifying the output vertices of G_1 with the inputs of S_1, identifying the output vertices of S_1 with the inputs of G_2 and deleting $6m - n$ outputs from G_2. Lastly we add a set E of n edges from the surviving inputs of G_1 to the surviving outputs of G_2. The construction of S is shown, perhaps more clearly, in figure 4.3.1. It is clear that the graph S has at most $72m + n + q$ edges. It remains only to check that it is an n-superconcentrator.

For some $r \leq n$ let X be a set of r inputs and Y a set of r outputs. We must show that there are r vertex-disjoint paths from X to Y. Let X be partitioned into two parts: X_0, those vertices that correspond through E to vertices in Y, and those vertices X_1 which do not correspond to vertices in Y.

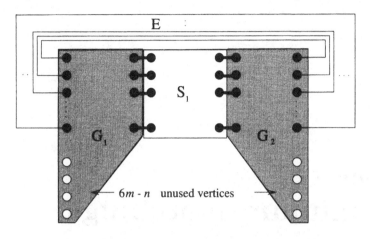

●━● denotes identification of vertices

Figure 4.3.1 The construction of an n-superconcentrator

Similarly let Y be partitioned as Y_0 and Y_1 respectively. There are certainly vertex-disjoint paths joining the vertices of X_0 to those in Y_0. Furthermore X_1, through E, corresponds to a set of vertices equinumerous to, but disjoint from, Y_1. Thus X_1 and Y_1 have a common cardinality k which is at most $n/2 \leq 3m$. Then, by Lemma 4.3.1, there are k edges joining the vertices of X_1 to some set X_1' of k outputs of G_1, and also k edges joining Y_1 to some set Y_1' of k inputs to G_2. Finally, since S_1 is a $4m$-superconcentrator there must be vertex-disjoint paths joining X_1' to Y_1'. The result follows.   □

For some results using superconcentrators in complexity theory we draw the reader's attention to the work of Tompa (1980) and Valiant (1976).

Exercise

4.3.1 ▽ Show that by counting more carefully in the proof of Theorem 4.3.2 it is possible to construct an n-superconcentrator with $40n$ edges.

Chapter 5
Maximum matchings

5.1 Properties of maximum matchings

A set of edges in a graph G is called a *matching* if no two edges have a common end vertex. A matching with the largest possible number of edges is called a *maximum matching*.

Many discrete problems can be formulated as problems about maximum matchings. Consider, for example, probably the most famous:

> *A set of boys each know several girls, is it possible for the boys each to marry a girl that he knows?*

This situation has a natural representation as the bipartite graph with bipartition (V_1, V_2), where V_1 is the set of boys, V_2 the set of girls and an edge between a boy and a girl represents that they know one another. The marriage problem is then the problem: does a maximum matching of G have $|V_1|$ edges?

Let M be a matching of a graph G. A vertex v is said to be *covered*, or *saturated* by M, if some edge of M is incident with v. We shall also call an unsaturated vertex *free*. A path or cycle is *alternating*, relative to M, if its edges are alternately in $E(G) \backslash M$ and M. A path is an *augmenting path* if it is an alternating path with free origin and terminus. Throughout this, and the following section, we shall identify a path P or a cycle C with the set of edges it embodies, and write $|P|$ and $|C|$ for the cardinalities of these sets of edges.

To appreciate the importance of augmenting paths in the structure of matchings consider the following two simple properties.

Property 5.1.1 *Let M be a matching and P an augmenting path relative to M. Then $M \triangle P$ is also a matching of G and $|M \triangle P| = |M| + 1$.* $\square$

Property 5.1.2 *Let M and N be matchings in G. Then each connected component of $G[M \triangle N]$ is one of the following:*

(1) an even cycle with edges alternately in $M \backslash N$ and $N \backslash M$, or

(2) a path whose edges are alternately in $M \backslash N$ and $N \backslash M$.   □

The next two results, which build on these observations, form the key to finding a maximum matching in a given graph (see Section 5.2).

Proposition 5.1.3 *Let M and N be matchings. If $|M| = r$, $|N| = s$ and $s > r$, then $G[M \triangle N]$ contains at least $s - r$ vertex-disjoint augmenting paths relative to M.*

Proof. Let the components of $G' = G[M \triangle N]$ be $C_1, C_2, \ldots, C_g$, and let $f(C_i) = |E(C_i) \cap N| - |E(C_i) \cap M|$. Then it follows from Property 5.1.2 that $f(C_i) \in \{-1, -0, 1\}$ for each $1 \le i \le g$ and $f(C_i) = 1$ if and only if C_i is an augmenting path relative to M. To complete the proof we need only observe that

$$\sum_{i=1}^{g} f(C_i) = |N \backslash M| - |M \backslash N| = |N| - |M| = s - r.$$

Hence there are at least $s - r$ components with $f(C_i) = 1$, and at least $s - r$ vertex-disjoint augmenting paths relative to M.   □

Combining Proposition 5.1.3 with Property 5.1.1 we can deduce the following theorem.

Theorem 5.1.4 (Berge (1957)) *M is a maximum matching if and only if there is no augmenting path relative to M.*   □

Indeed Proposition 5.1.3 and Property 5.1.2 also imply the following property.

Property 5.1.5 *If M is a matching of G then there exists a maximum matching M^* of G such that the set of vertices covered by M is also covered by M^*.*   □

Theorem 5.1.6 (Dulmage, Mendelsohn (1958b)) *Let G be a bipartite graph with bipartition (V_1, V_2) and M_1 and M_2 be two matchings in G. Then there is a matching $M \subseteq M_1 \cup M_2$ such that M covers all the vertices of V_1 covered by M_1 and all the vertices of V_2 covered by M_2.*

Proof. Let X_i (respectively Y_i) be the vertices of V_1 (of V_2) covered by M_i, $i = 1, 2$. Let $G_1, G_2, \ldots, G_k$ be the connected components of $G[M_1 \triangle M_2]$.

By Property 5.1.2 each G_i is an even cycle or a path. Let $M_{1i} = E(G_i) \cap M_1$ and $M_{2i} = E(G_i) \cap M_2$ $(1 \le i \le k)$ and define in each G_i a matching P_i:

$$P_i = \begin{cases} M_{1i} & \text{if } G_i \text{ is a cycle,} \\ M_{1i} & \text{if there is a vertex } v \in V(G_i) \cap (X_1 \backslash X_2), \\ M_{2i} & \text{if there is a vertex } v \in V(G_i) \cap (Y_2 \backslash Y_1). \end{cases}$$

It is not, then, difficult to check that $M = (M_1 \cap M_2) \cup P_1 \cup P_2 \cup \ldots \cup P_k$ is the required matching. $\qquad\square$

Alternating paths and cycles also provide a method to transform one maximum matching into another.

Theorem 5.1.7 *A maximum matching N of a graph G can be obtained from any other maximum matching M by a sequence of transfers along alternating cycles and paths of even length.*

Proof. By Property 5.1.1 and Theorem 5.1.4 every component of $G[M \triangle N]$ is either an alternating cycler or an alternating even path relative to M. Changing M in each component in turn will transform M into N. $\qquad\square$

A *perfect matching* of a graph G is a matching which covers every vertex of G. Clearly if a graph has two perfect matchings M and N then all components of $G[M \triangle N]$ are even cycles. Therefore, using Theorem 5.1.7, we can deduce the following result.

Corollary 5.1.8 *Let a graph G have a perfect matching M. Then any other perfect matching can be obtained from M by a sequence of transfers along alternating cycles relative to M.* $\qquad\square$

We discuss further properties of maximum and perfect matchings in Section 6.1.

Exercises

5.1.1 △ Show that in a connected graph G with q edges a maximum matching contains at most $\left\lfloor \dfrac{q+1}{2} \right\rfloor$ edges.

5.1.2 △ Show that a tree T has a perfect matching if and only if the graph $T - u$ has exactly one odd component for each vertex u of T.

5.1.3 △ Show that a connected, balanced, bipartite graph has a perfect matching if and only if its spectrum does not contain zero. (Longe-Higgins (1950))

5.1.4   Let G be a k-regular bipartite graph with $k \geq 3$, and e_1, e_2 be edges
        of G. Show that there exists a perfect matching of G using e_1 but
        not e_2 unless $G - \{e_1, e_2\}$ is not connected. (Häggkvist (1978))

5.1.5   Let G be a k-regular bipartite graph on $2n$ vertices, and B_1 be a
        matching of G. Furthermore, let $B_2 \subseteq E(G) \backslash B_1$ be a set of edges
        so that $k - |B_1| \geq \dfrac{n-1}{2}$ and $|B_2| \leq k - |B_1| - 1$. Show that G
        contains a perfect matching F such that $B_1 \subseteq F$ and $B_2 \cap F = \emptyset$.
        (Häggkvist (1978))

5.1.6   Let M_1 and M_2 be two distinct matchings of a graph G such that
        $|M_1| = r$, $|M_2| = s$ and $r \leq s$. Show that there exist two distinct
        matchings M_1' and M_2' such that

$$\left\lfloor \frac{s+r}{2} \right\rfloor \leq |M_1'| \leq |M_2'| \leq \left\lceil \frac{s+r}{2} \right\rceil.$$

5.1.7   Show that a bipartite graph G with at least one edge contains a
        matching with at least $\left\lfloor \dfrac{|E(G)|}{\Delta(G)} \right\rfloor$ edges.

5.2 Finding a maximum matching

In this section we shall deal with the practical question of how to actually
find a maximum matching in a bipartite graph efficiently. Since augmenting
paths will play a central role in the algorithms which we shall consider, let
us establish rather more about the structure of augmenting paths.

Lemma 5.2.1   *Let M be a matching with $|M| = r$ and suppose that the
cardinality of a maximum matching is s. Then there exists an augmenting
path relative to M of length at most $2\lfloor r/(s-r) \rfloor + 1$.*

Proof.   Let N be a maximum matching. Then by Proposition 5.1.3, $M \triangle N$
contains $s - r$ vertex-disjoint augmenting paths relative to M. Trivially
these paths contain at most r edges from M, so one of them must contain
at most $\lfloor r/(s-r) \rfloor$ edges from M and so at most $2\lfloor r/(s-r) \rfloor + 1$ edges
altogether.   □

An augmenting path P relative to a matching M is **shortest** relative to M
if P is of least cardinality among all augmenting paths relative to M.

Lemma 5.2.2   *Let M be a matching, P be a shortest augmenting path
relative to M, and Q be an augmenting path relative to $M \triangle P$. Then
$|Q| \geq |P| + |P \cap Q|$.*

Proof.   Let $N = M \triangle P \triangle Q$. Then N is a matching, and by Property 5.1.1
$|N| = |M| + 2$. So by Property 5.1.2 $M \triangle N$ contains two vertex-disjoint

augmenting paths relative to M, P_1 and P_2 say. Now, we certainly have $|M \triangle N| = |P \triangle Q| \geq |P_1| + |P_2| \geq 2|P|$, since P is a shortest augmenting path and the result follows by combining this inequality with the identity $|P \triangle Q| = |P| + |Q| - |P \cap Q|$. $\qquad \square$

Corollary 5.2.3 *Let P be a shortest augmenting path relative to a matching M, and Q be a shortest augmenting path relative to $M \triangle P$. Then if $|P| = |Q|$ the paths P and Q are vertex-disjoint and moreover Q is also a shortest augmenting path relative to M.*

Proof. Applying Lemma 5.2.2 we have $|P| = |Q| \geq |P| + |Q \cap P|$ and so $P \cap Q = \emptyset$. Thus P and Q are edge-disjoint. If they were to have a vertex v in common then they would also share the edge incident with v in $M \triangle P$, and so they are also vertex-disjoint. $\qquad \square$

Let G be a simple bipartite graph with bipartition (V_1, V_2) which has p vertices and q edges. Most algorithms so far developed to find a maximum matching start with a matching M and obtain a matching of greater cardinality, if possible, by identifying an augmenting path P relative to M and forming $M \triangle P$. The problem is to search for these paths efficiently. We shall consider two methods which implement this technique. The first is due to Hopcroft and Karp (1973) which gives an algorithm to find a maximum matching of G in $O(q\sqrt{p})$ operations. The second, due to Alt et al. (1991), uses some more recent techniques to use $O(p^{2.5}/\sqrt{\log p})$ operations, an improvement provided q is large. Corollary 5.2.3 is the key to both algorithms, and both break into a number of stages, at which some partial matching has been constructed and some way is sought to increase it. At stage i we have the matching M_i and we search for $\{R_1, R_2, \ldots, R_t\}$, a maximal set of vertex-disjoint, shortest augmenting paths, relative to M_i. The new matching for the next stage is then formed as

$$M_{i+1} = M_i \triangle R_1 \triangle R_2 \triangle \ldots \triangle R_t.$$

Our first task is to see how many stages such a process requires. This is an easy consequence of Lemma 5.2.1.

Proposition 5.2.4 *Let s be the cardinality of a maximum matching in a graph G, then to construct a maximum matching by the above process requires at most $2\lfloor \sqrt{s} \rfloor + 2$ stages.*

Proof. Let $N_0, N_1, \ldots, N_s$ be a sequence of matchings in G, where $N_0 = \emptyset$ and $N_{i+1} = N_i \triangle P_i$ and P_i is a shortest augmenting path relative to N_i for each $i = 1, \ldots, s - 1$. Then an upper bound for the number of stages we require can be given by bounding the number of distinct integers in the sequence $|P_0|, |P_1|, \ldots, |P_{s-1}|$. Let $r = \lfloor s - \sqrt{s} \rfloor$. Then since $|N_r| = r$ Lemma 5.2.1 gives

$$|P_r| \leq 2\lfloor r/(s-r) \rfloor \leq 2\lfloor \sqrt{s} \rfloor + 1.$$

Thus, for each $i \leq r$, P_i is one of the $\lfloor \sqrt{s} \rfloor + 1$ odd numbers less than or equal to $2\lfloor \sqrt{s} \rfloor + 1$. Certainly, the remaining $s - r = \lceil \sqrt{s} \rceil$ paths contribute at most $\lceil \sqrt{s} \rceil$ distinct integers to the collection, and so the total number is at most

$$\lfloor \sqrt{s} \rfloor + 1 + \lceil \sqrt{s} \rceil \leq 2\lfloor \sqrt{s} \rfloor + 2. \qquad \square$$

The problem now is to implement the search for shortest augmenting paths in an efficient way. This is where the two methods differ.

Let M_i be the matching of G at stage i. We define $\vec{G}_{M_i}$ to be the directed graph with the same vertex set as G, but with edge set

$$E(\vec{G}_{M_i}) = \{\overrightarrow{uv} : u \in V_1, v \in V_2 \text{ and } uv \in E(G) \backslash M_i\}$$
$$\cup \{\overrightarrow{vu} : u \in V_1, v \in V_2 \text{ and } uv \in M_i\}.$$

Observe then that directed paths in $\vec{G}_{M_i}$ from free vertices in V_1 to free vertices in V_2 are precisely augmenting paths relative to M_i. Hopcroft and Karp (1973) find shortest augmenting paths for each stage using this directed representation with a two step search.

Hopcroft and Karp's two step search
First step:

From $\vec{G}_{M_i}$ construct the subgraph $\hat{G}_{M_i}$ defined as follows: let L_0 be the set of free vertices in V_1 and define

$$E_i = \{\overrightarrow{uv} \in E(\vec{G}_{M_i}) : u \in L_i, v \notin L_0 \cup L_1 \cup \ldots \cup L_i\}, \quad i = 0, 1, 2, \ldots,$$
$$L_{i+1} = \{v \in V(\vec{G}_{M_i}) : \text{for some } u, \overrightarrow{uv} \in E_i\}, \quad i = 0, 1, 2, \ldots.$$

If now we define $i^* = \min\{i : L_i \cap \{\text{free vertices in } V_2\} \neq \emptyset\}$ then we can form a new graph $\hat{G}_{M_i}$ with

$$V(\hat{G}_{M_i}) = L_0 \cup L_1 \cup \ldots \cup L_{i^*-1} \cup \left(L_{i^*} \cap \{\text{free vertices in } V_2\} \right),$$
$$E(\hat{G}_{M_i}) = E_0 \cup E_1 \cup \ldots \cup E_{i^*-2}$$
$$\cup \left\{ \overrightarrow{uv} : u \in L_{i^*-1} \text{ and } v \in \{\text{free vertices in } V_2\} \right\}.$$

With this definition of the graph $\hat{G}_{M_i}$ directed paths from L_0 to L_{i^*} are precisely in one-to-one correspondence with shortest augmenting paths relative to M_i in G.

Second step:

We find a maximal set of vertex-disjoint paths from L_0 to L_{i^*}. For this we use a straightforward depth-first search. In the algorithm below we assume that each vertex v has a list $List(v)$ of the vertices in $N^+(v)$ (the out-neighbours of v) in some arbitrary order, and that the vertices of V_1 are also ordered.

Algorithm: Depth-first search for augmenting paths

      •**While** $V_1 \neq \emptyset$

           •**Let** v be the first vertex in V_1.

     (1)•**While** $List(v) \neq \emptyset$

           •**Let** u be the first element in $List(v)$.

           •**If** u already occurs in a constructed path **then** delete it from $List(v)$.

           •**Else** add u to *path* and **let** $v = u$.

        •**End while**

        •**If** v is in L_{i^*} **then** *path* is complete.

        •**Else if** v is in L_0 **then** delete v from the V_1.

           •**Else** 'retreat':

               •**Let** u be the previous vertex in *path*.

               •Delete v from $List(u)$ and **let** $v = u$.

               •**Goto** (1).

      •**End while**

Notice that at each visit to (1) a vertex has either just been added to the present path, deleted from a list or retreated from. Since all of these events can happen only once, the running time of this algorithm is bounded by $O(p + q)$. Thus, recalling that we need at most $\sqrt{p}$ stages to carry out this process we see that Hopcroft and Karp's algorithm runs in $O(q\sqrt{p})$ operations.

Alt, Blum, Melhorn and Paul's one step search

Using techniques to implement two step searches as a one step process Alt, Blum, Melhorn and Paul (1991) have given an improved method for finding augmenting paths. Their method begins with the empty matching and the graph $\vec{G} = \vec{G}_\emptyset$. Their algorithm creates 'layers' similar to $L_0, L_1, \ldots$ as it runs, but this is done in a rather different way. The algorithm maintains two labels $layer[v]$ and $free[v]$ for each vertex v. The distance label $layer[v]$ is initially 0 if $v \in V_2$ and 1 if $v \in V_1$. The other label $free[v]$ is a flag showing whether v is a free vertex.

At any given stage the vertices v with $layer[v] = i$ can then be thought of as L_i, but this set will change from step to step. Also required is the function $ce[v]$ which returns the vertex w if $\overrightarrow{vw}$ is an edge of $\vec{G}$ and $layer[v] = layer[w] + 1$, but otherwise returns the empty set.

Algorithm ABMP:
Single step construction of a maximum matching
> •Let $L = 1$ and $M = \emptyset$.
> •**While** $L \le \sqrt{p}\gamma$
>> •**While** there is a free vertex v with $layer[v] = L$
>> •Add v to *path*.
>> •**While** $path \neq \emptyset$
>>> •**Let** v be the last vertex of *path*.
>>> •**If** $layer[v] = 0$ and $free(v) = 1$ **then** the path is complete:
>>>> •$free[first\_vertex[path]] = 0$; $free[v] = 0$.
>>>> •Reverse the direction of all edges in *path*.
>>>> •**Let** $M = M \triangle E(path)$.
>>>> •**Let** $path = \emptyset$.
>>> •**Else If** $ce[v] \neq \emptyset$ **then** add $ce[v]$ to *path*
>>>> •**Else** $layer[v] = layer[v] + 2$; remove v from path
>> •**End while**
>> •**End while**
>> •**Let** $L = L + 2$.
> •**End while**

The algorithm runs by initially putting all vertices of V_1 in layer 1 and all vertices of V_2 in layer 0. Each loop of the algorithm then begins with a free vertex, v, in layer L, and tries to construct a directed path to a free vertex of V_2. This path is then used to update the matching M. At each step a search is made for an 'eligible' vertex to extend the path by calling the function $ce[v]$, and if no such vertex exists a retreat is made along the path until there is success. The gain over Hopcroft and Karp's algorithm is obtained by changing the layer labels on the graph $\vec{G}$ itself during the algorithm, rather than constructing a new $\hat{G}$ at each stage. To analyse the running time of this algorithm we need to make the following observations.

Lemma 5.2.5   *At all times during the execution of the algorithm* **ABMP** *the following hold:*

(1) *if* $\overrightarrow{uv} \in E(\vec{G})$ *layer*$[u] \leq$ *layer*$[v] + 1$,

(2) *layer*$[v]$ *is even if and only if* $v \in V_2$,

(3) *all free vertices of* V_1 *are in layers* L *and* $L + 2$. $\qquad\qquad\square$

It is then clear that each path which is completed is a shortest augmenting path relative to the present matching. Also when the algorithm stops there is no augmenting path relative to the present matching, M, which is shorter than $\gamma\sqrt{p}$, where γ is a parameter to be chosen a little later. Thus, if M^* is a maximal matching, $M \triangle M^*$ must contain $|M^*| - |M|$ vertex-disjoint augmenting paths by Proposition 5.1.3. Thus $(|M^*| - |M|)\gamma\sqrt{p} < p$ and $|M| > |M^*| - \sqrt{p}/\gamma$. Thus the matching M could be completed with at most $\sqrt{p}/\gamma$ phases of the Hopcroft-Karp algorithm with running time $O(q\sqrt{p}/\gamma)$.

As for the running time of this routine we can see from *(3)* that the maximum layer number is $\gamma\sqrt{p} + 2$, and so the total number of layer increases is at most $p^{1.5}\gamma$. Similarly the function $ce[v]$ is called at most $O(p^{1.5}\gamma)$ times. It is obvious that the function $ce[v]$ can be calculated in time $O(p)$ at each step, but by being rather more careful we can save a factor of $\log p$ (see Cheriyan, Hagerup and Mehlhorn (1990)). Let $L = [l_{iv}]$ be a $(\sqrt{p} + 2) \times p$ matrix with

$$l_{iv} = \begin{cases} 1 & \text{if } level[v] = i, \\ 0 & \text{otherwise,} \end{cases} \quad 1 \leq i \leq \sqrt{p} + 2 \text{ and } v \in V(\vec{G}),$$

and let $D = [d_{uv}]$ be a $p \times p$ matrix with

$$d_{uv} = \begin{cases} 1 & \text{if } \overrightarrow{uv} \in E(\vec{G}), \\ 0 & \text{otherwise,} \end{cases} \quad u, v \in V(\vec{G}).$$

The task of returning a vertex for $ce[v]$ requires only an examination of row v in D and row $i = level[v] - 1$ in L for a vertex w with $d_{vw} \cdot l_{(i-1)w} = 1$. The 'trick' is to store the entries of these binary matrices in blocks of length $\log p$. Each of these blocks can then be stored as one RAM word, and the processing of a block in the search can be done in constant time using elementary binary operations. In this way $ce[v]$ can be implemented in time $O(p/\log p)$. In total, then, the running time of the routine **ABMP** is $O(p^{2.5}\gamma/\log p)$.

Now, finally, we can combine these two algorithms to give the most efficient search for a maximal matching.

Theorem 5.2.6 *A maximum cardinality matching in a bipartite graph with p vertices and q edges can be computed in* $\mathrm{O}\left(\min\left\{q\sqrt{p}, p^{1.5}\sqrt{q/\log p}\right\}\right)$ *operations, which is at worst* $\mathrm{O}(p^{2.5}/\sqrt{\log p})$ *operations.*

Proof. The first of the two terms in the minimum is simply the running time of the Hopcroft-Karp algorithm. To obtain the second we use both algorithms together. Let $\gamma = \sqrt{q\log p}/p$. As we have seen we can use Alt, Blum, Mehlhorn and Paul's single step search to find a matching which is at most $\sqrt{p}/\gamma$ edges short of maximal, and then finally apply the Hopcroft-Karp algorithm to complete the process. This method runs in $\mathrm{O}(p^{2.5}\gamma/\log p + q\sqrt{p}/\gamma)$ operations. The result follows. $\qquad\square$

5.3 Maximum matchings in convex bipartite graphs

In this section we shall turn to constructing a maximum matching in convex bipartite graphs. This is an interesting special case, since it is precisely this structure which occurs in many practical problems, and in addition, the extra structure provides an algorithm for finding maximum matchings which is much simpler than the general method.

Example 1. *A certain product requires a machined part from a set V_1 and a second from a set V_2. Associated with each of the parts is its size $s(x)$. The two parts $u \in V_1$ and $v \in V_2$ can fit together provided $|s(u) - s(v)| < \varepsilon$, where ε is some permitted tolerance. Our objective is to fit the maximum number of parts together.*

This problem corresponds to finding a maximum matching in the bipartite graph G with bipartition (V_1, V_2) and edge set

$$E(G) = \{uv : u \in V_1, v \in V_2 \text{ and } |s(u) - s(v)| < \varepsilon\}.$$

What is more, if we order the parts V_2 in order of increasing size, this bipartite graph is convex.

The second example is in similar vein.

Example 2. *A set V_1 of skiers have each specified the smallest and largest skis they will accept from a set V_2 of skis. Our task is to allocate as many skis as possible.*

Then once again the obvious bipartite graph is convex, and the problem of assigning the skis as well as possible is that of finding a maximum matching.

Notice that in both of these examples the ordering on V_2 which makes the graph convex is clear. However, any given bipartite graph can be tested for possessing such an ordering and if possible rearranged in time $\mathrm{O}(|E(G)| +$

$|V(G)|$) using an algorithm due to Booth and Lueker (1976). Consequently it will be enough for our purpose to consider the following:

Let $V_1 = \{u_1, u_2, \ldots, u_n\}$, $V_2 = \{v_1, v_2, \ldots, v_m\}$ and G be a convex bipartite graph with bipartition (V_1, V_2). Also, for each u_i let the numbers

$$\mu_i = \min\{j : u_i v_j \in E(G), 1 \le j \le m\},$$
$$\tau_i = \max\{j : u_i v_j \in E(G), 1 \le j \le m\}$$

be specified. Then Glover (1967) showed that the following algorithm produces a maximum matching of G.

Glover's rule:

- **Begin** with the empty *matching*.

- **For** $j = 1, 2, \ldots, m$.

  - **If** v_j has a free neighbour in V_1 **then add** to the *matching* the edge $u_i v_j$ for which u_i is free and τ_i is as small as possible.

- **End for**

To see the correctness of this algorithm consider the following proposition.

Proposition 5.3.1 *Let G be a bipartite graph with bipartition (V_1, V_2). If $uv \in E(G)$, $u \in V_1, v \in V_2$ and $N(u) \subseteq N(w)$ for all $w \in N(v)$ then there is a maximum matching containing uv.*

Proof. Suppose that M is a maximum matching not containing the edge uv. If u is free then we may replace the edge of the matching incident with v with uv, and do similarly if v is free. This satisfies the proposition for this case. Suppose, therefore, that ux and vy are matched edges, for some $x \in V_2$ and $y \in V_1$. Since $x \in N(u) \subseteq N(y)$, it follows that we may replace ux, vy with uv and xy. The result follows. $\qquad\qquad\square$

Exercises

5.3.1   Show that the graph G in Example 1 is in fact doubly convex.

5.3.2   Verify the correctness of Glover's rule using Proposition 5.3.1.

5.3.3   Show that Glover's rule runs in time $O(|V_1| \cdot |V_2|)$.

5.4 Stable matchings

In this section we shall introduce a style of ranking for matchings, one of many ways to make some perfect matchings preferable to others. The idea of a *stable matching* is inspired by the search for an idyllic society of married couples. In such a society no man and woman would choose each other over their present partners if any – marriages would be stable. With this model in mind we pose the following graph theoretic problem.

Suppose that $G = K_{n,n}$ is the complete bipartite graph with bipartition (V_1, V_2) where $V_1 = \{u_1, u_2, \ldots, u_n\}$ and $V_2 = \{v_1, v_2, \ldots, v_n\}$. Let $L = [l_{ij}]$ and $R = [r_{ij}]$ be $n \times n$ matrices, with each row of L and column of R a permutation of the numbers 1 to n. Our aim is to use the two matrices L and R as rankings for the perfect matchings of G. Each vertex $u_i \in V_1$ has ordered the vertices in V_2 (these rankings give the entries of L) so that v_j is ranked l_{ij}th on u_i's list. Similarly the vertices of V_2 have ranked the vertices of V_1 to give the entries of the matrix R. Our aim is to find a matching of V_1 in which every vertex is as 'happy' as possible with its matched vertex.

We define a matching F to be *unstable* if there are vertices $u \in V_1$ and $v \in V_2$ which are not matched under F, but rank each other higher than the vertices with which they are matched under F. We say that a stable matching F is V_1-*optimal* if every vertex in V_1 is matched to a vertex ranked at least as high as any vertex with which it can be matched in a stable matching. A V_1-optimal stable matching, then, is analogous to stable marriages in which one sex is as happy as it could ever be. It is certainly not clear that we can guarantee anything other than unstable matchings in general, let alone V_1-optimal matchings, but in fact a V_1-optimal stable matching always exists for any given pair of ranking matrices L and R.

Theorem 5.4.1 (Gale, Shapley (1962)) *For any pair of ranking matrices L and R there exists a V_1-optimal, stable matching of $K_{n,n}$.*

Proof. To prove the theorem we shall give an algorithm which produces a V_1-optimal stable matching, and of course prove that this algorithm works. The algorithm is described in the language of marriage proposals, which we hope will clarify the process. We leave the reader to assign the sexes of V_1 and V_2.

Algorithm: Stable Matching

- Each vertex in V_1 proposes to its favourite vertex in V_2.
- Each vertex in V_2 provisionally gets engaged with its favourite, and rejects all others.
- **While** some vertex in V_2 has received no proposal.
    - Every presently rejected vertex in V_1 removes its favourite

vertex from the top of its list and proposes to the vertex in V_2 which is its new favourite.

- Each vertex in V_2 chooses its best offer from amongst the new proposals and possibly its present provisional match, provisionally gets engaged with the best and rejects all the others.

- **End while**

- Every vertex in V_2 'marries' (matches with) its 'betrothed'.

As no vertex of V_1 can propose to a vertex of V_2 more than once, at some point (in fact, in at most $n^2 - 2n + 2$ stages) each vertex will receive a proposal. Clearly this algorithm terminates. It remains only to observe that the perfect matching which the algorithm produces when it stops is V_1-optimal and stable.

It is easy to see that the matching is stable, for suppose that u and v are not matched, but u prefers v to its matched vertex. Then u must have proposed to v and been rejected, and v must prefer its final choice to u. There is no instability.

We prove the V_1-optimality by induction on the steps in the algorithm. We call a vertex $v \in V_2$ *impossible* for a vertex $u \in V_1$ if there is no stable matching in which u is matched with v. We shall show that, in the algorithm, vertices of V_1 are only rejected by impossible vertices of V_2.

Suppose that up to a certain stage each vertex has been rejected only by impossible vertices, and that now u is rejected by v in favour of U. We know that U prefers v to all other vertices, except for those that have previously rejected it. These, by assumption, were impossible vertices for U. Consider a hypothetical matching in which u and v are matched and U is matched to some other possible vertex. Then under such an arrangement U is matched to a vertex less desirable than v, and v to a vertex less desirable than U, and the matching is unstable. Thus v is impossible for u.

Our conclusion is that the algorithm rejects only matches with impossible vertices and the matching is both stable and V_1-optimal. $\square$

As a slight extension of this consider a problem concerning students and universities. There are m universities and the kth university is prepared to admit n_k students. Each of the n students and the universities has some list of preferences and we would like to find a stable matching.

This problem can be easily transformed to the stable matching problem on $K_{n,n}$ by making some simple observations. The first thing to notice is that we may assume that $n_1 + n_2 + \ldots + n_m = n$. We may achieve this by adding fictitious universities or fictitious students, whichever are needed, and making these fictitious entries as undesirable as possible. This done let us form the bipartite graph in which one colour class consists of the n students and the other consists of n vertices, n_k vertices representing the kth university (for each $k = 1, \ldots, m$) each equally preferable to the students. The problem of assigning students to universities is now the stable matching problem on $K_{n,n}$.

Of course, it is only natural to ask how the problem of finding stable matchings extends to general bipartite graphs. In fact, to deal with the general case, we can employ a trick, in the same vein as the one we used above, and reduce the problem to an instance of finding a stable matching on a complete bipartite graph. The idea is as follows. Let $G \subset K_{n,n}$ be a subgraph, and L and R be the ranking matrices. From G we construct an instance of the stable matching problem on $K_{n,n}$. We shall refer to this as the *completed system* relative to G. It remains only to specify the rankings. They are as follows. For each vertex in $u \in V_1$ the ranking of $N(u)$ remains as in L. We complete the ranking for u by ranking the vertices of V_2 outside $N(u)$ in some random order below the lowest ranked vertex of $N(u)$. Similarly for a vertex $v \in V_2$ we begin with $N(v)$ ordered as in R followed by the other vertices in V_1 in some random order. Using this trick we can prove the following result.

Theorem 5.4.2 *Let G be any bipartite graph with ranking matrices L and R. Then G has a stable matching which is a maximal matching in G.*

Proof. We begin by embedding G into a complete bipartite graph H, and extending the rankings in such a way that at each vertex those edges of G are preferable to the new edges. Then Theorem 5.4.1 implies that this complete system has a V_1-optimal stable matching M. Certainly if M is stable in H $M \cap E(G)$ must be stable in G. Thus it remains only to show that $M \cap E(G)$ is non-empty and that if it is not a maximal matching it can be extended to a maximal matching which is also stable.

To see that $M \cap E(G)$ is non-empty we need only follow the steps of the algorithm of Gale and Shapley. In the first round some vertex $u \in V_1$ will certainly be provisionally matched with a vertex $v \in V_2$ across an edge e of G. Now, since v ranks all new edges below e it will only change this provisional matching to match provisionally with some other edge of G. In other words some edge of G must survive to the final stable matching M.

To see that $M \cap E(G)$ can be extended to a maximal matching, consider an edge $uv \in E(G)$ which is incident with no edge of $M \cap E(G)$. Then if $(M \cap E(G)) \cup \{uv\}$ is unstable there must be some edge $xy \in M \cap E(G)$ and (without loss of generality) be some edge $xv \in E(G)$ so that x ranks v above y and v ranks x above u. But this would imply that M were itself unstable; a contradiction. Thus $(M \cap E(G)) \cup \{uv\}$ is stable, and $M \cap E(G)$ can be extended to a stable maximal matching of G. $\square$

Finally, we shall give a result which perhaps demonstrates an effect of complete inequality of the sexes (see Knuth (1976)).

Proposition 5.4.3 *If one stable matching is preferable to another from the point of view of* V_1, *then from the point of view of* V_2 *the second is preferable to the first.*

Proof. To prove this result we shall prove the following property:

If a stable matching M of a graph G contains the edge uv and another stable matching N contains edges ua and bv then either

    (1) u ranks a higher than v, and v ranks u higher than b, or

    (2) u ranks v higher than a, and v ranks b higher than u.

Let us introduce some notation. If a vertex a ranks b over c we shall write $a : b > c$. Now let $u = X_0$, $v = x_0$ and $a = x_1$ and suppose that u ranks a over v. Then in our new notation $X_0 : x_1 > x_0$. Let X_1 be the neighbour of x_1 in M. Then since M is stable $x_1 : X_1 > X_0$. Similarly let x_2 be the neighbour of X_1 in the matching N. The stability of N implies that $X_1 : x_2 > x_1$ and letting X_2 be the neighbour of x_2 in M we have $x_2 : X_2 > X_1$. Continuing in similar vein the matchings will be

$$M = \{X_0 x_0, X_1 x_1, X_2 x_2, \ldots\} \text{ and}$$
$$N = \{X_0 x_1, X_1 x_2, X_2 x_3, \ldots\}$$

and we have that $X_k : x_{k+1} > x_k$ and $x_{k+1} : X_{k+1} > X_k$ for each $k \geq 0$.

There are of course only finitely many edges in each matching so there must be some $j < k$ for which $X_j = X_k$. Let j be the smallest value with this property and k be the smallest value for which $X_k = X_j$. Then, by the construction we also have $x_j = x_k$. In fact we have that $j = 0$, as we shall see.

Suppose that $j \neq 0$. Then $X_{k-1} x_k = X_{k-1} x_j$ is an edge of N but this edge $X_{k-1} x_j$ must be the edge $X_{j-1} x_j$, contradicting the choice of j. Thus $j = 0$ and $X_{k-1} x_0$ is an edge of N. In other words $X_{k-1} = b$, $X_k = u$ and $x_k = v$. The result follows since, by construction, $x_k : X_k > X_{k-1}$. $\square$

Application

5.5 The Generalised Assignment Problem

The well known *Assignment Problem* is formulated as follows: there are n workers and n jobs. The ability of worker i to carry out job j is $w(ij)$. What is the optimal assignment of workers to jobs, that is the permutation $\pi \in S_n$ which maximises $\sum_{i=1}^{n} w(i\pi(i))$?

Sadly, in practice, it is rarely true that the assignment with maximum weight is also, in our eyes, a satisfactory solution. There are always hidden factors which cannot be entered into this basic model. Rather than simply finding the assignment with the maximum weight, then, it might be better to find perhaps the k best assignments to give some choice.

We can rephrase the problem as a problem on a bipartite graph. Let $G = K_{n,n}$ have bipartition (V_1, V_2) and let each edge $e \in E(G)$ be given a weight $w(e)$. The weight of a matching M is $w(M) = \sum_{e \in M} w(e)$. A matching with maximum weight is called an *optimal* matching. The problem is to find k distinct perfect matchings $M_1, \ldots, M_k$ such that $w(M_1) \geq w(M_2) \geq \ldots \geq w(M_k) \geq w(M)$ for every perfect matching $M \notin \{M_1, \ldots, M_k\}$. We call this problem the k *Best Perfect Matchings Problem*. We shall solve this problem first for the case $k = 1$ and then for the case $k = 2$. Finally we shall see that the general case can be solved using these two special cases.

Let us begin with the case $k = 1$ which corresponds to the Assignment Problem. We shall follow the treatment of Bondy and Murty (1976).

Let $f : V_1 \cup V_2 \longrightarrow \mathbb{R}$ be a function with the property that

$$f(u) + f(v) \geq w(uv) \quad \text{for each } u \in V_1, v \in V_2.$$

f is then a *feasible* vertex labelling, and $f(u)$ is the *label* of the vertex u. Happily, a simple example of a feasible labelling always exists, regardless of the weight function w, namely the function f_0:

$$f_0(x) = \begin{cases} \max\{w(xy) : y \in N(x)\} & \text{if } x \in V_1, \\ 0 & \text{if } x \in V_2. \end{cases}$$

For a given feasible vertex function f let E_f be the subset of edges uv for which $f(y) + f(v) = w(uv)$. The *equality graph* of f, G_f, is then the spanning subgraph of G with edge set E_f. The following theorem gives a link between the graph G_f and optimal matchings of G.

Theorem 5.5.1 *Let f be a feasible vertex labelling. If G_f contains a perfect matching M, then M is an optimal matching for G.*

Proof. First suppose that G_f has a perfect matching, M. Then since G_f is a spanning subgraph of G, M is also a perfect matching for G. Also, since each vertex of G is covered once and once only by M, and G_f is the equality graph of f, $w(M) = \sum_{e \in M} w(e) = \sum_{x \in V(G)} f(x)$. On the other hand, if N is any perfect matching in G then $w(N) = \sum_{e \in N} w(e) \leq \sum_{x \in V(G)} f(x)$. The result follows. $\square$

We now present an algorithm due to Kuhn (1955) and Munkres (1957), set on the complete bipartite graph $K_{n,n}$ with bipartition (V_1, V_2) and weight function w.

Optimal assignment: Kuhn and Munkres

- **Begin** with the feasible labelling $f = f_0$. Find the equality graph G_f and a maximum matching of G_f, M_0. **Let** $i = 0$.
- **While** M_i is not a perfect matching of G_f
  - **Let** u be a free vertex in V_1. **Let** $S = \{u\}$ and $T = \emptyset$.
  - (1) • **If** $N_{G_f}(S) = T$ **then**
    - Compute $d_f = \min\limits_{x \in S, y \notin T} \{f(x) + f(y) - w(xy)\}$.
    - Update the function f to

    $$f(x) = \begin{cases} f(x) - d_f & \text{if } x \in S, \\ f(x) + d_f & \text{if } x \in T, \\ f(x) & \text{otherwise}, \end{cases}$$

    and calculate the new G_f.
  - Select a vertex $y \in N_{G_f}(S) \setminus T$.
  - If y is saturated **then add** y to T, and the neighbour of y in M_i to S. **Goto** (1).
  - **Else** there is an augmenting path P in $G_f[S \cup T]$ joining u to y. Let $M_{i+1} = M_i \triangle E(P)$ and $i = i + 1$.
- **End while**
- **Output** M_i, an optimal matching.

The algorithm runs by using a two pronged attack to increase the size of the matching in G_f. The first is the familiar technique of searching for an augmenting path relative to the present matching. The second is by changing the function f itself. Notice that when the function is updated this has the effect of increasing the size of $N(S)$ to assist the search for an augmenting path.

Now to the running time of the algorithm. There can be at most n matchings M_i. For each of these matchings the set T can be increased only at most n times, and at each of these increases the function f may have to be updated, at a cost of $O(n^2)$ steps. Thus the algorithm terminates after at most $O(n^4)$ operations.

So we can find an optimal matching of $K_{n,n}$. Now we shall turn to finding the second best perfect matching using an algorithm of Murty (1968).

Second best perfect matching:

- Let $M_1 = \{e_1, e_2, \ldots, e_t\}$ be an optimal matching in G.
- For $i = 1, \ldots, t$.
  - Let H_i be the graph obtained from G by deleting all the end vertices of $\{e_j : j < i\}$ from G. We define a weight function for H_i by
$$w_i(e) = \begin{cases} w(e) & \text{if } e \neq e_i, \\ -\infty & \text{if } e = e_i. \end{cases}$$
  - Find an optimal matching N_i in H_i.
  - Let $N_i = N_i \cup \{e_j : j < i\}$.
- **End for**
- The second best matching M_2 is the matching with the greatest weight amongst $N_1, \ldots, N_t$.

The basic idea of the algorithm is to partition the solution space by sequentially forbidding the edges of the optimal matching one by one. If M_2 does not contain e_1 then it will be found as N_1, otherwise it does contain e_1 and will be one of the later matchings. If M_2 contains e_1 but not e_2 it will be found as N_2, etc.

The algorithm of Hamacher and Queyranne (1985) to find the k best perfect matchings uses a generalisation of this type of partitioning technique. We define $\Omega_{I,O}$ to be the set of all perfect matchings M which contain the edges of I, but do not contain the edges of O.

k Best Perfect Matchings: Hamacher, Queyranne

- Let $I_1 = O_1 = \emptyset$
- Find the best and second best matchings M_1 and N_1 for the graph G with weight function w.
- For $i = 2, \ldots k$
  - Let p be the index of the matching N_p with the maximum weight amongst $N_1, \ldots, N_{i-1}$.
  - Let $M_i = N_p$.

- Choose $e \in M_p \backslash N_p$.
- **Let** $I_i = I_p$ and $O_i = O_p \cup \{e\}$.
- **Let** $I_p = I_p \cup \{e\}$.
- Compute a new second best matching N_p in the new Ω_{I_p, O_p}.
- Compute the second best matching N_i in Ω_{I_i, O_i}.
- **If** either N_p or N_k does not exist **then** there are only $i-1$ matchings. **Stop.**
- **End for**

Before explaining the algorithm in full, perhaps we should first explain how to find the second best matching in this restricted solution space $\Omega_{I,O}$. This is done with the now familiar trick of deleting from G all the vertices incident with the edges I and setting the weight of the edges in O to be $-\infty$. Then the second best matching in this new setting will exactly correspond to the second best matching in $\Omega_{I,O}$.

Now to the algorithm itself. As we have mentioned the algorithm uses a partitioning technique a little similar to that of the second best matching algorithm. It should be observed that at each stage i the sets I_j and O_j $(j \leq i)$ have evolved in such a way that M_j and N_j are the best and second best matchings in Ω_{I_j, O_j}. Also, when I_p and O_p are updated M_p remains the best matching in Ω_{I_p, O_p}. If we further observe that the solution sets $\{\Omega_{I_j, O_j} : j \leq i\}$ form a partition of the total set of perfect matchings it can be seen that, when the new N_p is chosen, it is the best matching besides the $i-1$ best matchings so far, $M_1, \ldots, M_{i-1}$, and so is what we would like to call M_i.

Exercise

5.5.1   Given the graph $K_{n,n}$ and weight function $w : E(G) \longrightarrow \mathbb{N}$, show that the following algorithm constructs a perfect matching M_0 such that $\min_{e \in M_0} w(e) \geq \min_{e' \in M} w(e')$ for every perfect matching M of $K_{n,n}$. Find a perfect matching M_1 and delete from $K_{n,n}$ all edges e' with $w(e') \leq \min_{e \in M_1} w(e)$. If the obtained graph has a perfect matching M_2 then repeat the procedure with the matching M_2. Otherwise M_1 is the required matching. (Gross (1959))

Chapter 6

Expanding properties

This chapter is devoted to bipartite graphs which have the property that any subset of vertices, from one of the colour classes, has a neighbour set which is at least as large as itself, be it by some additive or multiplicative factor. We shall see that such graphs have a very useful structure, and a variety of applications.

6.1 Graphs with Hall's condition

We begin by considering bipartite graphs with bipartition (V_1, V_2) which satisfy the condition $|N(A)| \geq |A|$ for every $A \subseteq V_1$ (where $N(\emptyset) = \emptyset$). Graphs with this property arise very often in practice. Many practical problems can be reformulated as problems about finding a matching M such that every vertex in V_1 is incident with an edge in M. Such a matching is called a matching of V_1 into V_2. The 'marriage problem', which we have already considered in Section 5.1, is one of many examples.

It is clear that for us to hope to find a matching of V_1 into V_2 we must have $|N(A)| \geq |A|$ for every $A \subseteq V_1$. Remarkably, as the following result of Philip Hall (1935) shows, this condition is also sufficient.

Theorem 6.1.1 (Hall's Theorem) *Let G be a bipartite graph with bipartition (V_1, V_2). Then G has a matching of V_1 into V_2 if and only if $|N(A) \geq |A|$ for every $A \subseteq V_1$.*

(In light of this result the above condition is often called *Hall's condition*.)

Proof. First notice that, if there is a matching from V_1 into V_2, Hall's condition does indeed hold.

To prove the converse suppose that $|N(A)| \geq |A|$ for each $A \subseteq V_1$. We shall follow a proof of Halmos and Vaughn (1950),[†] proceeding by induction on the cardinality of V_1; if $|V_1| = 1$ the result is immediate.

[†]For an alternative proof of this theorem see Section 11.1

Case A. For every non-empty subset $A \subset V_1$ we have $|A| < |N(A)|$

Let $u \in V_1$ and $v \in V_2$ be adjacent vertices and $G' = G - \{u, v\}$. Then let $A \subseteq V_1 \backslash \{u\}$. If $A = \emptyset$ then $|A| = 0 = |N(A)|$. If $A \neq \emptyset$ then $|N_{G'}(A)| \geq |N_G(a)| - 1 \geq |A|$. Hence by the induction hypothesis there is a matching of $V_1 \backslash \{u\}$ into $V_2 \backslash \{v\}$ and so, with uv, a matching of V_1 into V_2 as required.

Case B. Some non-empty subset $A \subset V_1$ has $|A| = |N(A)|$

For this case we shall split G into two smaller subgraphs $G_1 = G[A \cup N(A)]$ and $G_2 = G - A - N(A)$ and show that each of these satisfies Hall's condition. First suppose that $X \subseteq A$. Then $N_G(X) \subseteq N_G(A)$ and so $N_{G_1}(X) = N_G(X)$. Thus by assumption $|N_{G_1}(X)| \geq |X|$. Now let $X \subseteq V_1 \backslash A$. Then $N_G(X \cup A) = N_{G_2}(X) \cup N_G(A)$ and therefore

$$|N_{G_2}(X)| = |N_G(X \cup A)| - |N_G(A)| \geq |X \cup A| - |N_G(A)|$$
$$= |X \cup A| - |A| = |X| \quad \text{since } X \cap A = \emptyset.$$

Thus applying the induction hypothesis to both G_1 and G_2 there must exist matchings M_1 of A into $N(A)$ in G_1 and M_2 of $V_1 \backslash A$ into $V_2 \backslash N(A)$ in G_2. The union $M_1 \cup M_2$ is then a matching of V_1 into V_2 as required. $\square$

Putting this theorem into the language of boys and girls we obtain an answer to the marriage problem:

Theorem 6.1.2 *It is possible to marry a group of n boys each to a girl that he knows if and only if every subset of k boys communally know at least k girls, for each $k = 1, \ldots, n$.* $\square$

Hall's Theorem has many simple consequences some of which we shall explore in the remainder of this section. We shall begin by considering graphs with perfect matchings.

Corollary 6.1.3 *A bipartite graph G with bipartition (V_1, V_2) has a perfect matching if and only if $|V_1| = |V_2|$ and $|N(A)| \geq |A|$ for each $A \subset V_1$.* $\square$

Interestingly Hall's Theorem itself follows from Corollary 6.1.3 and so they are equivalent. Actually, this result is a reformulation of the following result of Frobenius (1917) on determinants: if all the terms of a determinant of order n are zero, then there is a positive integer r such that all elements which have one of a certain set of r rows and one of a set of $n - r + 1$ columns in common are zero.

Corollary 6.1.4 (König (1914, 1916)) *Every regular bipartite graph has a perfect matching.*

Proof. Let G be a k-regular bipartite graph with bipartition (V_1, V_2). It is easy to see that $|V_1| = |V_2|$, we need only establish Hall's condition.

Consider $A \subseteq V_1$ and the subgraph $H = G[A \cup N(A)]$. Then counting the number of edges in H we have $k|A| = |E(H)| \leq k|N(A)|$. Thus $|A| \leq |N(A)|$ and G has a perfect matching.[†]   □

Indeed in Section 9.3 we shall show that a k-regular, n by n bipartite graph has at least $n!(k/n)^n$ perfect matchings.

A bipartite graph G is called *elementary* if it is connected and each edge is contained in some perfect matching. The next theorem, obtained by Hetyei (1964) and independently by Little, Grant and Holton (1975), gives criteria for a graph to be elementary.

Theorem 6.1.5 *The following three statements are equivalent for a bipartite graph G with bipartition (V_1, V_2):*

(1) *G is an elementary bipartite graph,*

(2) *$G = K_2$ or $|V(G)| \geq 4$ and for each pair $x \in V_1$, $y \in V_2$ the graph $G - \{x, y\}$ has a perfect matching,*

(3) *$|V_1| = |V_2|$ and $|N(X)| \geq |X| + 1$ for every non-empty proper subset $X \subset V_1$.*

Proof. *(1) implies (3)*

G has a perfect matching, so $|V_1| = |V_2|$. Now suppose that $\emptyset \neq X \subset V_1$ is a subset of vertices with $|N(X)| \leq |X|$. Then $|N(X)| = |X|$ since G has a perfect matching. Certainly, since G is connected, $G[X \cup N(X)]$ is joined to the rest of G by an edge e, and e must have its ends in $N(X)$ and $V_1 \backslash X$. Also, any perfect matching contains precisely $|X|$ edges leaving X and these edges cover all the vertices of $N(X)$. But then e cannot belong to any perfect matching of G, giving the required contradiction.

(3) implies (2)

We must show that $G - \{x, y\}$ has a perfect matching whenever $x \in V_1$ and $y \in V_2$. Since $|V_1| = |V_2|$ it suffices to show that there is a matching in $G - \{x, y\}$ covering $V_1 \backslash \{x\}$. In other words we must show that for each subset $X \subset V_1 \backslash \{x\}$ we have $|N_{G-\{x,y\}}(X)| \geq |X|$. This is obvious if $X = \emptyset$, and otherwise we have $|N_{G-\{x,y\}}(X)| \geq |N_G(X)| - 1 \geq |X|$.

(2) implies (1)

Suppose G is disconnected. Then choose a component of G, G_1, so that $|V(G_1) \cap V_1| \leq |V(G_1) \cap V_2|$. Let $x \in V(G_1) \cap V_1$ and $y \in V_2 \backslash V(G_1)$. Then

---

[†]Actually in 1887 Martinetti stated a result in geometric language which is equivalent to Corollary 6.1.4 for simple regular bipartite graphs. However, the first published proof, also in geometric language, appears in the doctoral dissertation of Steinitz (1894). In his words he considered simple regular bipartite graphs as 'configurations'. We thank Harald Gropp for pointing this out.

$G - \{x, y\}$ cannot have a perfect matching, since the vertices $V(G_1) \cap V_2$ are adjacent to only $|V(G_1) \cap V_1| - 1 < |V(G_1) \cap V_2|$ vertices; a contradiction.
□

Using this result we can give a more structural characterisation of elementary bipartite graphs.

Theorem 6.1.6 (Hetyei (1964)) *A bipartite graph G is elementary if and only if it can be written in the form $G = G_0 \cup P^1 \cup P^2 \cup \ldots \cup P^k$, where $G_0 = K_2$, and each P^i is a path of odd length which joins two vertices of $G_0 \cup P^1 \cup \ldots \cup P^{i-1}$ from different colour classes, but has no other vertices in common.*

Proof. First let us show that graphs of the described form are indeed elementary. They are certainly bipartite and connected. We shall use induction on k to also show that every edge lies in a perfect matching.

Suppose that $G' = G_0 \cup P^1 \cup \ldots \cup P^{k-1}$ is elementary, and let x and y be the end vertices of P^k. Observe that every perfect matching of G' extends to a perfect matching of G, simply by adding every second edge of P^k starting with the second edge. Thus, every edge of G' as well as every second edge of P^k belongs to some perfect matching of G.

On the other hand, by Theorem 6.1.5, $G' - \{x, y\}$ has a perfect matching. To this matching, this time starting with the first edge of P^k, we can add every second edge of P^k, to give a perfect matching of G. Thus every edge of G lies in a perfect matching.

Conversely, suppose that G is an elementary bipartite graph. Let F be a perfect matching of G, and let G_0 be the graph formed by one of the edges of this matching. We shall choose the paths $P^1, P^2, \ldots, P^k$ one by one so that they have the required properties. Indeed, we shall impose the additional requirement that each path is also an alternating path with respect to F.

Suppose that the paths $P^1, \ldots, P^i$ are already chosen and $G' = G_0 \cup P^1 \cup \ldots \cup P^i \neq G$. Let e be an edge not in G', but having one end in G'. G is connected, so such an edge certainly exists. Of course we can find a perfect matching containing e. Let this matching be F_1. The additional requirement that all of the paths are alternating paths with respect to F gives us that $E(G') \cap F$ is a perfect matching for G', and what is more $F_0 = F - E(G')$ is disjoint from G'. Certainly then $F_1 \cup F_0$ will contain a path ending with e. This path, P^{i+1}, has both ends in G' and has no other vertex in $V(G')$. For reasons of parity, P^{i+1} also has odd length and connects two vertices in different colour classes of G'. Finally, by definition, the new path alternates with respect to F. Thus $G_0 \cup P^1 \cup \ldots \cup P^{i+1}$ is a graph of the required form and, provided edges remain, we can repeat this procedure to find another path extending the union. The result follows.
□

In the previous chapter we considered properties of maximum matchings. Using Hall's Theorem we can now give some further results.

Corollary 6.1.7   *Every bipartite graph G has a maximum matching which covers all the vertices of maximum degree.*

Proof.   Consider a $\Delta(G)$-regular graph H which contains G as an induced subgraph (the existence of such a graph follows from Proposition 2.2.7). By Corollary 6.1.4 H has a perfect matching, so G has a matching which covers all the vertices of maximum degree. Finally, by Property 5.1.5 G also has a maximum matching which covers all the vertices of maximum degree. $\square$

Corollary 6.1.8   (Ore (1955)) *A bipartite graph G with bipartition (V_1, V_2) has a matching of cardinality $r \leq |V_1|$ if and only if*

$$|N(A)| \geq |A| + r - |V_1| \text{ for any } A \subseteq V_1.$$

Proof.   Let $V_1 = \{x_1, x_2, \ldots, x_n\}$ and $V_2 = \{y_1, y_2, \ldots, y_m\}$. Consider a bipartite graph H obtained by adding $n - r$ new vertices $y_{m+1}, \ldots, y_{m+n-r}$ to V_2 and joining each vertex of V_1 to all the new vertices.
By Hall's Theorem V_1 can be matched into $V_2' = \{y_1, \ldots, y_{m+n-r}\}$ if and only if for each $A \subseteq V_1$

$$|A| \leq |N_H(A)| = |N_G(A)| + n - r.$$

Thus, since each matching in H contains at most $n - r$ edges not in $E(G)$ the result follows. $\square$

Corollary 6.1.9   *Let G be a bipartite graph with bipartition (V_1, V_2). Then the number of edges in a maximum matching of G equals*

$$|V_1| + \min_{A \subseteq V_1} \{|N(A)| - |A|\}. \tag{6.1.1}$$

Proof.   Let $m(G)$ be the cardinality of a maximum matching of G. Corollary 6.1.8 implies that

$$m(G) \leq |V_1| + \min_{\emptyset \neq A \subseteq V_1} \{|N(A)| - |A|\}. \tag{6.1.2}$$

Since G has no matching with $1 + m(G)$ edges Corollary 6.1.8 also gives that

$$|N(A_1)| < |A_1| + 1 + m(G) - |V_1|$$

for some $A_1 \subset V_1$ and so

$$m(G) > |V_1| + \min_{A \subseteq V_1} \{|N(A)| - |A|\} - 1. \tag{6.1.3}$$

Thus combining (6.1.2) and (6.1.3) the result follows. $\square$

To conclude this section we return to the existence of perfect matchings, but this time for general graphs. Rather pleasingly the following famous theorem of Tutte (1947) can be deduced from Hall's Theorem. For a subset of vertices S of a graph G we denote by $c_G(S)$ the number of components of $G - S$ which have an odd number of vertices (odd components).

Theorem 6.1.10 (Tutte's Theorem) *Let G be a simple non-trivial graph. Then G has a perfect matching if and only if $|S| \geq c_G(S)$ for every $S \subset V(G)$.*

Proof. We follow the proof of Anderson (1971).

Necessity being trivial we need only prove the sufficiency of the condition. Setting $S = \emptyset$ we see that $|V(G)|$ must be even. Furthermore, we note that for each $S \subset V(G)$ the number $c_G(S)$ and $|S|$ are of the same parity, since $|V(G)|$ is even.

We proceed by induction on n where $|V(G)| = 2n$. The case $n = 1$ is trivial so we shall proceed to the induction step, which breaks into two cases.

Case A.

Suppose that $c_G(S) < |S|$ for all subsets S with $2 \leq |S| \leq 2n - 1$. Then $c_G(S) \leq |S| - 2$ because $c_G(S)$ and $|S|$ are of the same parity. Choose any edge $uv \in E(G)$, and consider $G' = G - \{u, v\}$. For any set $T \subset V(G')$ we must have $c_{G'}(T) \leq |T|$ for otherwise we have

$$c_G(T \cup \{u.v\}) = c_{G'}(T) > |T| = |T \cup \{u, v\}| - 2$$

and so $c_G(T \cup \{u, v\}) \geq |T \cup \{u, v\}|$ contradicting our assumptions. Thus, by induction, G' and hence G has a perfect matching.

Case B.

Suppose instead that there is a set S such that $c_G(S) = |S| \geq 2$. Let S be a such a subset with maximal cardinality and $|S| = k$. Further, let $G_1, ..., G_k$ denote the odd components of the graph $G' = G - S$. These are the only components of G', for if G_0 were a component of G' with even number of vertices then there is a vertex $u_0 \in V(G_0)$ such that $c_G(S \cup \{u_0\}) = |S \cup \{u_0\}|$, which contradicts the maximum property of S.

Our first step will be show that each of components $G_1, ..., G_k$ can be paired with a vertex of S to which it is connected in G. To do this we apply Hall's Theorem. We know that this pairing can be accomplished unless there are some k odd components which are only connected in G to a subset $T \subset S$ of $l < k$ vertices. But then we have $c_G(T) \geq k > |T|$ which is impossible. Thus Hall's Theorem ensures that we can take one vertex u_i from each odd component G_i and pair it with a vertex v_i of S.

For $i = 1, ..., k$ consider the graph $H_i = G_i - u_i$. If H_i contains some subset T of vertices such that $c_{H_i}(T) > |T|$, then $c_{H_i}(T) \geq |T| + 2$, so that

$$c_G(T \cup S \cup \{u_i\}) = c_G(S) + c_{H_i}(T) - 1 \geq |S| + |T| + 1 = |T \cup S \cup \{u_i\}|$$

which contradicts the maximum property of S. Thus, $c_{H_i}(T) \geq |T|$ for each $T \subset V(H_i)$. Therefore, by induction, H_i has a perfect matching for each $i = 1, ..., k$. These matchings together with edges $u_1 v_1, ..., u_k v_k$ produce a perfect matching of G.   □

Hall's Theorem and Tutte's Theorem are in fact equivalent. They are but two of a family of fundamental results in graph theory and combinatorics which are all equivalent to each other. For a fuller discussion of this topic we direct the reader to Jacobs (1969).

Exercises

6.1.1 △ Let G be a regular bipartite graph with bipartition (V_1, V_2). Show that if there exists a set $A \subseteq V_1$ with $|N(A)| < |A|$ then there exists a set $B \subseteq V_2$ with $|N(B)| < |B|$.

6.1.2 ▽ Let G be a bipartite graph with bipartition (V_1, V_2) where $|V_1| \geq |V_2|$ and $d_G(x) = d$ for each $x \in V_1$. Show that there exists $X \subseteq V_1$ so that the induced subgraph $G[X \cup N_G(X)]$ has a perfect matching.

6.1.3   Let G be a bipartite graph with bipartition (V_1, V_2). Show that G satisfies Hall's condition if $d_G(x) \geq d_G(y)$ for each edge xy with $x \in V_1$ and $y \in V_2$.

6.1.4   Deduce Hall's Theorem from Corollary 6.1.3.

6.1.5   Suppose that Hall's condition is satisfied in the marriage problem, and that each of the n boys knows at least t girls. Show that the marriages can be arranged in at least $t!$ ways if $t \leq n$ and in $t!/(t-n)!$ ways if $t > n$. (M. Hall (1948))

6.1.6 △ Show that every elementary bipartite graph on at least three vertices is 2-connected.

6.1.7 ▽ Show that an elementary bipartite graph G with p vertices and q edges has at least $q - p + 2$ distinct perfect matchings.

6.1.8 △ Find a polynomial algorithm for recognising an elementary bipartite graph.

6.1.9 ▽ Berge (1978) defines a graph G to be *regularisable* if a regular graph may be obtained from G by 'multiplying' each edge $e \in E(G)$ by a positive integer $k(e)$, where by multiplying we mean replacing e with $k(e)$ parallel edges connecting the ends of e. Show that a bipartite graph is regularisable if and only if it is elementary.

6.1.10 ▽ A simple graph G is called *n-extendable* $(n \leq (|V(G)| - 2)/2)$ if it has a perfect matching and every matching of cardinality n is a subset of some perfect matching of G. Prove that every n-extendable graph is also $(n-1)$-extendable. (Plummer (1980))

6.1.11   Show that a connected, bipartite graph G with bipartition (V_1, V_2) is n-extendable if and only if $|V_1| = |V_2|$ and $N(A)| \geq |A| + n$ for every $A \subset V_1$ with $|A| \leq |V_1| - n$.

6.1.12   Let G be an n-extendable bipartite graph with bipartition (V_1, V_2). Prove that $G + xy$ is n-extendable, for each pair of non-adjacent vertices $x \in V_1$, $y \in V_2$. (Yu (1992))

6.1.13 ▽ Prove that a k-regular, simple bipartite graph on m vertices is n-extendable if and only if $n \leq 2k - m$. (Brualdi, Csima (1986))

6.2 Expanding graphs

As we have seen, a graph G with bipartition (V_1, V_2) which satisfies Hall's condition must have a matching of V_1 into V_2. It is natural to ask whether a stronger condition might give more powerful matching theorems, or some other useful structure.

We define a graph to be *k-expanding* if $|N(X)| \geq |X| + k$ for any non-empty set $X \subseteq V_1$. Of course the case $k = 0$ is dealt with by Hall's Theorem. Our interest will be when $k \geq 1$.

Theorem 6.2.1 (Lovász (1970a), Las Vergnas (1970)) *A bipartite graph G, with bipartition (V_1, V_2), is k-expanding if and only if G has a spanning subgraph H which is itself k-expanding and for which $d_H(u) = k + 1$ for every vertex $u \in V_1$.*

Proof. It is clear that if H exists G must also be k-expanding. Thus we need only prove the necessity. Consider a subgraph H which is k-expanding and has the minimum number of edges. Choose an arbitrary vertex $u \in V_1$ and let $N_H(u) = \{v_1, \ldots, v_d\}$. Since H is k-expanding $d \geq k + 1$, and also since H is minimal, $G - uv_i$ is $(k-1)$-expanding, for $i = 1, \ldots, d$. Thus the graphs $H_i = H - uv_i$ must each contain a set $\emptyset \neq X_i \subseteq V_1$ with $|N_{H_i}(X_i)| = |X_i| + k - 1$, for all $i = 1, \ldots, d$. Then $|N_H(X_i,)| = |X_i| + k$ and $x \in X_i$.

Now consider two non-empty subsets $A, B \subseteq V_1$ for which $|N_H(A)| = |A| + k$, $|N_H(B)| = |B| + k$, and $A \cap B \neq \emptyset$. Then

$$|N_H(A \cup B)| + |N_H(A \cap B)| \leq |N_H(A) \cup N_H(B)| + |N_H(A) \cap N_H(B)|$$
$$= |N_H(A)| + |N_H(B)| = |A| + |B| + 2k.$$
$$(6.2.1)$$

On the other hand, since $A \cap B \neq \emptyset$

$$|N_H(A \cup B)| \geq |A \cup B| + k \quad \text{and} \quad |N_H(A \cap B)| \geq |A \cap B| + k$$

and so

$$|N_H(A \cup B)| + |N_H(A \cap B)| \geq |A \cup B| + |A \cap B| + 2k$$
$$= |A| + |B| + 2k. \tag{6.2.2}$$

Hence we have equality throughout (6.2.1) and (6.2.2), and, in particular, $|N_H(A \cap B)| = |A \cap B| + k$.

Returning to the X_i we see that since $x \in X_0 = X_1 \cap \ldots \cap X_d$ by the above we must have $|N_H(X_0)| = |X_0| + k$, and so we need only show that $X_0 = \{x\}$ to complete the proof. But this follows easily, since otherwise

$$|N_H(X_0)| = |N_H(x)| + |N_H(X_0 \backslash \{x\})| \geq k + 1 + |X_0| + k - 1 = |X_0| + 2k$$

giving a contradiction.   □

Corollary 6.2.2 *A bipartite graph with bipartition (V_1, V_2) is 1-expanding if and only if it contains a spanning forest F such that $d_F(u) = 2$ for every vertex $u \in V_1$*

Proof. Plainly, such a forest is a 1-expanding bipartite graph. On the other hand, by Theorem 6.2.1, a 1-expanding graph contains a subgraph F for which $d_F(u) = 2$ for every $u \in V_1$. Suppose that F contains a cycle C. Then $|N_F(V(C) \cap V_1)| = |V(C) \cap V_1|$, contradicting F being 1-expanding. Thus F is a forest and we are done. □

Observe that Corollary 6.2.2 implies that the graph $bip(T)$ is 1-expanding for each tree T. Moreover, any connected 1-expanding graph which is minimal with respect to edge deletion is $bip(T)$ for some tree T.

A graph G has a matching of V_1 into V_2 if and only if G is 0-expanding, but what can be said about matchings if G is k-expanding and $k \geq 1$? As an easy corollary of Hall's Theorem we have

Property 6.2.3 *Let G be a k-expanding bipartite graph, with bipartition (V_1, V_2), and U be any subset of k vertices from V_2. Then if there is a matching of U into V_1, this matching can be extended to a matching of V_1 into V_2.*
                                                                              □

In general, the property that every matching of k edges can be extended to a matching of V_1 into V_2 is not sufficient to ensure that the graph is k-expanding; but in the case $k = 1$ we have

Property 6.2.4 *A connected bipartite graph G with bipartition (V_1, V_2) is 1-expanding if and only if every edge is contained in a matching of V_1 into V_2.*
                                                                              □

The proof of Property 6.2.4 is essentially identical to that of Theorem 6.1.5. In the case when $k = |V_2| - |V_1|$ we can also say rather more.

Proposition 6.2.5 *Let G be a bipartite graph with bipartition (V_1, V_2) where $m = |V_1| < |V_2| = n$. Then G is $(n - m)$-expanding if and only if for every subset $U \subset V_2$ of m vertices there is a matching of V_1 into U.*

Proof. Let G be $(n - m)$-expanding and $U \subset V_2$ be some subset of m vertices. Then for any subset $X \subset V_1$ we have $|N_G(X) \cap U| \geq |X| + n - m - (n - m) = |X|$, and so by Hall's Theorem there is a matching of V_1 into U.

Conversely suppose that there is a matching of V_1 into any subset of m vertices, and yet there is some subset $\emptyset \neq X \subseteq V_1$ with $|N_G(X)| \leq |X| + n - m - 1$. Then we may choose subsets $U_1 \subset N_G(X)$ with $|U_1| = |X| - 1$ and $U_2 \subset V_2 \backslash N_G(X)$ with $|U_2| = m - |X| + 1$. However, then $U = U_1 \cup U_2$ is a set of m vertices so that the graph $H = G[V_1 \cup U]$ does not have a perfect matching, since $|N_H(X)| = |N_G(X) \cap U| = |X| - 1$; a contradiction. $\square$

In other words Proposition 6.2.5 means that an $(n - m)$-expanding bipartite graph is an (n, m)-concentrator (see Section 4.3) with input set V_2 and output set V_1.

Property 6.2.4 gives an easy method to recognise if a graph is 1-expanding. Conveniently, we can recognise if a graph is k-expanding by trying to find matchings in some auxiliary graphs. The next theorem, a variation of a result of Blum et al. (1981), which appears in the book of Lovász and Plummer (1986), deals with exactly this and shows that there is always a polynomial algorithm.

Given a vertex $x \in V_1$ we denote by $G(x, k)$ the graph obtained by adding k new vertices to V_1 each joined to the vertices of $N_G(x)$.

Theorem 6.2.6 *Let $k \geq 1$ and G be a bipartite graph with bipartition (V_1, V_2) where $|V_1| < |V_2|$. Then G is k-expanding if and only if for every $x \in V_1$ the graph $G(x, k)$ has a matching of cardinality $k + |V_1|$.*

Proof. Let G be k-expanding and let $x \in V_1$, and $x_1, \ldots, x_k$ be the new vertices to be added to V_1 in the graph $H = G(x, k)$. Then for every subset $A \subseteq V_1$ we have $|N_H(A)| = |N_G(A)| \geq |A| + k \geq |A|$ and for every $A \subseteq V_1 \cup \{x_1, \ldots, x_k\}$ for which $A \cap \{x_1, \ldots, x_k\} \neq \emptyset$ we have

$$|N_H(A)| = |N_G(V_1 \cap A \cup \{x\})| \geq |V_1 \cap A \cup \{x\}| + k$$
$$\geq |V_1 \cap A| + k \geq |A|.$$

Thus, by Hall's Theorem $G(x, k)$ has a matching of cardinality $|V_1| + k$.

On the other hand, suppose that k is an integer satisfying the conditions of the theorem. Then let $\emptyset \neq A \subseteq V_1$, and $x \in A$. Consider the graph $H = G(x, k)$, with new vertices $x_1, \ldots, x_k$. Then by the hypothesis H has a matching of $V_1 \cup \{x_1, \ldots, x_k\}$ into V_2. Thus

$$|N_G(A)| = |N_H(A \cup \{x_1, \ldots, x_k\})| \geq |A \cup \{x_1, \ldots, x_k\}| = |A| + k. \quad \square$$

Exercises

6.2.1 △ Construct a bipartite graph which is not k-expanding, in which every matching of cardinality k can be extended to a matching of V_1 into V_2.

6.2.2   Is it true in a k-expanding bipartite graph G that for every set of k vertices $U \subseteq V_2$ there is a matching of V_1 into V_2 which covers U?

6.2.3   Show that for each pair of positive integers p and Δ ($p \geq 1 + \Delta$) there exists an $(n - m)$-expanding graph G with $p = n + m$ vertices and maximum degree Δ.

6.2.4   Let $\sigma(G) = \min_{\emptyset \neq A \subseteq V_1}(|N(A)| - |A|)$. A set $A \subseteq V_1$ is called σ-*critical* if $\sigma(G) = |N(A)| - |A|$ and a σ-critical set is called maximal if it is not a proper subset of a σ-critical set.

(a)   Show that if $\sigma(G) < 0$ then any two σ-critical sets have a non-empty intersection, and G has a unique maximal σ-critical set.

(b)   Show that if $\sigma(G) \geq 0$ then any two maximal σ-critical sets are disjoint.

(c)   Construct bipartite graphs G and H with $\sigma(G) > 0$ and $\sigma(H) < 0$.

6.2.5 ▽ Let G be a k-expanding bipartite graph where $d_G(x) = k + 1$ for each $x \in V_1$. Show that V_1 can be partitioned into maximal σ-critical sets. (Suškov (1975))

6.3 Expanders

In the previous section we considered an additive condition on the size of neighbourhoods which gave rise to the definition of k-expanding graphs. In this section we shall insist on a multiplicative condition. We define an $(n, \gamma, \Delta, \lambda)$-*expander* to be an n by n bipartite graph with maximum degree Δ, and bipartition (V_1, V_2) such that if $X \subset V_1$ and $|X| \leq \gamma |V_1|$ then $|N(X)| \geq |X| + \lambda |X|(1 - |X|/n)$. We call an $(n, 1, \Delta, \lambda)$-expander simply an (n, Δ, λ)-expander. Let us begin by showing the existence of such graphs.

Proposition 6.3.1 *For any $n, \lambda \geq 1$ and $\gamma < 1$ with $\gamma\lambda < 1$ there is an*
$$(n, \gamma, \Delta, \lambda - 1)\text{-expander for any } \Delta \geq \frac{1 + \log_2 \lambda + (\lambda + 1)\log_2 e}{-\log_2(\lambda\gamma)} + \lambda + 1.$$

Proof. To prove the theorem we shall consider the set of all Δ-regular n by n bipartite graphs $G_{n,\Delta}$, and estimate the proportion of these which are not $(n, \gamma, \Delta, \lambda-1)$-expanders; these we call **bad** graphs. The result will follow by showing that the proportion of bad graphs is strictly less than 1, provided Δ is as stated, and so there must be at least one $(n, \gamma, \Delta, \lambda - 1)$-expander.

Let us begin by considering a way to generate the members of $G_{n,\Delta}$. To do this begin with the vertex sets V_1 and V_2. Replace each vertex in V_1 and V_2 with a set of Δ vertices; this gives two new vertex sets V_1' and V_2', each of Δn vertices. Now let G' be the bipartite graph which has bipartition (V_1', V_2') whose edges consist of a perfect matching from V_1' to V_2', and let G be the graph obtained from G' by re-identifying each of the sets of Δ vertices with a single vertex, to give a bipartite graph with bipartition (V_1, V_2). Then $G \in G_{n,\Delta}$ as required. This done, counting the proportion of bad graphs is relatively easy.

If G is bad then there must be some subset $U \subset V_1$ with $|U| = z \leq \gamma n$ whose neighbourhood has cardinality at most $(1 + (\lambda - 1)(1 - z/n))z \leq \lambda z$. Let g_z be the proportion of graphs which have such a subset; then, by using the above construction we see that

$$g_z = \binom{n}{z}\binom{n}{\lfloor \lambda z \rfloor}\frac{\binom{\lfloor \lambda z \rfloor \Delta}{\Delta z}}{\binom{n\Delta}{z\Delta}}. \tag{6.3.1}$$

By using the approximation that $\binom{n}{z} \leq (en/z)^z$ in (6.3.1) we see that

$$g_z \leq \left(\frac{en}{z}\right)^z \left(\frac{en}{\lfloor \lambda z \rfloor}\right)^{\lfloor \lambda z \rfloor} \left(\frac{\lfloor \lambda z \rfloor}{n}\right)^{z\Delta} \leq \left(e^{\lambda+1}\frac{\lfloor \lambda z \rfloor^{\Delta-\lambda}}{zn^{\Delta-\lambda-1}}\right)^z.$$

Then, since $\Delta \geq \lambda + 1$ and $z \leq \gamma n$

$$g_z \leq \left(e^{\lambda+1}\left(\frac{\lambda z}{n}\right)^{\Delta-\lambda-1}\lambda\right)^z \leq \left(e^{\lambda+1}(\gamma\lambda)^{\Delta-\lambda-1}\lambda\right)^z.$$

When $\Delta \geq \dfrac{1 + \log_2 \lambda + (\lambda + 1)\log_2 e}{-\log_2(\lambda\gamma)} + \lambda + 1$ we see that $g_z \leq 2^{-z}$ and so the proportion of bad graphs in $G_{n,\Delta}$ is

$$\sum_{z=1}^{\lfloor \gamma n \rfloor} g_z \leq \sum_{z=1}^{\lfloor \gamma n \rfloor} 2^{-z} < 1. \qquad \square$$

Although this result shows that expanders certainly do exist, they are rather difficult to construct explicitly. It was Margulis (1973) who first succeeded in giving an explicit construction of an infinite family of expanders.

Theorem 6.3.2 (Margulis (1973)) *For $m \geq 4$, let V_1 and V_2 be two disjoint copies of $\mathbb{Z}_m \times \mathbb{Z}_m$ (where $\mathbb{Z}_m$ is the set of integers modulo m), and define a bipartite graph G_m with bipartition (V_1, V_2) by joining $(x, y) \in V_1$ to (x, y), $(x + 1, y)$, $(x, y + 1)$, $(x, x + y)$ and $(-y, x)$ in V_2. Then G_n is an $(m^2, 1/\lambda, 5, \lambda)$-expander, for some constant $\lambda > 0$.* $\qquad \square$

Often it is exactly explicit expanders that are required for applications, and so it would be very convenient to have some technique to test the expansion properties of a given graph. To do this precisely is not really feasible – even the problem of deciding if a graph is an $(n, 1/2, \Delta, 0)$-expander is co-NP-complete (see Blum et al. (1981)). However, in the remainder of this section we shall give some tools to provide some bounds; results which show that the expansion properties of a graph are intimately connected with the eigenvalues of certain associated matrices.

Proposition 6.3.3 (Tanner (1984)) *Let G be a simple n by n, r-regular bipartite graph, with bipartition (V_1, V_2) and with biadjacency matrix $\mathbf{B}(G)$. Let the spectrum of the matrix $\mathbf{B}(G)\mathbf{B}(G)^{\mathrm{T}}$ be $\lambda_1 \geq \lambda_2 \geq \ldots \geq \lambda_n$ with $\lambda_1 > \lambda_2$. Then if $X \subset V_1$, and $\alpha = |X|/n$ we have*

$$|N(X)| \geq \left(\frac{r^2}{\alpha(r^2 - \lambda_2) + \lambda_2} \right) |X|.$$

Proof. Let $V_1 = \{u_1, \ldots, u_n\}$ and $V_2 = \{v_1, \ldots, v_n\}$ and let $\mathbf{B} = \mathbf{B}(G)$. Let the row vector $\mathbf{x}$ be the characteristic vector of some set $X \subset V_1$ of size $|X| = \alpha n$. That is $\mathbf{x} = (x_i)_{1 \leq i \leq n}$ and

$$x_i = \begin{cases} 1 & \text{if } u_i \in X, \\ 0 & \text{otherwise.} \end{cases}$$

Let $||\mathbf{x}||$ denote the norm of $\mathbf{x}$, $||\mathbf{x}|| = \left(\sum_{i=1}^n x_i^2 \right)^{1/2}$. Then $\mathbf{x}\mathbf{x}^{\mathrm{T}} = ||\mathbf{x}||^2 = \alpha n$. Let $Y = N(X)$ and $\mathbf{y}$ be its characteristic row vector. Our task is to bound $|Y|$.

Observe that if $N(u_i) \cap N(u_j) = \emptyset$ for all $i \neq j$ then $\mathbf{x}\mathbf{B} = \mathbf{y}$,

$$\mathbf{x}\mathbf{B}\mathbf{B}^{\mathrm{T}}\mathbf{x}^{\mathrm{T}} = \alpha n r = \mathbf{x}\mathbf{B}(1, \ldots, 1)^{\mathrm{T}},$$

and $|Y| = ||\mathbf{x}\mathbf{B}||^2 = \alpha n r$. It can easily be seen that any intersections between the neighbourhoods will cause $\mathbf{x}\mathbf{B}\mathbf{B}^{\mathrm{T}}\mathbf{x}^{\mathrm{T}}$ to increase. We shall bound $|Y|$ below by bounding $\mathbf{x}\mathbf{B}\mathbf{B}^{\mathrm{T}}\mathbf{x}^{\mathrm{T}}$ above.

Let $\mathbf{z} = \mathbf{x}\mathbf{B}$. By the convexity of the function $f(x) = x^2$ we may use Jensen's inequality to obtain

$$||\mathbf{z}||^2 = \sum_{j=1}^m z_j^2 \geq \left(\sum_{j=1}^m z_j/|Y| \right)^2 |Y| = (\alpha n r)^2/|Y|.$$

Thus

$$|Y| \geq \frac{(\alpha n r)^2}{||\mathbf{z}||^2}.$$

Recall that since $\mathbf{BB}^T$ is a real symmetric non-negative definite matrix (i.e. $\mathbf{xBB}^T\mathbf{x}^T \geq 0$ for all $\mathbf{x}$), it is diagonalisable, has real non-negative eigenvalues, and orthogonal eigenvectors, $\mathbf{e}_1, \ldots, \mathbf{e}_n$. We may express $\mathbf{x}$ in terms of these eigenvectors as $\mathbf{x} = \sum_{i=1}^n \gamma_i \mathbf{e}_i$. Then $\mathbf{xBB}^T = \sum_{i=1}^n \lambda_i \gamma_i \mathbf{e}_i$ and $||\mathbf{z}||^2 = \sum_{i=1}^n \lambda_i \gamma_i^2$, since the eigenvectors are orthogonal.

Since G is r-regular its maximum eigenvalue is r (see Exercise 2.3.5) and so $\lambda_1 = r^2$. Furthermore, we may take $\mathbf{e}_1 = (1, \ldots, 1)/\sqrt{n}$, without loss of generality. This implies that $\gamma_1 = \mathbf{xe}_1^T = \alpha\sqrt{n}$. Thus

$$||\mathbf{z}||^2 = \lambda_1\gamma_1^2 + \sum_{i=2}^n \lambda_i\gamma_i^2 = \lambda_1\alpha^2 n + \sum_{i=2}^n \lambda_i\gamma_i^2$$

$$\leq \lambda_1\alpha^2 n + \lambda_2 \sum_{i=2}^n \gamma_i^2 = \alpha^2 n(\lambda_1 - \lambda_2) + \lambda_2||\mathbf{x}||^2$$

$$\leq \alpha^2 n(r^2 - \lambda_2) + \lambda_2\alpha n.$$

The result follows. □

Given a simple, r-regular, n by n bipartite graph G we define the matrix $\mathbf{Q}(G) = r\mathbf{I} - \mathbf{A}(G)$, where $\mathbf{I}$ is the identity matrix of order $2n$ and $\mathbf{A}(G)$ is the adjacency matrix of G. Let $\xi(G)$ be the second smallest eigenvalue in the spectrum of $\mathbf{Q}(G)$. The following result shows that we need only calculate $\xi(G)$ to assess the expanding properties of G.

Theorem 6.3.4 (Alon (1986)) *Let G be a simple, r-regular, n by n bipartite graph. Then G is an (n, r, λ)-expander, where*

$$\lambda = \frac{2r\xi(G) - \xi(G)^2}{r^2}.$$

Proof. Let $\mathbf{B} = \mathbf{B}(G)$ be the biadjacency matrix of G and $\mathbf{A} = \mathbf{A}(G)$ be the adjacency matrix of G. Then since $r\mathbf{I} - \mathbf{Q} = \mathbf{A}(G)$ it is easy to see that

$$(r\mathbf{I} - \mathbf{Q})^2 = \begin{bmatrix} \mathbf{BB}^T & 0 \\ 0 & \mathbf{B}^T\mathbf{B} \end{bmatrix}.$$

From the definition of $\mathbf{Q}(G)$ we have that the two largest eigenvalues of $r\mathbf{I} - \mathbf{Q}$ are r and $r - \xi(G)$. The symmetry of the spectrum of G (see Proposition 2.3.3) also implies that the two smallest eigenvalues of $r\mathbf{I} - \mathbf{Q}$ are $-r$ and $-(r - \xi(G))$. Thus the two largest eigenvalues of $\mathbf{BB}^T$ are r^2 and $(r - \xi(G))^2$. Observe that if $\xi(G) = 0$ then the statement of the theorem amounts to establishing that G satisfies Hall's Theorem, which follows easily. On the other hand if $\xi(G) > 0$ we may apply Proposition 6.3.3 to see that any set $X \subset V_1$ of size $|X| = \alpha n$ must have a neighbourhood at least as large as

$$|N(X)| \geq \frac{r^2}{\alpha(r^2 - (r - \xi(G))^2) + (r - \xi(G))^2}|X|$$

$$= \left(1 + \frac{(2r\xi(G) - \xi(G)^2)(1 - \alpha)}{r^2 - (2r\xi(G) - \xi(G)^2)(1 - \alpha)}\right)|X|$$

$$\geq \left(1 + \frac{(2r\xi(G) - \xi(G)^2)}{r^2}\left(1 - \frac{|X|}{n}\right)\right)|X|. \qquad \square$$

Of course, besides having a good test for the expansive properties of graphs, we should also wish to have explicit constructions of some classes of sparse, yet highly expanding graphs. One such example arises from bipartite graphs associated with finite projective geometries. To close this section we shall give this construction.

Let $d,q > 2$ be integers. Then a *d-dimensional finite projective geometry* of order q consists of a finite set of *points* and a finite set of *hyperplanes*, which have the properties that

> *(1)* every d points lie on one and only one common hyperplane,
>
> *(2)* every d hyperplanes pass through one and only one common point,
>
> *(3)* every hyperplane must pass through precisely $k = (q^d - 1)(q - 1)$ points, and every point must lie on precisely k hyperplanes.

It is well known that such a geometry always exists if q is a prime power. Furthermore any finite projective geometry of dimension d and order q must have $n = (q^{d+1} - 1)/(q - 1)$ points and n hyperplanes, and each two points must lie on $\lambda = (q^{d-1} - 1)/(q - 1)$ common hyperplanes. We refer the reader to M. Hall's book (1967) for basic facts about finite projective geometries, and proofs of the above, although this information will suffice for our purposes.

Let V_1 and V_2 be the sets of points and hyperplanes of a finite projective geometry of dimension d and order q. Let $G(d,q)$ be the bipartite graph with bipartition (V_1, V_2) in which $p \in V_1$ is joined to $h \in V_2$ if the point p is incident with the hyperplane h. We shall show that $G(d,q)$ is in general a highly expanding graph.

Proposition 6.3.5 *Let $n = (q^{d+1} - 1)/(q - 1)$, and $k = (q^d - 1)/(q - 1)$. Then $G(d,q)$ is a k-regular n by n bipartite graph, where for any $X \subset V_1$*

$$|N(X)| \geq n - \frac{n^{1+1/d}}{|X|}.$$

Proof. The first properties follow directly from the definition of $G(d, q)$ and basic properties of finite projective geometries. The bound on the neighbourhood of X follows from Proposition 6.3.3. To see this we must first establish the appropriate eigenvalue. Let $\mathbf{B} = \mathbf{B}(G(d, q))$.

Due to the structure of $G(d, q)$ every row of $\mathbf{B}$ contains precisely k 1's and every two rows contain λ common 1's. Thus it is an easy task to see that $\mathbf{BB}^{\mathrm{T}} = \lambda \mathbf{J} + (k - \lambda)\mathbf{I}$, where $\lambda = (q^{d-1} - 1)/(q - 1)$, $\mathbf{I}$ is the $n \times n$ identity matrix and $\mathbf{J}$ the $n \times n$ matrix of all 1's. It follows that the eigenvalues of $\mathbf{BB}^{\mathrm{T}}$ are $\lambda_1 = k^2$ and $\lambda_2 = \ldots = \lambda_n = k - \lambda$ and applying Proposition 6.3.3 with $|X| = \alpha n$ gives the bound we require,

$$
\begin{aligned}
|N(X)| &\geq \frac{k^2}{\alpha(k^2 - (k - \lambda)) + (k - \lambda)}|X| \\
&= \frac{(n-1)^2 \alpha n}{\alpha(n^2 - n(q+1)) + n(q-1) + 1} \\
&= n - \frac{(n - |X|)(n(q-1) + 1)}{n(q-1) + 1 + (n - q - 1)|X|} \\
&\geq n - \frac{n(n(q-1) + 1)}{(n - q - 1)|X|} \geq n - \frac{nq}{|X|} \geq n - \frac{n^{1+1/d}}{|X|}. \qquad \square
\end{aligned}
$$

Exercise

6.3.1 Prove that for any $n \geq 1$, $\lambda \geq 1$ and $\gamma < 1$ such that $\gamma\lambda < 1$ there is a Δ-regular, n by n bipartite graph, in which $|N(X)| \geq \lambda|X|$ for any $X \subset V(G)$ with $|X| \leq \gamma n$ for some suitable Δ which depends only on γ and λ.

Application

6.4 Expanders and sorters

Suppose that we have some collection of n objects which we must sort by using a sequence of pairwise comparisons: is a less than b? Normally we will think of the objects as being the numbers $1, 2, \ldots, n$, but this type of task occurs in innumerable different guises, and so it is very natural to ask how quickly such a sorting process can be carried out. It is well known that any successful algorithm which sorts n objects must use at least $\mathrm{O}(n \log_2 n)$ comparisons, and what is more, there are algorithms which achieve this theoretical lower bound (see Knuth (1973)). However, in these days of parallel computation there is the opportunity to carry out a number of independent comparisons as one single operation. This suggests an alternative computational model, in which we carry out independent comparisons in a number of parallel steps or *rounds*, and between each of these rounds have some form of processing of the gathered information. Since it is only possible to carry out $n/2$ independent comparisons in each round we must use at least $C \log_2 n$ rounds. The aim of this section is to show that indeed this sorting process can be carried out in $C \log_2 n$ rounds; a result of Ajtai, Komlós and Szemerédi (1983) which utilises the structure of a special class of expanding graphs. Actually we shall deal with an improved version of their original proof due to Paterson (1990).

Suppose that we had a *halving algorithm* which, given a set of elements, sorted them into two (as near as possible) equally sized sets L and R, so that every element of L is less than every element of R. Then, by recursively applying the halving algorithm first to the whole set, then to each of L and R to produce four sets, and so on, we could create a binary tree-like structure of depth $\log_2 n$ in which the n elements would be sorted correctly at the leaves. Since we could carry out each of the halvings on any level of the tree in parallel, we would sort the entire collection of elements in $\log_2 n$ rounds.

Our basic building blocks will be comparison algorithms with a particularly simple form. Let $F_1, \ldots, F_l$ be matchings in the complete graph K_m with vertex set $\{1, \ldots, m\}$. Then we can define a *comparator network* (which we denote by $\mathcal{C}[F_1, \ldots, F_l; m])$ which reorders the elements $\{x_1, \ldots, x_m\}$ according to the following algorithm:

- Place the elements into registers marked $1, \ldots, m$.

- Round 1: Compare the elements in registers joined by an edge in F_1, taking the edges one by one; if the larger numbered register contains the smaller element then swap the contents of the registers, otherwise do nothing.

- Round 2: Compare the elements using the edges in F_2, interchanging the contents of registers where necessary.

$\vdots$

- Round l: Compare the elements using the edges of F_l.

Then if we regard the initial ordering of $x_1, \ldots, x_m$ as a permutation $\pi \in S_m$, then each $\mathcal{C}[F_1, \ldots, F_l; m]$ can be regarded as a function of permutations.

We say that a permutation $\pi \in S_m$ (for some $m \geq 1$) is *ε-halved* if for every $K = \{1, \ldots, k\}$, with $k \leq m/2$, the inequality $\pi(i) \leq m/2$ holds for all but at most $\varepsilon|K|$ members i of K, and similarly that for every $K' = \{k, k+1, \ldots, m\}$, $k \geq m/2$, the inequality $\pi(i) \geq m/2$ for all but at most $\varepsilon|K|$ members i of K'. A network of comparisons is an *ε-halver* if it accepts an arbitrary permutation π and produces an ε-halved permutation. If we base the edge sets $F_1, \ldots, F_l$ on the edges of a certain type of expanding bipartite graph then we can build an ε-halver. The proof of the following lemma is almost identical to that of Proposition 6.3.1.

Lemma 6.4.1 *For every $\varepsilon > 0$ there exists an $l = l(\varepsilon)$ so that for every $m \in \mathbb{N}$ there is an $\lfloor m/2 \rfloor$ by $\lceil m/2 \rceil$ bipartite graph G, with bipartition (V_1, V_2), and maximum degree l, for which if $X \subset V_1$ and $|X| \leq \varepsilon|V_1|$, or $X \subset V_2$ and $|X| \leq \varepsilon|V_2|$, we have $|N(X)| \geq (1 - \varepsilon)|X|/\varepsilon$.* ☐

Lemma 6.4.2 *For every $\varepsilon > 0$ there is a constant $l = l(\varepsilon)$ such that for every $m \in \mathbb{N}$ there is an ε-halver of the form $C[F_1, \ldots, F_l; m]$.*

Proof. Let G_m be a bipartite graph with colour classes $V_1 = \{1, \ldots, \lfloor m/2 \rfloor\}$ and $V_2 = \{\lfloor m/2 \rfloor + 1, \ldots, m\}$ with the properties given in Lemma 6.4.1, and let $F_1, \ldots, F_l$ be a partition of the edge set of G_m into matchings. We claim that $C[F_1, \ldots, F_l; m]$ is an ε-halver.

First notice that if $i \in V_1$ and $j \in V_2$ with $ij \in E(G_m)$, then at the end of the algorithm the contents of register i will be less than the contents of register j. This will be true after the edge occurs in the algorithm, and, after this, the contents of register i can only decrease, whilst the contents of register j can only increase.

Suppose that we have some $K = \{1, \ldots, k\}$, $k \leq m/2$, and some subset $X \subset K$ with $|X| > \varepsilon|K|$ so that the elements of X end up in the registers forming some subset $Y \subset V_2$. Then we have that $|N(Y)| \geq (1 - \varepsilon)k$, by the properties of G_m, and also, by the above, each register in $N(Y)$ has a content which is at most as large as the content of its neighbour in Y, in other words at most k. But this is impossible, since $|N(Y) \cup Y| > k$. Thus any such X can have at most $\varepsilon|K|$ elements.

The argument for K' goes similarly. ☐

We call a comparator network a $(\lambda, \varepsilon, \varepsilon_0)$-*separator* (on m inputs) if it returns a partition of its m input values into four parts FL, CL, CR and FR of sizes $\lambda m/2$, $(1 - \lambda)m/2$, $(1 - \lambda)m/2$ and $\lambda m/2$ respectively. The subset FL (the 'far left') has the property that for any $k \leq \lambda m/2$ it contains all but at most εk of the least k elements in the true order of its inputs. Also, for any $k \leq m/2$ the CL ('centre left') and FL together contain all but $\varepsilon_0 k$ of the least k elements in the true order of the m inputs. The subsets FR and CR have the analogous properties on the 'far' and 'centre' right.

Using several ε_0-halvers we can construct a $(\lambda, \varepsilon, \varepsilon_0)$-separator which will be sufficient for our purposes. If we apply an ε_0-halver to the m inputs of the separator, and then apply it again to each of the output $m/2$ sets, and then repeatedly apply ε_0-halvers to the extreme $m/4$ sets etc. (see figure 6.4.1), then in p such stages the resulting network is a $(2^{-p+1}, p\varepsilon_0, \varepsilon_0)$-separator with the extreme sets forming FL and FR and the centre sets together

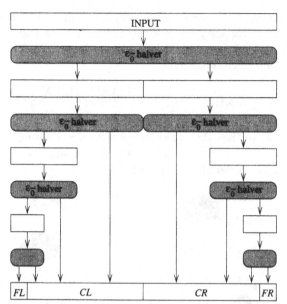

Figure 6.4.1 The construction of a $(1/8, 4\varepsilon_0, \varepsilon_0)$-separator from ε_0-halvers.

forming CL and CR. It will be enough for us to take $p = 4$, $\lambda = 1/8$, $\varepsilon_0 = 1/72$ and $\varepsilon = 1/18$.

In the remainder of this section we shall describe how these basic building blocks are used in Paterson's parallel sorting algorithm. However, before this, it is worth observing that the construction of our $(\lambda, \varepsilon, \varepsilon_0)$-separators depends solely on the existence of a certain type of expanding bipartite graph, and ingenious use of the properties of such graphs.

The actual sorting algorithm is structured just as our naïve approach was, as a binary tree, with $(\lambda, \varepsilon, \varepsilon_0)$-separators playing the parts of the halving algorithm. Associated with each vertex of the tree is a *bag* which contains a number of elements constrained by the bag's *capacity*. The algorithm works in *stages*. At each stage, every full bag is approximately halved by a $(\lambda, p\varepsilon_0, \varepsilon_0)$-separator. At each such vertex the elements which were put into FL and FR are sent back up the tree to that vertex's parent bag, whilst the elements in CL and CR are respectively sent down the tree to the bags of that vertex's left and right children, respectively. We begin at stage 1 with all the n elements in the bag at the root – level 0. Thus at odd numbered stages the bags on odd levels are empty, and some of the bags on the even levels are full; at even stages the roles are reversed. Our aim is to vary the capacities of the bags in such a way that the elements are gradually squeezed down the tree by carefully controlling the capacities of the bags.

We let the capacity of each bag at level d be $r(t) \cdot A^d$, for some function $r(t)$

of the stage number t. Consider a bag which is empty at the beginning of some stage and has capacity b (see figure 6.4.2). At the end of the stage it will be filled to its new capacity νb by elements it receives from its children and its parent. Thus we must have

$$\nu b = 2\lambda b A + \frac{(1-\lambda)b}{2A} = \left(\frac{4\lambda A^2 + (1-\lambda)}{2A}\right) b.$$

We require $\nu < 1$, so that the capacities of the bags diminish at each stage. We shall take $A = 3$ and $\nu = 43/48$. Then the capacity of a bag at level d and stage t will be $cn\nu^t A^d$ for some constant c.

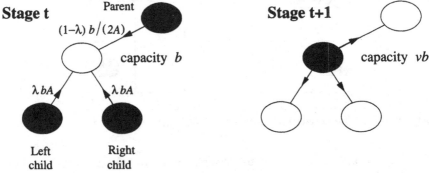

Figure 6.4.2 Reduction of bag capacities at each stage

Now that we have control over the bag capacities we turn to keeping track of how well the algorithm is sorting. The 2^d bags at each level d are associated with a partition of the n elements into intervals in the linear order; each interval being the elements which should lie in that bag if all the halvings were perfect. Using this sequence of partitions we can define the *strangeness* of an element at a given stage as a measure of how far from its correct position it is. The strangeness is calculated according to the following rules:

(1) all elements in the root bag have strangeness 0;

(2) the strangeness of an element, if non-zero, decreases by one if that element is sent back to its bag's parent;

(3) the strangeness of an element increases by one if it is sent to the wrong child bag.

For any bag B and integer $j > 0$ we define $S_j(B)$ to be the proportion of elements in B which have strangeness j or more. We shall show that the $S_j(B)$'s remain bounded throughout the algorithm; indeed that

$$S_j(B) < \mu \delta^{j-1} = s_j \tag{6.4.1}$$

for all B, $j \geq 1$, and at every stage, with $\mu = 1/36$ and $\delta = 1/40$.

Suppose that (6.4.1) held after the previous stage at every bag. Consider a bag B, with new capacity νb and old capacities of its parent and children b/A and bA respectively. Then when $j > 1$, the elements of strangeness at least j find their way into B either as elements which had strangeness $j + 1$ and came from its children or as elements which had strangeness $j - 1$ and were wrongly sent down from its parent. With $\mu \leq \lambda/2$ we ensure that the FL and FR parts of each separator are sufficiently large to accommodate the elements of positive strangeness, and so only at most a proportion ε of these are sent down to B from its parent. Thus the new strangeness of B, $S'_j(B)$, must satisfy

$$S'_j(B)\nu b < 2bA\mu\delta^j + \frac{\varepsilon b}{A}\mu\delta^{j-2} \quad \text{for } j > 1.$$

Thus we can ensure that (6.4.1) holds at this new stage by choosing our parameters such that $2A^2\delta^2 + \varepsilon \leq \nu A\delta$, an inequality which is satisfied by our choices.

It remains only to see that $S_1(B)$ remains bounded, but this is a little more complicated. The term from elements of strangeness at least 2 which are sent up from B's children appears just as before, but the term from B's parent is a little different.

If it were true that when B's parent bag is halved it contains equal numbers of elements which should be sent to B and elements which should be sent to B's sibling bag, C, then the only strange elements which would appear in B would be there because of the mistakes of the separator. Unfortunately we cannot guarantee this good behaviour, and in general the parent bag may have far too many elements which should go into B's sibling's bag C. Thus to estimate $S_1(B)$ we must estimate how large this surplus can be. Let V be the set of elements which would have strangeness zero in bag C. Then the 'natural' place for these elements to lie is filling all the bags on even layers of the subtree rooted at C, as well as half of B's parent, one eighth of B's great-grandparent, and so on. In reality things are not this way and the elements of V have been displaced from their natural space.

The bag of a child of C may contain up to s_2bA elements from outside V. A bag two levels down may contain s_4bA^3 elements from outside V and so on. Thus the total number of elements intruding into V's natural space in levels below that of B is at most

$$2s_2bA + 8s_4bA^3 + 32s_6bA^5 + \ldots < 2\frac{\mu\delta bA}{1 - 4\delta^2A^2}.$$

In the area above B's parent there may be no elements from V at all. This space amounts to

$$\frac{b}{8A^3} + \frac{b}{32A^5} + \ldots = \frac{b}{8A^3 - 2A}.$$

Thus taking into consideration the up to $s_1 b/A$ strangers in B's parent, even if the halving of B's parent were done perfectly, some excess of elements from V and other strangers would spill across into B. This surplus is certainly bounded by the sum of the three contributions given above, together with a further term of $b\varepsilon_0/(2A)$ from the separator error. Thus finally we have

$$S_1'(B)\nu b < \left[2\frac{\mu\delta A}{1 - 4\delta^2 A^2} + \frac{1}{8A^3 - 2A} + \frac{\mu}{A} + \frac{\varepsilon_0}{2A} \right] b$$

which with our choice of parameters is less than $\mu\nu b$ as required.

There are various technical details which still remain unresolved in the above analysis. But these problems can be resolved, and, for want of space, and to avoid clouding the major structure of the algorithm, we invite the interested reader to consult Paterson's and Ajtai, Komlós and Szemerédi's papers. Instead we shall turn to the final stages of the algorithm.

At an odd stage, when all the odd level bags are empty, the number of elements at level $2j$ which are in the wrong half of the tree is bounded above by $rA^{2j}s_{2j} = rA^{2j}\mu\delta^{2j-1} \le rA^2\mu\delta$ since $A\delta \le 1$. Thus the number of elements which are in the wrong half of the tree is less than 1, i.e. 0, by the time r has reduced to $1/(A^2\mu\delta) = 160$.

At this time the number of elements in the root bag is at most 160, and so we can split this bag precisely in constant time, and what is more we can split the binary tree into two new binary trees rooted at the children of the old root. In similar fashion, from this stage on, we can make such a 'tree split' at regular bounded intervals, resulting in a rapidly growing forest of independent sorting trees. Finally, once these trees reduce to a conveniently small size, we can complete the ordering of the n elements by sorting in each remaining tree by some simple sorting algorithm, in constant time. It remains to estimate the number of stages required. Notice that bags at level $\log_2 n$ initially have capacity $\Theta(nA^{\log_2 n})$. In the final stage these bags have capacity $\Theta(1)$, since they are within a bounded number of levels of bags of bounded capacity. Thus the number k of stages must satisfy $nA^{\log_2 n}\nu^k = \Theta(1)$ and so $k = \log_2 n \left(\frac{\log_2 2A}{-\log_2 \nu} \right) + O(1) < 17\log_2 n + O(1)$. This is the number of stages in the algorithm, but not the number of rounds of simple comparisons, since our stages require separators built from ε_0-halvers which themselves entail some bounded number of rounds. With very careful analysis and choice of parameters, Paterson (1990) has managed to form a network which requires at most $6100\log_2 n$ rounds.

Chapter 7
Subgraphs with restricted degrees

7.1 (g,f)-factors

$\mathbf{L}$et G be a bipartite graph and let $f(x)$ and $g(x)$ be integer valued functions on the vertex set $V(G)$ such that $0 \leq g(x) \leq f(x) \leq d_G(x)$ for each vertex $x \in V(G)$. Then a spanning subgraph $F \subseteq G$ is called a (g,f)-*factor* of G if $g(x) \leq d_F(x) \leq f(x)$ for each vertex $x \in V(G)$. For instance, if $f(x) \equiv g(x) \equiv 1$, then a (g,f)-factor is simply the graph induced by a perfect matching in G. Indeed (g,f)-factors can be thought of as one generalisation of matchings. Accordingly, it is not surprising that we define a trail to be alternating relative to a (g,f)-factor F if its edges are alternately in $E(F)$ and in $E(G)\backslash E(F)$. We also define $f(S) = \displaystyle\sum_{s \in S} f(s)$ for any subset $S \subseteq V(G)$.

There are many characterisations of graphs with (g,f)-factors, see for example Lovász (1970b). Here we give one of these characterisations for bipartite graphs.

Theorem 7.1.1 (Heinrich et al. (1990)) *A bipartite graph G has a (g,f)-factor if and only if for every set $S \subseteq V(G)$*

$$\sum_{z \notin S} \max \left\{0, g(z) - d_{G-S}(z)\right\} \leq f(S). \qquad (7.1.1)$$

Proof. The case when $g(x) \equiv 0$ is trivial. Suppose, then, that $g(x) \not\equiv 0$ and that G has a (g,f)-factor F. Then for any set S and any $z \notin S$ for which $d_{G-S}(z) < g(z)$ there must be at least $g(z) - d_{G-S}(z)$ edges in F from z to

S. Thus $\sum_{z \notin S} \max\{0, g(z) - d_{G-S}(z)\}$ is at most the number of edges in F from $V(G)\backslash S$ to S, which is at most $f(S)$ as required.

Conversely, suppose that G has no (g, f)-factor and yet (7.1.1) is satisfied. Let F be a $(0, f)$-factor which minimises the quantity

$$\mu(F) = \sum_{z \in V(G)} \max\{0, g(z) - d_F(z)\}.$$

Let $Z = \{z \in V(G) : d_F(z) < g(z)\}$. Then, by the minimality of $\mu(F)$ we can assume that each edge joining two vertices of Z belongs to F.

Let S be the set of vertices not in Z which are the terminal vertices of alternating paths, of odd length, starting in Z. Similarly, let T be the set of such terminal vertices of even alternating paths starting in Z.

Property 1 *Let e be an edge with ends x and y. Then*

(1) if $e \in E(G)\backslash E(F)$ and $x \in T \cup Z$ then $y \in S$,

(2) if $e \in E(F)$ and $x \in S$ then $y \in T \cup Z$.

Property 2 $d_F(z) = f(z)$ *for each* $z \in S$.

Proof. Let $u_1 e_1 u_2 e_2 \ldots e_{2k-1} u_{2k}$ be an alternating path relative to F where $e_1 \in E(G)\backslash E(F)$, $u_1 \in Z$ and $u_{2k} \in S$. If $d_F(u_{2k}) < f(u_{2k})$ then the $(0, f)$-factor F' with $E(F') = (E(F)\backslash\{e_2, e_4, \ldots, e_{2k-2}\}) \cup \{e_1, e_3, \ldots, e_{2k-1}\}$ satisfies $\mu(F') < \mu(F)$; a contradiction. $\square$

Also, using a similar proof we have the following two properties.

Property 3 $d_F(z) = g(z)$ *for each* $z \in T$.

Property 4 $S \cap T = \emptyset$.

Proof. Suppose the contrary and let $x \in S \cap T$. Then, by definition, there are vertices $z_1, z_2 \in Z$ with z_1 connected to x by an alternating path of even length, and z_2 connected to x by an alternating path of odd length. Then z_1 is connected to z_2 by an alternating path of odd length and, just as in the proof of Property 2, we can use this path to obtain a $(0, f)$-factor contradicting the minimality of F. $\square$

It remains only to combine these properties. From Property 1 we obtain $d_{G-S}(z) = d_F(z) - |E_G(S, z)| \leq g(z)$ for each $z \in T \cup Z$. Therefore,

$$\sum_{z \notin S} \max\{0, g(z) - d_{G-S}(z)\} \geq \sum_{z \in T \cup Z} \max\{0, g(z) - d_{G-S}(z)\}$$

$$= \sum_{z \in Z} (g(z) - d_F(z)) + \sum_{z \in T \cup Z} |E_G(S, z)|. \tag{7.1.2}$$

The result now follows by observing that Properties 1 and 2 give $f(S) = \sum_{z \in T \cup Z} |E_G(S, z)|$, and so since $Z \neq \emptyset$, (7.1.2) strictly exceeds $f(S)$, which contradicts (7.1.1). $\qquad\square$

The above proof also provides an effective method to find a (g, f)-factor, if such exists. The key is to find a $(0, f)$-factor F minimising $\mu(F)$. This can be done, by using Properties 2 and 3 in the proof above, in a manner similar to the Hopcroft-Karp algorithm of Section 5.2. If, for any vertex x, $d_F(x) < g(x)$ the graph G has no (g, f)-factor. Otherwise F itself is a (g, f)-factor of G. Hell and Kirkpatrick (1993) showed that this method can be realised in $O(\sqrt{g(|V(G)|)}|E(G)|)$ operations.

We can also use Theorem 7.1.1 to give a useful existence criterion.

Corollary 7.1.2 *A bipartite graph G has a (g, f)-factor if*

$$\frac{g(y)}{d_G(y)} \leq \frac{f(x)}{d_G(x)}$$

for each pair of vertices $x, y \in V(G)$.

Proof. Let $W = \{z \notin S : g(z) \geq d_{G-S}(z)\}$, and observe that for any set $S \subseteq V(G)$ we have $\sum_{z \in W} (d_G(z) - d_{G-S}(z)) \leq d_G(S) = \sum_{z \in S} d_G(z)$. We shall now verify the condition of Theorem 7.1.1.

$$f(S) - \sum_{z \notin S} \max \left\{0, g(z) - d_{G-S}(z)\right\}$$

$$= f(S) - \sum_{z \in W} (g(z) - d_{G-S}(z))$$

$$\geq \left(\frac{f(S)}{d_G(S)}\right) \sum_{z \in W} (d_G(z) - d_{G-S}(z)) - \sum_{z \in W} (g(z) - d_{G-S}(z))$$

$$= \frac{1}{d_G(S)} \sum_{z \in W} \left((f(S)d_G(z) - d_G(S)g(z)) + (d_G(S) - f(S)) d_{G-S}(z) \right) \geq 0.$$

$\qquad\square$

Corollary 7.1.3   *Let G be a bipartite graph and a and b be positive integers with $a < b \leq \Delta(G)$ and $a/b \leq \delta(G)/\Delta(G)$. Then G has a subgraph H with $a \leq d_H(x) \leq b$ for each vertex $x \in V(G)$.* $\qquad\square$

In fact the last result is true not only for a bipartite graph but also for an arbitrary graph G (see Kano and Saito (1983)).

Exercises

7.1.1   Let $g(x)$ and $f(x)$ be integral functions on the vertex set $V(G)$ of a graph G and $g(x) \leq f(x) \leq d_G(x)$ for each vertex x. Define functions $\bar{g}(x) = d_G(x) - g(x)$ and $\bar{f}(x) = d_G(x) - f(x)$. Prove that if F is a (g, f)-factor of G then the spanning subgraph F' with $E(F') = E(G) \backslash E(F)$ is an $(\bar{f}, \bar{g})$-factor of G.

7.1.2   Let G be a bipartite graph and F a (g, f)-factor of G. Also, let P be an alternating trail relative to F, connecting two distinct vertices x and y. Prove that, if $e \notin E(F)$ and e' are the first and last edges of P, there is an alternating path relative to F also connecting x to y, whose first edge is also not in $E(F)$ and whose last edge is in $E(F)$ if and only if $e' \in E(F)$.

7.1.3   Let G be a bipartite graph and $f(x)$ be an integer valued function on $V(G)$. A $(0, f)$-factor of G with a maximum number of edges is called a *maximum* $(0, f)$-*factor* of G. Prove that a $(0, f)$-factor F is maximum if and only if there is no alternating path relative to F with end vertices x and y, $d_F(x) < f(x)$ and $d_F(y) < f(y)$ and end edges $e, e' \notin E(F)$.

7.1.4   Let G be a tree on p vertices and $f(x)$ an integral function on $V(G)$. Show that a maximum $(0, f)$-factor of G can be found using $O(p)$ operations.

7.1.5 $\triangledown$ Show that for any $p \geq 9$ there exists a forest on p vertices having at least $(10^{p/5})/2$ maximum $(0, 2)$-factors. (Skupien (1986))

7.2 Subgraphs with given degrees

$\mathbf{L}$et G be a graph and $f(x)$ an integer valued function on the set $V(G)$ such that $0 \leq f(x) \leq d_G(x)$ for each $x \in V(G)$. A spanning subgraph F is called an f-*factor* of G if $d_F(x) = f(x)$ for each vertex $x \in V(G)$. Thus in the notation of the previous section an f-factor is an (f, f)-factor, and Theorem 7.1.1 gives the following condition for existence.

Corollary 7.2.1   *A bipartite graph G has an f-factor if and only if for each $S \subseteq V(G)$*

$$\sum_{z \notin S} \max\{0, f(z) - d_{G-S}(z)\} \leq f(S). \qquad (7.2.1)$$

Condition (7.2.1) has a 'two part' form, since S can contain vertices from either colour class. The next theorem is a 'one part' condition for the existence of an f-factor.

Theorem 7.2.2 (Ore (1956)) *Let G be a bipartite graph with bipartition (V_1, V_2), then G has an f-factor if and only if $f(V_1) = f(V_2)$ and for any set $U \subseteq V_1$*

$$f(U) \leq \sum_{y \in V_2} \min \left\{ f(y), |E_G(U, y)| \right\}. \tag{7.2.2}$$

Proof. The necessity of condition (7.2.2) is evident. Thus we need only turn our attention to the sufficiency. For this suppose that (7.2.2) holds for any $U \subseteq V_1$. We shall show that (7.2.2) is equivalent to the condition

$$f(U) \leq f(W) + |E_G(U, V_2 \backslash W)| \quad \text{for any } U \subseteq V_1 \text{ and } W \subseteq V_2. \tag{7.2.3}$$

To this end, put $W_0 = \{ y \in V_2 : f(y) \leq |E_G(U, y)| \}$. Then

$$\sum_{y \in V_2} \min \left\{ f(y), |E_G(U, y)| \right\} = f(W_0) + |E_G(U, V_2 \backslash W_0)|. \tag{7.2.4}$$

Moreover, for any $W \subseteq V_2$

$$f(W_0) + |E_G(U, V_2 \backslash W_0)| \leq f(W) + |E_G(U, V_2 \backslash W)|. \tag{7.2.5}$$

Then (7.2.4) and (7.2.5) imply that (7.2.2) is equivalent to (7.2.3).

By exchanging $V_1 \backslash U$ for U and $V_2 \backslash W$ for W in (7.2.3), and recalling that $f(V_1) = f(V_2)$, we obtain

$$f(W) \leq f(U) + |E_G(V_1 \backslash U, W)| \quad \text{for any } U \subseteq V_1 \text{ and } W \subseteq V_2. \tag{7.2.6}$$

Then, by the same arguments as above we see that (7.2.6) is equivalent to

$$f(W) \leq \sum_{x \in V_1} \min \left\{ f(x), |E_G(x, W)| \right\} \tag{7.2.7}$$

for each $W \subseteq V_2$.

Finally we shall show that together (7.2.2) and (7.2.7) imply (7.2.1). Let $S = U' \cup W'$ where $U' \subseteq V_1$ and $W' \subseteq V_2$. Put $U = V_1 \backslash U'$ and $W = V_2 \backslash W'$. Then

$$\sum_{z \notin S} \max\{0, f(z) - d_{G-S}(z)\} = \sum_{z \notin S} f(z) - \sum_{z \notin S} \min\{f(z), d_{G-S}(z)\}$$

$$= \left(f(U) - \sum_{y \in V_2} \min\{f(y), |E_G(U, y)|\} \right) + \sum_{y \in W'} \min\{f(y), |E_G(U, y)|\}$$

$$+ \left(f(W) - \sum_{x \in V_1} \min\{f(x), |E_G(x, W)|\} \right) + \sum_{x \in U'} \min\{f(x), |E_G(x, W)|\}.$$

Now applying (7.2.2) and (7.2.7) we deduce that

$$\sum_{z \notin S} \max\{0, f(z) - d_{G-S}(z)\} \le f(U') + f(W') = f(S).$$

Hence (7.2.2) implies (7.2.1) and by Corollary 7.2.1 G has an f-factor. □

Just as for matchings, we can transform any f-factor into any other by a series of transformations, and indeed these transformations are based on alternating cycles. The following property follows easily from the assumption that G is bipartite.

Property 7.2.3 *Let F be an f-factor of a bipartite graph G and P an alternating closed trail relative to F. Then P is the edge-disjoint union of alternating cycles relative to F.*

Theorem 7.2.4 *Let G be a bipartite graph with an f-factor F. Then any f-factor of G can be obtained from F by a sequence of transfers along alternating cycles relative to F.*

Proof. Let $V(G) = \{v_1, v_2, \ldots, v_p\}$ and $E(v_i)$ be the set of edges incident with v_i. For each $i = 1, \ldots, p$ we define the two disjoint sets

$$A_i = \{a_i^e : e \in E(v_i)\},$$
$$B_i = \{b_i^k : 1 \le k \le d_G(v_i) - f(v_i)\},$$

which allow us to define bipartite graph G_0 as follows: G_0 has vertex set $V(G_0) = \bigcup_{i=1}^{p}(A_i \cup B_i)$ and for each i we join each vertex of A_i to each vertex of B_i. Furthermore, a_i^e is joined to a_j^e if and only if v_i and v_j are the ends of an edge $e \in E(G)$. It is now not difficult to check that every perfect matching of G_0 corresponds exactly to an f-factor of G (by contracting the sets $A_i \cup B_i$ to the single vertex v_i).

Now let F and F^* be two f-factors of G. Each corresponds to a perfect matching in G_0, M and M^* respectively. By Corollary 5.1.8 M can be transformed into M^* by a sequence of transfers along alternating cycles relative to M. Each of these cycles, after contraction, corresponds to a closed, alternating trail relative to F, and so, by Property 7.2.3, this trail is itself an edge-disjoint union of alternating cycles relative to F. The result follows. □

In a chordal graph f-factors can be transformed using rather simpler transformations, called **switches**. A switch is a transfer along an alternating cycle of length 4. The following result appears in Berge (1970).

Proposition 7.2.5 *Let G be a chordal bipartite graph having an f-factor. Then any f-factor of G can be obtained from any other by a sequence of switches.*

Proof. As we have already shown in Theorem 7.2.4 any f-factor can be transformed into any other by a sequence of transfers along alternating cycles. It suffices then to show that any transfer along an alternating cycle can be regarded as a sequence of switches.

Let $C = a_1a_2\dots a_{2k}a_1$ be an alternating cycle relative to F. The assertion is trivial for $k = 2$. We shall proceed by induction on k and suppose that it has already been shown that any transfer along a cycle of length less than $2k$ can be regarded as a sequence of switches.

Since G is a chordal graph C has a chord splitting it into two alternating cycles both of which are shorter than $2k$ and at least as long as 4. Thus, by induction, transfers along each can be regarded as a sequence of switches relative to F. The result follows.   □

Clearly an f-factor is an (f, f)-factor, and so can be constructed using the direct polynomial algorithm given in the previous section. However, there is also a more indirect method. First we construct a graph G_0, from our graph G, as in the proof of Theorem 7.2.4. Then, using methods from Section 5.2 we find a maximum matching M in G_0. If M is a perfect matching then G has an f-factor, otherwise it does not.

Of particular interest is the problem of existence and construction of an f-factor in the complete bipartite graph $K_{n,m}$.

Let $V_1 = \{x_1,\dots,x_n\}$ and $V_2 = \{y_1,\dots,y_m\}$ be two disjoint sets, and (R, S) be a pair of sequences of non-negative integers, where $R = (r_1, r_2,\dots,r_n)$ and $S = (s_1, s_2,\dots,s_m)$. The pair (R, S) is called *non-increasing* if both R and S are non-increasing sequences, and is called *graphical* if there exists a simple bipartite graph G with bipartition (V_1, V_2) such that $d_G(x_i) = r_i$, for $i = 1,\dots,n$, and $d_G(y_j) = s_j$ for $j = 1,\dots,m$. The graph G is then called a *realisation* of (R, S). We denote by $\mathcal{U}(R, S)$ the set of all simple bipartite graphs G with bipartition (V_1, V_2) which are realisations of (R, S).

Property 7.2.6 *If $\mathcal{U}(R, S) \neq \emptyset$ then each $H \in \mathcal{U}(R, S)$ is an f-factor of the graph $K_{n,m}$ with bipartition (V_1, V_2), where $f(x_i) = r_i, 1 \leq i \leq n$ and $f(y_j) = s_j, 1 \leq j \leq m$.*   □

Since $K_{n,m}$ is a chordal bipartite graph, Property 7.2.6 together with Proposition 7.2.5 give us the following result.

Theorem 7.2.7 *If $\mathcal{U}(R, S) \neq \emptyset$ then any graph $H \in \mathcal{U}(R, S)$ can be obtained from any other in the set by a finite sequence of switches.*   □

Property 7.2.6 and Theorem 7.2.2 also give rise to a direct criterion for a pair (R, S) to be graphical. This result was obtained independently by Gale (1957) and Ryser (1957).

Theorem 7.2.8 *The non-increasing pair (R, S) is graphical if and only if*

$$\sum_{i=1}^{n} r_i = \sum_{j=1}^{m} s_j \text{ and } \sum_{i=1}^{k} r_i \leq \sum_{j=1}^{m} \min\{s_j, k\} \text{ for each } k = 1, \ldots, n. \qquad \square$$

The next theorem also gives a criterion for a pair (R, S) to be graphical. It is a variant of a result obtained by Havel (1955) and independently by Hakimi (1962).

Theorem 7.2.9 *A non-increasing pair (R, S) is graphical if and only if*

$$r_1 \leq m, \sum_{i=1}^{n} r_i = \sum_{j=1}^{m} s_j \text{ and the pair } (R', S') \text{ is also graphical, where}$$

$$R' = (0, r_2, \ldots, r_n) \text{ and } S' = (s_1 - 1, \ldots, s_{r_1} - 1, s_{r_1+1}, \ldots, s_m).$$

Proof. First suppose that the pair (R', S') is graphical, with a realisation H. Then from H we can easily create a realisation of (R, S): simply join an isolated vertex of V_1 to those vertices in V_2 with degrees $s_1 - 1, \ldots, s_{r_1} - 1$ and the resulting graph will be a realisation of (R, S).

Conversely, suppose that (R, S) is graphical and that $G \in \mathcal{U}(R, S)$. The conditions on r_1 and the sums of the sequences are evident. It remains to show that (R', S') is graphical.

If x_1 is adjacent to $y_1, y_2, \ldots, y_{r_1}$ then we have a realisation of (R', S') by deleting from G the edges $x_1 y_1, x_1 y_2, \ldots, x_1 y_{r_1}$. Thus suppose that there is some vertex in $\{y_1, \ldots, y_{r_1}\}$ which is not adjacent to x_1 and let y_i be the vertex with the smallest index with this property. Then, of course, x_1 is adjacent to some vertex y_j with $j > r_1$. By definition $d_G(y_j) \leq d_G(y_i)$ and so there exists a vertex x_l adjacent to y_i but not to y_j. By deleting from G the edges $x_1 y_j, x_l y_i$ and adding the edges $x_1 y_i, x_l y_j$ we obtain a switching of G which is a realisation of (R, S) in which x_1 is adjacent to $y_1, \ldots, y_i$. Thus by repeating this process, if necessary, we can reduce this case to the former, and find a realisation of (R', S'). $\qquad \square$

The proof of this theorem also gives us an effective way to construct a realisation, should one exist. The procedure is to begin by joining x_1 to $y_1, \ldots, y_{r_1}$. Now consider the pair (R', S'), as in the theorem, with the sequence R' ordered in non-increasing order. Join x_2 to the first r'_1 y_i's for which $s_i > 0$. If (R, S) is graphical then repeating this procedure will build a realisation of (R, S).

Exercises

7.2.1 △ A spanning k-regular subgraph is called a *k-factor*. Using Theorem 7.2.2 prove that a bipartite graph G with bipartition (V_1, V_2) has a k-factor if and only if $|V_1| = |V_2|$ and $\sum_{y \in V_2} \min \{k, |E_G(U, y)|\} \geq k|U|$ for any $U \subseteq V_1$.

7.2.2 ▽ Let G be a simple n by n bipartite graph. Show that G has an $\lfloor n/4 \rfloor$-factor if $\delta(G) \geq n/2$.

7.2.3 ☐ Show that a connected, regular bipartite graph on p vertices has a connected k-factor, where $k \leq \sqrt{p}$. (Häggkvist (1978))

7.2.4   Let (R, S) be a graphical pair in which every integer is positive. Prove that the set $\mathcal{U}(R, S)$ contains a graph with a 1-factor if and only if $n = m$ and the pair (R', S') is graphical, where $R' = (r_1 - 1, \ldots, r_n - 1)$ and $S' = (s_1 - 1, \ldots, s_m - 1)$.

7.2.5 ▽ Let (R, S) be a graphical pair and let ρ_{min}, ρ_{max}, respectively, be the minimum and maximum numbers t for which there is a graph in $\mathcal{U}(R, S)$ with a maximum matching of cardinality t.

(a)   Prove that for each ρ between ρ_{min} and ρ_{max} there is a graph in $\mathcal{U}(R, S)$ with a maximum matching of cardinality ρ. (Ryser (1963))

(b)   Let $\mathcal{U}_1(R, S)$ be the set of graphs from $\mathcal{U}(R, S)$ with a maximum matching of cardinality ρ_{max}. Prove that any graph $G \in \mathcal{U}_1(R, S)$ can be obtained from any other by a sequence of switches, in such a way that all intermediate graphs are also in $\mathcal{U}_1(R, S)$. (Tarakanov (1993))

7.2.6   Let $R = (r_1, \ldots, r_n)$ and $S = (s_1, \ldots, s_m)$ be two non-increasing sequences of non-negative integers. Show that (R, S) is graphical if $r_1 \leq m$, $s_1 - s_m \leq 1$ and $\sum_{i=1}^{n} r_i = \sum_{j=1}^{m} s_j$.

7.2.7   Let (R, S) be a non-increasing pair, and let q_l denote the number of elements of S which equal l. Prove that the pair (R, S) is graphical if and only if $\sum_{i=1}^{n} r_i = \sum_{j=1}^{m} s_j$ and $\sum_{i=1}^{k}(r_i + iq_{k-i}) \leq km$ for any $k = 1, 2, \ldots, n$. (I. Zverovich, V. Zverovich (1992))

7.2.8   Let G be a connected m by n bipartite graph with $m \leq n$. Show that if $n - m \geq \Delta(G) - 1$ then $n - m$ is the minimum number of vertices which must be added to G to construct a $\Delta(G)$-regular graph, containing G as a subgraph. (Alavi et al. (1986))

7.2.9 △ Let D be a simple directed graph, and for each $v \in V(D)$ let $r_1(v)$ and $r_2(v)$ be non-negative integers. Consider $G(D)$, the bipartite graph with bipartition (V_1, V_2), where $V_1 = \{v' : v \in V(D)\}$ and $V_2 = \{v'' : v \in V(D)\}$, in which $N_{G(D)}(v') = \{u'' : u \in N_D^+(v)\}$, for each $v' \in V_1$ and $N_{G(D)}(v'') = \{u' : u \in N_D^-(v)\}$, for each $v'' \in V_2$. Put $f(v') = r_1(v)$ and $f(v'') = r_2(v)$ for each $v \in V(D)$, and show that D has a subgraph H with $d_H^+(v) = r_1(v)$ and $d_H^-(v) = r_2(v)$ for each $v \in V(D)$ if and only if G has an f-factor.

7.3 2-factors and Hamilton cycles

In this section we shall more carefully consider 2-factors of bipartite graphs. Our main interest in 2-factors is motivated by a search for a connected 2-factor – what is more familiarly known as a *Hamilton cycle*. In other words a Hamilton cycle is a cycle containing all the vertices of the graph. A graph which contains a Hamilton cycle is called *hamiltonian*.

Example 7.3.1 *Can a knight tour a chess board, so that it visits each square precisely once, and finally returns to its starting square?*

This problem, which dates from nineth century India (see Murray (1913)), is precisely equivalent to the problem of the existence of a Hamilton cycle in a bipartite graph – the graph with bipartition (V_1, V_2), where V_1 is the set of white squares on a chessboard, V_2 is the set of black squares, and in which two vertices are joined by an edge if a knight can move between their corresponding squares. In fact, there are Hamilton cycles in this graph; one such cycle produces the closed knight-tour shown in figure 7.3.1. Other examples of applications of Hamilton cycles can be found in Section 11.2.

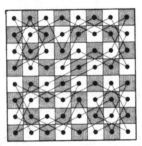

Figure 7.3.1 A knight-tour

Since 2-factors are actually f-factors when $f \equiv 2$, the results of Section 7.2 imply that the existence of a 2-factor can be recognised in polynomial time. Indeed, applying Theorem 7.2.2 to this particular case, we have the following corollary.

Corollary 7.3.2 *A bipartite graph G with bipartition (V_1, V_2) has a 2-factor if and only if $|V_1| = |V_2|$ and $2|U| \leq \sum_{y \in V_2} \min\{2, |E_G(U, y)|\}$ for any $U \subseteq V_1$.*

In contrast with the problem of existence of a 2-factor, the problem of deciding whether a bipartite graph has a Hamilton cycle is NP-complete (see Krishnamoorthy (1975)). The approaches to problems of existence of 2-factors and Hamilton cycles have been from rather different points of view. For example if the graph G is hamiltonian then it must have the property that the removal of any subset of vertices S can leave no more than $|S|$ components in $G - S$. Graphs with this property are called *tough*; not all tough graphs are hamiltonian, but it is known that every tough bipartite graph, on at least three vertices, has a 2-factor (see Katerinis (1987)).

Another example of such a comparison pertains to expanding properties of a graph. Woodall (1973) has shown that an arbitrary graph G is hamiltonian if $|N(S)| \geq 3|S|/2$ for any $S \subset V(G)$ with $N(S) \neq V(G)$. If G is bipartite then the 'one sided' restriction of this condition, although it is not enough to guarantee a Hamilton cycle, is sufficient for a 2-factor.

Theorem 7.3.3 (Katerinis (1987)) *A bipartite graph G with bipartition (V_1, V_2) has a 2-factor if*

(1) $|V_1| = |V_2| \geq 2$, and

(2) $|N(X)| \geq 3|X|/2$ for every subset $X \subset V_1$ with $N(X) \neq V_2$.

Proof. Suppose that G satisfies the conditions of the theorem and yet does not have a 2-factor. Then by Corollary 7.3.2 there exists $U \subseteq V_1$ such that $2|U| > \sum_{y \in V_2} \min\{2, |E_G(U, y)|\}$. Let $R_i = \{y \in V_2 : |E_G(U, y)| = i\}$ and $r_i = |R_i|$, $i \geq 1$. Then the last inequality is equivalent to

$$2|U| > r_1 + 2(r_2 + \ldots + r_{\Delta(G)}).   \tag{7.3.1}$$

Let us choose U such that it is minimal with respect to (7.3.1). Now suppose that there exists $x \in U$ with

$$|N(x) \cap (R_1 \cup R_2)| \geq 2.   \tag{7.3.2}$$

Let $U_0 = U \backslash \{x\}$, $R_i' = \{y \in V_2 : |E_G(U_0, y)| = i\}$ and $r_i' = |R_i'|$, for $i \geq 1$. Then (7.3.2) gives $r_1' + 2(r_2' + \ldots + r_{\Delta(G)}') \leq r_1 + 2(r_2 + \ldots + r_{\Delta(G)}) - 2$ and U_0 contradicts the minimality of U. Thus $|N(x) \cap (R_1 \cup R_2)| \leq 1$ for each $x \in U$, and

$$|U| \geq |E_G(R_1 \cup R_2, U)| = r_1 + 2r_2.   \tag{7.3.3}$$

Observe that R_1 cannot be empty. For, if $R_1 = \emptyset$ then (7.3.1) implies that $|U| > |N(U)|$, and so by *(2)* $N(U) = V_2$ and $|V_1| > |V_2|$, which contradicts *(1)*. Thus, combining (7.3.1) and (7.3.3) we have $3|U| > 2|N(U)|$ and a vertex $v \in R_1$. By *(2)*, $N(U) = V_2$ and so $|U| \geq \lfloor \frac{1}{3}|V_2| \rfloor + 1$. But if u is the element of U to which v is adjacent in G then by defining $U_0 = U \backslash \{u\}$, we have $|U_0| = |U| - 1 \geq \lfloor 2|V_2|/3 \rfloor$, but $N(U_0) \neq V_2$, so we once again contradict *(2)*. Hence G has a 2-factor. $\square$

In the case when G is the product of two bipartite graphs, we can sometimes establish the existence of a 2-factor or Hamilton cycle simply from properties of the factors. For example, if either of the factors has a 2-factor then so does G. The next result is an example of when we can establish the existence of a Hamilton cycle.

Theorem 7.3.4 (Paulraja (1993)) *If G is a 2-connected, simple, bipartite graph with $\Delta(G) \leq 3$, then $G \times K_2$ is hamiltonian.*

Proof. Let x be a vertex of degree 3, which is adjacent to two other vertices y and z also of degree 3. Then it is easy to check that at least one of the edges xy and yz can be deleted without destroying the 2-connectivity of the graph. By repeating such deletions, we can construct a 2-connected, spanning subgraph H which has a collection of edges F so that $H[F]$ is a disjoint union of paths and $H - F$ is 2-regular.

Let H' be the graph obtained from H by replacing each of the edges of F by a pair of parallel edges. Then every vertex in H' has degree 2 or 4, and so H' can be represented as an edge-disjoint union of cycles. Since G is bipartite, all of these cycles are even, and so we can colour the edges of each cycle alternately red and blue. We shall use this colouring of the edges to define a closed trail P by employing the rule that, on reaching a vertex, if there is another unused edge incident with that vertex with the same colour as the edge just traversed, then that edge must be the next edge in the trail. It is easy to see that H' has such a trail.

We can now exhibit the Hamilton cycle in $G \times K_2$. Regard $G \times K_2$ as two copies G_1 and G_2 of G joined by a perfect matching. The Hamilton cycle is obtained by traversing edges of G in the sequence of the edges of P, but if the edge of H' is blue we use the corresponding edge in G_1 and if the edge is red we use the edge in G_2; where necessary we use a matching edge to transfer between G_1 and G_2. Since each of the sets of blue and red edges in H' forms edge-disjoint paths which use all the vertices of H' (and so of G), it is easy to see that the cycle defined in this way is hamiltonian. $\qquad \square$

Bondy and Chvátal (1976) proposed a unified constructive method for investigating whether a graph is hamiltonian, by introducing a concept of *closure*. Although their theory implies results about graphs in general we shall restrict our comments to the point of view of bipartite graphs.

Firstly, let us observe that the problem of deciding whether a graph G is hamiltonian is equivalent to the problem of deciding whether the 'simplified' graph G', obtained from G by deleting all but one of the edges which join each pair of adjacent vertices, is hamiltonian. Furthermore, every hamiltonian bipartite graph must be balanced. In light of these two observations, we need only consider balanced, simple bipartite graphs. Bondy and Chvátal's technique is based on the following lemma.

Lemma 7.3.5 *Let G be a simple n by n bipartite graph with bipartition (V_1, V_2), and let $u \in V_1$ and $v \in V_2$ be non-adjacent vertices so that $d_G(u) + d_G(v) \geq n + 1$. Then G is hamiltonian if and only if $G + uv$ is hamiltonian.*

Proof. If G is hamiltonian then $G + uv$ certainly is too. Consider then the case when $G + uv$ is hamiltonian, but G is not. Then in G there exists a (u, v)-path $P = u_1 v_1 u_2 v_2 \ldots u_n v_n$ containing all the vertices of G where $u_1 = u$ and $v_1 = v$.

Since $d_G(u) + d_G(v) \geq n + 1$ there must be some i, $2 \leq i \leq n - 1$, so that both uv_i and $u_i v$ are edges of G. But then we have a Hamilton cycle $u_1 v_1 \ldots u_i v_n u_{n-1} v_{n-1} \ldots v_i u_1$ in G, giving the required contradiction.   □

The *bipartite closure* of an n by n bipartite graph G is the graph obtained from G by recursively joining pairs of non-adjacent vertices from different colour classes whose degree sum is at least $n + 1$. We say *the* graph, since it is not difficult to prove that the bipartite closure of G is unique, and we shall denote it by $c(G)$. Now from Lemma 7.3.5 and the definition of the closure, we immediately have the following results.

Theorem 7.3.6 (Bondy, Chvátal (1976)) *A simple bipartite graph G is hamiltonian if and only if its bipartite closure $c(G)$ is hamiltonian.*   □

Corollary 7.3.7 *A simple bipartite graph G is hamiltonian if its bipartite closure is $K_{n,n}$.*   □

Of course, if at any stage of the closure construction a Hamilton cycle can be recognised, then there is no need to continue the closure operation to the end. Conversely, under some circumstances the absence of a Hamilton cycle can also be recognised at early stages.

Proposition 7.3.8 (Asratian, Khachatrian (1988)) *Let G be a simple bipartite graph with bipartition (V_1, V_2), where $|V_1| = |V_2| = n \geq 2$. If there is a pair of non-adjacent vertices $x \in V_1$ and $y \in V_2$ such that $d_G(x) + d_G(y) \geq n + 1$ and $d_G(x, y) > 3$ then G is not hamiltonian.*

Proof. Let x and y be vertices as in the hypothesis, with $d_G(x) = k$ and $d_G(y) = t$, and let $S_1 = V_1 \backslash N(y)$ and $S_2 = V_2 \backslash N(x)$. Then

$$|S_1 \cup S_2| = 2n - (k + t) \leq n - 1.$$

Recall that $d_G(x, y) > 3$, thus there is no edge with one end in $N(x)$ and the other in $N(y)$. In other words, $G - (S_1 \cup S_2)$ consists of $k + t$ isolated vertices. Since $k + t \geq n + 1$, G is therefore not tough, and so is not hamiltonian.   □

Using Corollary 7.3.7 it is possible to deduce some explicit conditions for a bipartite graph to be hamiltonian.

Corollary 7.3.9 (Moon, Moser (1963)) *Let G be a simple bipartite graph with bipartition (V_1, V_2), where $|V_1| = |V_2| = n \geq 2$. If $d(x) + d(y) \geq n+1$ for each pair of non-adjacent vertices $x \in V_1$ and $y \in V_2$, then G is hamiltonian.*

Proof. It is simple to see that $c(G) = K_{n,n}$ and so G is hamiltonian by Corollary 7.3.7. $\qquad\square$

Corollary 7.3.10 (Chvátal (1972)) *Let G be a simple bipartite graph with bipartition (V_1, V_2), where $V_1 = \{u_1, \ldots, u_n\}$ and $V_2 = \{v_1, \ldots, v_n\}$, $n \geq 2$, and let the vertices be ordered so that*

$$d(u_1) \leq d(u_2) \leq \ldots \leq d(u_n), \quad \text{and} \quad d(v_1) \leq d(v_2) \leq \ldots \leq d(v_n).$$

If $d(u_k) \leq k < n$ implies that $d(v_{n-k}) \geq n - k + 1$, for $k = 1, \ldots, n$, then G is hamiltonian.

Proof. Once again, as we shall see, the closure of G is $K_{n,n}$. Let us denote the degree of a vertex v in $c(G)$ by $d'(v)$.

Suppose that $c(G) \neq K_{n,n}$. Then there must be vertices $u \in V_1$ and $v \in V_2$, which are not adjacent in $c(G)$. Certainly $d'(u) + d'(v) \leq n$ or u and v would already be joined. Let us choose this pair so that $d'(u) + d'(v)$ is as large as possible.

Let U_1 be the set of vertices in V_1 which are not adjacent to v in $c(G)$ and U_2 be the set of vertices which are not adjacent to u. Then, by definition we have $|U_1| = n - d'(v)$ and $|U_2| = n - d'(u)$. What is more, by the choice of u and v, no vertex of U_1 can have degree more than $d'(u)$, and no vertex of U_2 has degree more than $d'(v)$. Thus if we set $k = d'(u)$, V_1 must have at least k vertices of degree at most k, and V_2 must contain at least $n - k$ vertices of degree at most $n - k$. Since this must also be true of G, this implies that $d(u_k) \leq k$, but $d(v_{n-k}) \leq n - k$, which is a contradiction. $\qquad\square$

Corollary 7.3.11 *Let G be a simple bipartite graph with bipartition (V_1, V_2), where $|V_1| = |V_2| = n \geq 2$. If $|E(G)| \geq n^2 - n + 2$ then G is hamiltonian.*

Proof. Consider G as a subgraph of the complete bipartite graph $K_{n,n}$ with bipartition (V_1, V_2). Let $G \neq K_{n,n}$ and $E_0 = E(K_{n,n}) \backslash E(G)$. Then

$$|E_0| = n^2 - |E(G)| \leq n^2 - (n^2 - n + 2) = n - 2. \qquad (7.3.4)$$

Now consider a pair of non-adjacent vertices in G, $x \in V_1$, $y \in V_2$. Then clearly $|E_0| \geq (n - d(x)) + (n - d(y)) - 1$. This together with (7.3.4) implies that $d(x) + d(y) \geq n + 1$ and so, by Corollary 7.3.9, G is hamiltonian. $\qquad\square$

In fact, graphs which satisfy the conditions of this last result have much stronger properties.

Proposition 7.3.12 (Amar, Fournier, Germa (1989)) *Let G be a simple n by n bipartite graph with at least $n^2 - n + 2$ edges. Then for any integers $s, n_1, n_2, \ldots, n_s$ with $s \geq 1$ $n_i \geq 2$, for $i = 1, \ldots, s$, and $n_1 + \ldots + n_s = n$, there exist s disjoint cycles of length $2n_1, \ldots, 2n_s$.*

Proof. Since the case $s = 1$ is the conclusion of Corollary 7.3.11, we shall assume that $s \geq 2$, and of course we have that G has a Hamilton cycle $C = x_1 x_2 \ldots x_{2n-1} x_{2n}$. For $0 \leq r \leq n - 1$ we consider the system of pairs of vertices

$$M_r = \left\{ (x_{p+r}, x_{2n-p+1+r}) : 1 \leq p \leq n \right\}.$$

The sets $M_1, M_2, \ldots, M_{n-1}$ are clearly disjoint, and since $|E(G)| \geq n^2 - n + 2$ there must be some r for which every pair in M_r is an edge in G. Using these chord edges, and the Hamilton cycle C, it is now easy to construct the s disjoint cycles as required.   □

Rather than a bound on the total number of edges, we could instead appeal to a condition on the minimum degree. The following corollary follows immediately from Corollary 7.3.9.

Corollary 7.3.13 *Let G be a simple n by n bipartite graph. If $\delta(G) \geq (n + 1)/2$ then G is hamiltonian.*   □

It is not difficult to see that if G is 2-connected and regular then the bound on the minimum degree can be relaxed to $n/2$. However, Häggkvist (1976b) conjectured that this bound may be lowered even to $n/3$ which would be sharp. This is in fact essentially the case.

Theorem 7.3.14 (Jackson, Li (1994)) *Let G be a simple, 2-connected, k-regular bipartite graph on $2n$ vertices, where $k \geq \frac{n+19}{3}$. Then G is hamiltonian.*   □

It should be noticed that all the above results apply only to graphs which have a large edge density and small diameter, whilst also using only global information about the graph. Asratian and Khachatrian (1988, 1990, 1991) showed that by using the local structure of the graph, an improved method of closure can also apply to graphs with small edge density, and large diameter. For bipartite graphs some of their results are given in Exercise 7.3.7.

To close this section we shall briefly consider the special case of cubic bipartite graphs. If such a graph has a Hamilton cycle then that cycle cannot be unique. Indeed there must be an even number of Hamilton cycles (see Bosák (1967)). The problem of recognising a hamiltonian graph is still NP-complete even for planar, cubic bipartite graphs (see Akiyama, Nishizeki and Saito (1980)). Also, there are constructions of simple, 3-connected, cubic bipartite graphs which are not hamiltonian (see Horton (1982) for example). However, Barnette (1969) conjectured that every planar, simple, 3-connected, cubic, bipartite graph is hamiltonian. This conjecture is known to be true for graphs on at most 66 vertices (see Holton, Manvel and McKay (1985)).

Exercises

7.3.1 ▽ Construct a tough bipartite graph which is not hamiltonian.

7.3.2   Show that the graph $K_{n,n}$ has $\lfloor n/2 \rfloor$ edge-disjoint Hamilton cycles.

7.3.3 ▽ Show that every simple, bipartite hamiltonian graph with minimum degree d has at least $2^{1-d}d!$ Hamilton cycles. (Thomassen (1996))

7.3.4 ▽ Let G be a simple, n by n bipartite graph, with bipartition (V_1, V_2), where $d(u) + d(v) \geq n + 2$ for each pair of vertices $u \in V_1$, $v \in V_2$. Show that for any positive integers n_1, n_2, where $n_1, n_2 \geq 2$ and $n_1 + n_2 = n$, G contains two disjoint cycles of lengths $2n_1$ and $2n_2$. (Amar (1986))

7.3.5   Prove that a balanced bipartite graph G has a unique bipartite closure.

7.3.6   Let G be a simple, n by n bipartite graph where $d(u) + d(v) \geq n + 1$ for each pair of vertices u, v with $d(u, v) = 3$. Is it true that $c(G) = K_{n,n}$?

7.3.7 ▽ Let G be a simple, n by n bipartite graph. A pair of vertices u, v is called *hamiltonian stable* in G if $d(u, v) = 3$ and $d(u) + d(v) \geq 2 + |N(u) \cup N_2(v)|$. Furthermore, a graph H is called a *local closure* of G if it is obtained from G by recursively joining hamiltonian stable (in the graph at that stage) pairs of vertices, until no such pair exists. Show that

(a)   if u, v are hamiltonian stable, then G is hamiltonian if and only if $G + uv$ is hamiltonian,

(b)   a local closure of G is hamiltonian if and only if G is hamiltonian,

(c)   if $c(G) = K_{n,n}$ then $K_{n,n}$ is also the unique local closure of G,

(d) $K_{n,n}$ is a local closure of the graph shown in figure 7.3.2. (Asratian, Khachatrian (1988))

7.3.8   A *Hamilton path* of a graph G is a path containing all the vertices of G. Is it true that any simple, n by n bipartite graph G for which $d(u) + d(v) \geq n$ for every pair of vertices in different colour classes has a Hamilton path?

7.3.9 ▽   Let G be a simple, n by n bipartite graph which has a perfect matching $M = \{u_1v_1,\ldots,u_nv_n\}$. Show that G has a Hamilton path containing M if $d(u_i) + d(v_i) \geq n + 1$ for each $i = 1,\ldots,n$. (Häggkvist (1977))

7.3.10   Let $P_n = v_1v_2\ldots v_n$ and $P_m = u_1u_2\ldots u_m$ be two paths.

(a)   Show that the graph $P_n \times P_m$ is hamiltonian if and only if $m \cdot n$ is even.

(b)   Suppose that $m \cdot n$ is odd. The vertex $(v_i, u_j) \in P_n \times P_m$ is called even (odd) if the number $i + j$ is even (odd). Show that a Hamilton path between two vertices of $P_n \times P_m$ exists if and only if both vertices are even. (Itai, Papadimitrou, Szwarefiter (1982))

7.3.11   Show that an arbitrary $2r$-regular graph has a 2-factor. (Petersen (1891))

7.3.12 ▽   Let G be a simple bipartite graph with bipartition (V_1, V_2) with $|V_1| = |V_2| \geq 2$. Let $\alpha_{bip}(G)$ denote the maximum cardinality of a set of vertices X so that no two vertices of X are adjacent and $|V_1 \cap X| = |V_2 \cap X|$. Show that if $\alpha_{bip}(G) \leq \delta$ then $\kappa(G) = \delta(G)$. Moreover, prove that G is hamiltonian if and only if it is not isomorphic to one of the graphs G_1 and G_2 shown in figure 7.3.3. (Favaron, Mago, Ordaz (1993))

7.3.13   Let D be a simple directed graph. Let $G(D)$ be the bipartite graph with bipartition (V_1, V_2), where $V_1 = \{v' : v \in V(D)\}$ and $V_2 = \{v'' : v \in V(D)\}$. Let $N_{G(D)}(v') = \{u'' : u \in N_D^+(v)\} \cup \{v''\}$, for each $v' \in V_1$ and $N_{G(D)}(v'') = \{u' : u \in N_D^-(v)\} \cup \{v'\}$, for each $v'' \in V_2$. Show that there is a directed cycle in D which passes through each vertex of D precisely once if and only if $G(D)$ has a Hamilton cycle which uses every edge $v'v''$, for each $v' \in V_1$.

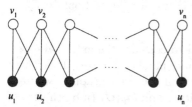

Figure 7.3.2

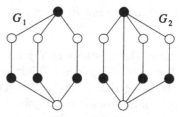

Figure 7.3.3

7.4 T-joins

So far in this chapter we have sought subgraphs in which the degree of each vertex has been restricted. This section will be no different, but here we will concern ourselves only with the parity of the degree at each vertex.

Let G be a graph and $T \subseteq V(G)$ be a subset of vertices. A set $F \subseteq E(G)$ is called a *T-join* if the number of edges in F incident with each $x \in V(G)$ is odd if $x \in T$ and even if $x \in V(G)\backslash T$. Throughout this section it will be convenient to identify paths and cycles with the sets of edges which they embody.

Proposition 7.4.1 *A connected graph G possesses a T-join if and only if T contains an even number of vertices.*

Proof. It is trivial to see that if a T-join exists then $|T|$ is even. Conversely, suppose that $|T|$ is even. Beginning with $F = \emptyset$ we do the following: if there are distinct vertices a and b from T for which $d_{G[F]}(a)$ and $d_{G[F]}(b)$ are even, then let P be an (a, b)-path in G. We update F, to $F \triangle P$. Then we see that, with the resulting F, $d_{G[F]}(a)$ and $d_{G[F]}(b)$ are now odd, but all other vertices have the same parities as earlier. Since $|T|$ is even, after $|T|/2$ steps we will obtain a T-join of G. $\qquad\square$

The above proof contains one algorithm to construct a T-join of G. Now we shall consider a method of transforming one T-join into another. Let F be a T-join and C a cycle. We will say that a T-join F' is obtained from F by a transformation along the cycle C if $F' = F \triangle C$.

Proposition 7.4.2 *Let F_1 and F_2 be two distinct T-joins of G. Then F_2 can be obtained from F_1 by a sequence of transformations along cycles.*

Proof. Clearly, the degree of each vertex of $G[F_1 \triangle F_2]$ is even, and $G[F_1 \triangle F_2]$ contains a cycle C_1. The deletion of the edges of C_1 gives a spanning subgraph G_1 in which each vertex still has even degree. If G_1 has no edges, then we have represented $G[F_1 \triangle F_2]$ as a union of edge-disjoint cycles. Otherwise, we can find a cycle C_2 in G_1 and by repeating the argument, a new graph G_2 obtained from G_1 by deleting the edges of C_2. By continuing this process we can represent $G[F_1 \triangle F_2]$ as a union of edge-disjoint cycles. By transforming F_1 along each of these cycles in turn we will obtain F_2. $\qquad\square$

Given the existence of T-joins it is natural to consider the extremal cases. A *minimum T-join* is a T-join with a minimum number of edges. It is worth observing that to find a minimum T-join in an arbitrary graph G we need only find a minimum T-join in the bipartite graph $bip(G)$ (n.b. the set of vertices T remains the same).

Property 7.4.3 *Every minimum T-join induces a forest.*

Proof. Let F be a minimum T-join. Then $G[F]$ can contain no cycle, since the edges of a cycle can be deleted without changing the parity of the vertices. Thus $G[F]$ is a forest.   □

Property 7.4.4 (Kwan (1960)) *A T-join F is minimum if and only if $|F \cap C| \le |C \setminus F|$ for every cycle C of G.*

Proof. If $|F \cap C| > |C \setminus F|$ for some cycle C, then the T-join $F \triangle C$ contains fewer edges than F; a contradiction.

Conversely, suppose that $|F \cap C| \le |C \setminus F|$ for every cycle C, but that there is some T-join G' with $|F'| < |F|$. Clearly, the degree of each vertex of $G[F \triangle F']$ is even. Using the same argument as we used in the proof of Proposition 7.4.2, we see that $G[F \triangle F']$ is a union of edge-disjoint cycles. Since $|F'| < |F|$, we have $|F' \cap C| < |F \cap C|$ for one of these cycles C, and so $|C \setminus F| < |F \cap C|$; a contradiction.   □

Let $T \subset V(G)$, and $a \in V(G)$, then let $T^a = \begin{cases} T \cup \{a\} & \text{if } a \notin T, \\ T \setminus \{a\} & \text{if } a \in T. \end{cases}$

If a and b are vertices, we write $T^{a,b}$ for $(T^a)^b$. With this definition we have the following property.

Property 7.4.5 *If F is a minimum T-join of G, and a, b are two distinct vertices of G, then there exist a minimum $T^{a,b}$-join F' of G and an (a, b)-path P such that $F = F' \triangle P$.*

Proof. Let F'' be any minimum $T^{a,b}$-join. Then in $G[F \triangle F'']$ all vertices except a and b have even degree. If we successively delete cycles, as we did in the proof of Proposition 7.4.2, we eventually obtain an (a, b)-path. So $G[F \triangle F'']$ is an edge-disjoint union of cycles $C_1, \ldots, C_k$ and an (a, b)-path P. Moreover, $F' = F'' \triangle (C_1 \cup \ldots \cup C_k)$ is also a minimum $T^{a,b}$-join and $F = F' \triangle P$.   □

Below we shall formulate and prove a duality result for T-joins in bipartite graphs, but first some definitions. Let X be a non-empty subset of vertices of a simple graph G such that $|X \cap T|$ is odd. Then the set of edges with only one end in X is called a T-cut. Let

$$\nu(G, T) = \max\{|C| : C \text{ is a family of disjoint } T\text{-cuts}\}.$$

Theorem 7.4.6 (Seymour (1981)) *Let G be a bipartite graph and a subset T contain an even number of vertices. Then the cardinality of a minimum T-join equals $\nu(G, T)$.*

Proof. We shall follow a short proof of Sebö (1987). If $T = \emptyset$ then the result is evident. Thus we shall assume that $T \ne \emptyset$.

Let vertices a and b of G be chosen so that the cardinality of a minimum $T^{a,b}$-join is minimal. Let F be a minimum T-join, then by Property 7.4.5 there are a minimum $T^{a,b}$-join F' and an (a,b)-path P so that $F = F' \triangle P$. Suppose that some edge $e \in F'$ has b as one end and b' as the other. Then $F' \backslash \{e\}$ is a $T^{a,b'}$-join with fewer edges than F', which contradicts the choice of a and b. Thus no edge of F' can have b as an end. Since $d_P(b) = 1$ we can conclude that $d_{G[F]}(b) = 1$ and that $b \in T$.

Let G^* be the bipartite graph obtained by deleting b from G and identifying all the vertices of $N_G(b)$ as a new vertex b^*. Observe that in the new graph we retain the old 'names' of the edges. Set

$$T^* = \begin{cases} T \backslash (\{b\} \cup N_G(b)) & \text{if } |T \cap N_G(b)| \text{ is odd,} \\ \{b^*\} \cup T \backslash (\{b\} \cup N_G(b)) & \text{otherwise.} \end{cases}$$

Let $E_G(b)$ denote the set of edges $E_G(b, V(G) \backslash \{b\})$. We shall show that $F^* = F \backslash E_G(b)$ is a minimum T^*-join of G^*.

Suppose that F^* is not a minimum T^*-join of G^*. Then by Property 7.4.4 there is some cycle K in G^* for which $|K \cap F^*| > |K \backslash F^*|$. Moreover, since G^* is bipartite $|K \cap F^*| \geq |K \backslash F^*| + 2$. The cycle K must use the new vertex b^* or we contradict the minimality of F. Indeed, K must correspond to an (x_1, x_2)-path in G, for some $x_1, x_2 \in N(b)$. The minimality of F and Property 7.4.4 now imply that $|(K \cup \{bx_1, bx_2\}) \cap F| \leq |(K \cup \{bx_1, bx_2\}) \backslash F|$, and so we have equality in both of the last two inequalities. This equality gives that neither bx_1 nor bx_2 can be an edge of F. But then the equality also gives that $T = F \triangle (K \cup \{bx_1, bx_2\})$ is a minimum T-join with $d_{G[T]}(b) = 3$; a contradiction.

To obtain the result we need now only observe that $|F \backslash E_G(b)| = |F| - 1$, that $E_G(b)$ is a T-cut of G containing no edges of G^*, and that any T^*-cut of G^* is a T-cut of G, when its edges are considered with their old 'names' in G. Thus by repeating the construction above, until we have the empty graph, we prove the desired equality. $\square$

Edmonds and Johnson (1973) gave a polynomial algorithm which finds a minimum T-join on an arbitrary weighted graph. For an unweighted bipartite graph, there is a polynomial algorithm of Sebö (1986) to find a minimum T-join and maximum family of T-cuts.

Having considered minimum T-joins, it is natural to be curious about maximum T-joins. Interestingly, this maximisation problem can be reduced to a minimisation problem: let T' be the set of all $x \in V(G)$ so that either $x \in T$ and $d_G(x)$ is even, or $x \notin T$ and $d_G(x)$ is odd. Then it is simple to check that if F is a minimum T'-join of G then $E(G) \backslash F$ is a maximum T-join. Thus the maximisation problem can also be solved in polynomial time.

A very closely connected maximisation problem is to find a spanning subgraph H of a graph G with the maximum number of edges so that every vertex in H has even degree. Let T be the set of vertices of G with odd degree. Then if F is a minimum T-join for G, $H = G \backslash F$ is the required subgraph, which can of course be found in polynomial time. We might also demand that H be connected. In contrast, the problem now becomes very difficult even for cubic bipartite graphs. For it is easily seen that such a spanning subgraph of a cubic graph is also a Hamilton cycle, but the problem of deciding whether a cubic bipartite graph is hamiltonian is NP-complete (see Akiyama, Nishizeki and Saito (1980)).

To end this section we shall note two applications of T-joins which are concerned with probably the most famous graph property – being *eulerian*. We say that a closed trail in a graph which contains all the vertices and edges is an *Euler trail*; a graph containing an Euler trail is called an *eulerian graph*. It is well known (see Euler (1736)) that a non-trivial connected graph G is eulerian if and only if it is connected and each vertex has even degree. We can now formulate two rather natural applications of T-joins.

Problem 1. Let G be a connected non-eulerian graph. Find an eulerian graph H with the minimum number of edges which contains G as a subgraph.

This problem is trivial to solve if we are allowed the luxury of adding 'new' vertices and 'new' edges – that is edges which do not have both their ends among the vertices of G (see Exercise 7.4.6). However, if we are restricted to only multiplying existing edges, the problem becomes much more interesting. In the latter case, the problem can be solved by using a minimum T-join F, where T is the set of vertices of G with odd degree. If we replace each edge of F by two parallel edges, then the resulting graph is the required eulerian graph.

Problem 2. (The Chinese Postman Problem) A postman on his daily route must pass down every street in a small town and then return to the post office. What route should he choose to minimise the length of his route?

Consider the graph G obtained from a map of the town by placing a vertex at each corner, and an edge for each street. To represent the distances we weight each edge with an integer weight function $w : E(G) \longrightarrow \mathbb{N}$. Now if G happens to be eulerian, then any Euler trail is a solution to the Chinese Postman Problem. But what if G is not eulerian? In that case we can mimic a postman reusing a street by adding an appropriate edge in parallel, which has the same weight as the original. The problem then becomes the weighted version of Problem 1, but still can be solved by finding an unweighted T-join as follows: let G' be the graph obtained from G, by replacing each edge $e \in E(G)$ by a path of length $w(e)$, and let T be the set of vertices of odd degree in G. If we now find a minimum T-join F in G', then those paths joining the vertices of G in F correspond to the edges of G which must be duplicated.

Exercises

7.4.1 △ Let G be a graph on $2n$ vertices. Show that G has a perfect matching if and only if a minimum $V(G)$-join contains n edges.

7.4.2 △ Let F be a T-join and Q be a T-cut of a graph G. Show that $F \cap Q \neq \emptyset$.

7.4.3 Show that any T-join in a graph G contains at least $\nu(G, T)$ edges. Construct a non-bipartite graph G which shows that the generalisation of Theorem 7.4.6 to graphs is false.

7.4.4 ▽ Let G be a cubic graph, and $G' = bip(G)$. By considering an appropriate T-join of G', show that G is the edge-disjoint union of copies of $K_{1,3}$ and unicyclic graphs.

7.4.5 ▽ Let G be a tree on p vertices. Show that there is a set S of at least $2\lfloor (n+1)/3 \rfloor$ vertices so that every vertex of $G[S]$ has odd degree. (Radcliffe, Scott (1995))

7.4.6 Let G be a bipartite graph and H be a bipartite graph with as few edges as possible which contains G as a subgraph and in which each vertex has even degree. Show that $|E(H) \backslash E(G)|$ equals the number of vertices of G with odd degree.

7.4.7 Let $\mathcal{W}(G)$ denote the set of all T-joins of a graph G, with $T = \emptyset$. Show that $\mathcal{W}(G)$ is a vector space over the Galois field $GF(2)$ with respect to the operation $\triangle$, the symmetric difference.

7.4.8 ▽ Let G be a k-regular bipartite graph on $2n$ vertices, $k \geq 3$. Prove that the vector space $\mathcal{W}(G)$ has a basis consisting of 2-factors. (Tutte (1971) for $k = 3$, Horton (1982) for $k \geq 4$)

Application

7.5 Isomer problems in chemistry

In recent times methods of graph theory have been being increasingly used in the areas of theoretical and experimental chemistry (see for example the book of Trinajstić (1983)). In the main these applications have been based around representing the molecular structure of a chemical by a graph, the vertices being atoms or some submolecule, the edges bonds. This approach allows chemists to generate theoretical models in which to give heuristic answers to a variety of questions. It is often the case that several chemical substances have the same chemical formula but different molecular structures.

These are called *isomers*, and are especially important in organic chemistry (a molecule of butane, C_4H_{10}, and its isomer are shown in figure 7.5.1).

Three problems about isomers arise very naturally. Firstly how many isomers of a given substance can possibly exist. Secondly, a technique to find new isomers from an existing one. Thirdly, chemists also need some effective method of recognising whether two chemical structures are the same; in terms of graphs this is an isomorphism problem. The second problem perhaps needs a little more explanation. If the chemist has some theoretical method to transform one graph to others with the same degree sequence, these graphs will be candidates for possible new isomers, which he can test experimentally.

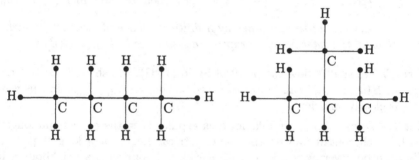

Figure 7.5.1 — Butane   Isobutane

Large classes of organic molecules have a representation as trees, for example saturated hydrocarbons, C_nH_{2n+2}. In this section we shall completely solve all the above three problems for molecules which can represented by trees.

For $p > 1$ let $H(d_1, d_2, \ldots, d_p)$ denote the set of all distinct trees with vertex set $\{v_1, v_2, \ldots, v_p\}$ and with degrees $d_T(v_i) = d_i$, $1 \leq i \leq p$. A tree is called a *caterpillar* if the removal of all its pendant vertices results in a path.

Proposition 7.5.1 *Let $d_1 \geq d_2 \geq \ldots \geq d_p$ be positive integers. Then $H(d_1, \ldots, d_p) \neq \emptyset$ if and only if $\sum_{i=1}^{p} d_i = 2(p-1)$.*

Proof. If $T \in H(d_1, \ldots, d_p)$ then certainly, by Proposition 2.2.1, $|E(T)| = p - 1$ and $\sum_{i=1}^{p} d_i = 2|E(T)| = 2(p-1)$.

Conversely, suppose that $d_1, d_2, \ldots, d_p$ satisfy the conditions of the theorem. If $p = 1$ then $d_1 = 0$ and $H(0) = K_1$; if $p = 2$ then $d_1 = d_2 = 1$ and $\{K_2\} = H(1,1)$. Otherwise, $p \geq 3$ and let k be an index so that $d_k > 1$, but $d_{k+1} = 1$. We shall build a tree with these degrees. Let the vertices $v_1, v_2, \ldots, v_{k+1}$ form a path. To these we join the other vertices $v_{k+2}, \ldots, v_p$ in order by joining the first $d_1 - 1$ vertices to v_1 and then consecutively the next $d_i - 2$ to each vertex v_i, for $i = 2, \ldots, k$. It is clear that this tree will have the required vertex degrees. □

We shall denote the tree which we constructed in the proof of Proposition 7.5.1 by $T_{d_1, d_2, \ldots, d_p}$. It is a caterpillar in which the vertices of degree greater than 1 are ordered in non-increasing order along the longest path. $T_{d_1, d_2, \ldots, d_p}$ is an *ordered caterpillar* and may be considered as the canonical tree realisation of the sequence $d_1, d_2, \ldots, d_p$.

Theorem 7.5.2 (Colbourn (1977)) *Let $d_1 \geq d_2 \geq \ldots \geq d_p$ be positive integers with $p > 2$ and $\sum_{i=1}^{p} d_i = 2(p-1)$. Then any tree in $H(d_1, \ldots, d_p)$ can be transformed to any other by a finite sequence of switches in such a way that all the intermediate graphs are also trees in $H(d_1, \ldots, d_p)$.*

Proof. We shall follow the proof of Syslo (1982). We shall show that any tree in $H(d_1, \ldots, d_p)$ can be transformed to the tree $T_{d_1, d_2, \ldots, d_p}$ in this way; this is clearly sufficient.

Let $T \in H(d_1, \ldots, d_p)$. If T is not a caterpillar then there exists no longest path which contains all non-pendant vertices. Let P be a longest path in T and v be a vertex not in P with $d_T(v) > 1$, but with a neighbour u in P. Let xy be an edge of P with $d_T(y) = 1$. We now perform a switch on T, removing the edges uv and xy and adding uy and vx. The resulting tree has fewer vertices of degree greater than 1 outside its longest path then T. By repeating this process we eventually arrive at a tree, none of whose non-pendant vertices lie outside any longest path. It remains only to make this an ordered caterpillar. To this end we need only observe that the transposition of two adjacent non-pendant vertices can also be interpreted as a switch (see figure 7.5.2). The result follows. □

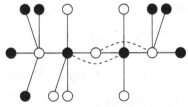

Figure 7.5.2

The next result, which appears in the book of Berge (1970), enumerates the number of trees with a given degree sequence.

Theorem 7.5.3 *Let* $d_1, \ldots, d_p$ *be positive integers with* $\sum_{i=1}^{p} d_i = 2(p-1)$.

Then $|H(d_1, d_2, \ldots, d_p)| = \dfrac{(p-2)!}{(d_1-1)!(d_2-1)!\ldots(d_p-1)!}$.

Proof. The proof is by induction on p. The result is trivial for the case $p = 2$. Suppose now that the hypothesis holds for all sequences of at most k integers, and let $d_1 \geq d_2 \geq \ldots \geq d_p$ $(p = k+1)$ be non-negative integers satisfying $\sum_{i=1}^{p} d_i = 2(p-1)$. Then by Proposition 7.5.1 $H(p, d_1, \ldots, d_p) \neq \emptyset$ and so $d_p = 1$. We can partition the trees $H(d_1, \ldots, d_p)$ according to the neighbour of v_p. It is then clear that

$$|H(d_1, \ldots, d_p)| = \sum_{i:d_i>1} |H(d_1, \ldots, d_i - 1, d_{i+1}, \ldots, d_{p-1})|.$$

Then applying the induction hypothesis we have

$$|H(d_1, \ldots, d_p)| = \sum_{i:d_i>1} |H(d_1, \ldots, d_i - 1, d_{i+1}, \ldots, d_{p-1})|$$

$$\dot{=} \sum_{i:d_i>1} \frac{(p-3)!}{(d_1-1)!(d_2-1)!\ldots(d_i-2)!\ldots(d_{p-1}-1)!}$$

$$= \frac{(p-3)!}{(d_1-1)!\ldots(d_{p-1}-1)!} \sum_{i=1}^{p}(d_i - 1)$$

$$= \frac{(p-2)!}{(d_1-1)!(d_2-1)!\ldots(d_{p-1}-1)!}$$

$$= \frac{(p-2)!}{(d_1-1)!(d_2-1)!\ldots(d_p-1)!}.$$

Hence the result follows by induction.   □

In Section 3.2 we proved that the centre of each tree consists of either one vertex or two adjacent vertices. Using this result we can determine whether or not any two given trees T and T' are isomorphic. The algorithm which we shall describe appears in the book of Busacker and Saaty (1965) and is due to Edmonds. It consists of three stages. First we divide the vertex sets of each tree into special subsets, and find the centres of the trees. If the centres of T and T' consist of different numbers of vertices then it is clear that T and T' cannot be isomorphic. If each centre consists of a pair of adjacent vertices, then by subdividing the edges which join these vertices we reduce the problem to the case when both trees have single central vertices. We shall consider only this case below. In the second stage we inductively construct codes for the vertices of each tree. Finally, to decide the isomorphism problem we compare only the codes given to the central vertices – if they are the same then T and T' are isomorphic, otherwise they are not. If T and T' are each on p vertices, then each of the three stages can be performed in $O(p)$ operations.

We begin with the first stage. For T, we inductively define vertex subsets $X_0, X_1, \ldots$ by letting X_0 be the set of all pendant vertices of T, and if $X_0, \ldots, X_{i-1}$ have already been defined, X_i to be the set of pendant vertices of $T - (X_0 \cup \ldots \cup X_i)$. Then if r is the radius of T, X_r will consist of the central vertex. Similarly we construct subsets $X'_0, \ldots, X'_k$ for T', where $k = r(T')$. Once again it is clear that if $k \neq r$ the two trees cannot be isomorphic. Thus we may assume that $k = r$.

The coding procedure in the second stage is the same for each tree; we shall describe it only for T. For each vertex $x \in T$ we will define a code $k_T(x)$ which will be a sequence of numbers. For this we need to impose an ordering on sequences of integers. We say of two sequences A and B, $A = a_1, a_2, \ldots, a_s$ and $B = b_1, b_2, \ldots, b_t$, that $A \triangleright B$ if $A = B$ or if for $i = \min\{j : a_j \neq b_j\}$ we have $a_i > b_i$.

For every vertex $x \in X_0$ we put $k_T(x) = 1$. Now suppose that we have already defined codes for the vertices of $X_0, \ldots, X_{i-1}$. Consider a vertex $x \in X_i$, and its neighbours $y_1, \ldots, y_t$ in X_{i-1}. We shall assume that the vertices $y_1, \ldots, y_t$ are ordered so that $k_T(y_1) \triangleright k_T(y_2) \triangleright \ldots \triangleright k_T(y_t)$. We put the code $k_T(x)$ to be the sequence $l + 1, k_T(y_1), \ldots, k_T(y_t)$, where l is the length of the sequence $k_T(y_1), \ldots, k_T(y_t)$. For examples of such a coding see figure 7.5.3 .

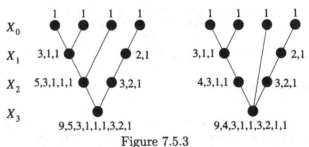

Figure 7.5.3

Property 7.5.4 *Let x be a vertex in X_i. Then the first number in $k_T(x)$ is the length of the sequence $k_T(x)$, and the code $k_T(x)$ uniquely specifies the codes of the neighbours of x in X_{i-1}.*

Proof. Suppose that $k_T(x)$ is the sequence $a_0, a_1, \ldots, a_s$. That a_0 is the length of the sequence follows easily from the construction. To recover the code words of the neighbours of x, notice that the second entry a_1 must be the length of the codes of x's first neighbour y_1 and this code word will be $a_1, \ldots, a_{a_1}$. Continuing in like vein, a_{a_1+1} will be the length of the code of x's second neighbour y_2 which will have code $a_{a_1+1}, \ldots, a_{a_1+a_{a_1}+1}$ and so on. In this way we can reconstruct all the codes of the neighbours of x in X_{i-1}. □

For the third stage it remains only to show that a comparison of the codes given to the central vertices does indeed decide the isomorphism problem for T and T'.

Theorem 7.5.5   *The trees T and T' are isomorphic if and only if their central vertices receive the same codes.*

Proof. Suppose first that T and T' are isomorphic, with isomorphism $\phi : V(T) \longrightarrow V(T')$. We shall show that $k_T(x) = k_{T'}(\phi(x))$ for each vertex $x \in T$. In particular, then, the central vertices of T and T' receive the same codes. If $x \in X_0$, the assertion is trivial. Suppose that the hypothesis is true for vertices of $X_0, X_1, \ldots, X_{i-1}$. Consider a vertex $x \in X_i$ and its neighbours $y_1, \ldots, y_t$ in X_{i-1}. Then the vertex $\phi(x)$ also has t neighbours in X'_{i-1}, $\phi(y_1), \ldots, \phi(y_t)$. By assumption $k_T(y_j) = k_{T'}(\phi(y_j))$, for each $j = 1, \ldots, t$, and so by construction $k_T(x) = k_{T'}(\phi(x))$. The claim follows by induction.

Conversely, suppose that the central vertices have the same codes. By construction, if $i \neq r$, then each vertex $x \in X_i$ has a unique neighbour $y(x)$ in X_{i+1}. We shall denote the component of the graph $T - y(x)$ which contains x by T_x. We shall prove that for each pair of vertices $x \in T$ and $x' \in T'$ which have the same codes, the trees T_x and $T'_{x'}$ are isomorphic. For vertices of X_0 this is true, since only pendant vertices of T and T' have code 1. Suppose then that the hypothesis is true for vertices from $X_0, \ldots, X_{i-1}$. Consider a vertex $x \in X_i$ ($i > 1$) and a vertex $x' \in X_j$ which has the same code as x. Let $y_1, \ldots, y_t$ be the neighbours of x in X_{i-1}, and $y'_1, \ldots, y'_s$ be the neighbours of x' in X'_{j-1}, where $k_T(y_1) \triangleright k_T(y_2) \triangleright \ldots \triangleright k_T(y_t)$ and $k_{T'}(y'_1) \triangleright \ldots \triangleright k_{T'}(y'_s)$. Since $k_T(x) = k_{T'}(x')$, Property 7.5.4 ensures that we can uniquely reconstruct the code words of $y_1, \ldots, y_t$ from $k_T(x)$ and $y'_1, \ldots, y'_s$ from $k_{T'}(x)$, and so t must equal s, and furthermore $k_T(y_j) = k_{T'}(y'_j)$, for $j = 1, \ldots, t$. Thus by assumption T_{y_i} is isomorphic to $T'_{y'_i}$, for each $i = 1, \ldots, t$, $i = j$ and T_x is isomorphic to $T_{x'}$. □

For example this theorem confirms that the two trees shown in figure 7.5.3 are not isomorphic.

Exercises

7.5.1   Prove the theorem of Cayley (1889) which states that the number of trees with the vertex set $\{v_1, \ldots, v_p\}$ is p^{p-2}.

7.5.2   Show that the number of trees T on vertex set $\{v_1, \ldots, v_p\}$ and
          with $d_T(v_1) = \ldots = d_T(v_k) = 1$ is $(p - k)^{p-2}$.

7.5.3   Show that the number of trees with vertex set $\{v_1, \ldots, v_p\}$ which
          contain a given edge is $2p^{p-3}$.

7.5.4   Show that all graphs with degree sequence $d_1, \ldots, d_p$ ($d_1 \geq d_2 \geq$
          $\ldots \geq d_p > 0$) are trees if and only if $p = 2$ and $d_1 = d_2 = 1$, or
          $p > 2$, $d_1 + d_2 = p$ and $d_3 = \ldots = d_p = 1$.

7.5.5   Show that the algorithm for deciding whether two given trees on p
          vertices are isomorphic can be performed in $O(p)$ operations.

Chapter 8
Edge colourings

8.1 Edge colourings and timetables

An *edge colouring* of a graph G with the *colours* $\alpha_1, \ldots, \alpha_t$ is an assignment of the colours to the edges of G, one colour to each edge. Such a colouring is called *proper* if no pair of adjacent edges receive the same colour. More formally, an edge colouring of G with colours from the set $\mathcal{C} = \{\alpha_1, \ldots, \alpha_t\}$ is a mapping $f : E(G) \longrightarrow \mathcal{C}$. If $f(e) = \alpha_k$ then we say that the edge e is *coloured* α_k. The set of edges of colour α_k we denote by $M(f, \alpha_k)$. Then $M(f, \alpha_k)$ is a matching for every $k = 1, \ldots, t$ if and only if f is a proper edge colouring. Usually we shall think of $\mathcal{C} = \{1, 2, \ldots, t\}$. We shall call an edge colouring with these colours $1, \ldots, t$ simply a *t-colouring* and call a path P *(k, l)-coloured* if its edges are alternately coloured with the colours k and l. A (k, l)-coloured path P is called *maximal* if there is no other (k, l)-coloured path P' with $V(P) \subset V(P')$.

Of course, if we have too few colours in the set $\mathcal{C}$ then there is no valid proper edge colouring. We call the minimum t for which there exists a proper t-colouring of G the *chromatic index* of G, which we denote by $\chi'(G)$. It is clear that for any graph G the chromatic index $\chi'(G) \geq \Delta(G)$. Actually, Vizing (1964) showed that for a simple graph G, $\chi'(G) \leq \Delta(G) + 1$. However, the problem of deciding whether $\chi'(G) = \Delta(G)$ is NP-complete even for simple graphs (see Holyer (1981)). Happily, the situation for bipartite graphs is much clearer, as the next theorem, due to König (1916), shows.

Theorem 8.1.1 (König's Colouring Theorem) *For any bipartite graph G* $\chi'(G) = \Delta(G)$.

Proof. As we have mentioned $\chi'(G) \geq \Delta(G)$. We shall describe an algorithm which produces a proper edge colouring with colours from $\mathcal{C} = \{1, \ldots, \Delta(G)\}$.

Let e be an edge which has so far not been coloured, and u and v be the ends of e. If there is a colour $t \in \mathcal{C}$ for which there is no edge coloured t adjacent to e, then use colour t to colour e. Otherwise, since we have $\Delta(G)$ colours, there is a colour $t_v \in \mathcal{C}$ which is not used to colour an edge incident with u, and a colour $t_u \in \mathcal{C}$, $t_u \neq t_v$, which is not used to colour an edge incident with v. The maximal (t_u, t_v)-coloured path P which originates at u cannot pass through v, otherwise $E(P) \cup \{e\}$ forms an odd cycle in G which contradicts G being bipartite. Thus if we interchange the two colours t_u and t_v on P, the colour t_u will no longer be used on an edge adjacent to either vertex, and we can colour e with t_u. In this way we can produce a proper edge colouring of the entire graph. $\square$

This proof certainly provides a polynomial algorithm to construct a proper $\Delta(G)$-colouring. In fact there are faster algorithms to produce such a colouring, for instance the algorithm of Gabow and Kariv (1982).

In the remainder of this section we shall consider various applications of König's Colouring Theorem. We begin with perhaps the most well known – the simplest of timetabling problems.

We have n classes $C_1, \dots, C_n$, m teachers $T_1, \dots, T_m$ and an $n \times m$ matrix $\mathbf{B} = (b_{ij})$ where b_{ij} is the number of one hour lectures which teacher T_j is required to teach to class C_i. A timetable of *length* t, corresponding to the matrix $\mathbf{B}$, is an $n \times t$ array $\mathbf{S} = (s_{ih})$ satisfying the following three conditions:

(1) each entry of S is either one of the members of the set $\{T_1, \dots, T_m\}$ or the empty symbol;

(2) the symbol T_j occurs precisely b_{ij} times in the ith row of $\mathbf{S}$, for $j = 1, \dots, m$;

(3) in each column of $\mathbf{S}$ all non-empty symbols are different.

The problem is to determine whether there exists a timetable of length t corresponding to the matrix B. For instance figure 8.1.1 shows a timetable $\mathbf{S}$ of length 4 corresponding to the matrix $\mathbf{B}$.

$$\mathbf{B} = \begin{pmatrix} 1111 \\ 1200 \\ 1011 \end{pmatrix} \quad \mathbf{S} = $$

Hour / Class	1	2	3	4
C_1	T_1	T_2	T_3	T_4
C_2	T_2		T_1	T_2
C_3	T_3	T_4		T_1

Figure 8.1.1 A requirement matrix $\mathbf{B}$ and a corresponding
timetable $\mathbf{S}$ of length 4

Perhaps it would be helpful at this stage to explain the interpretation of the timetable **S**. If s_{ih} is empty then the class C_i has a free lesson in the hth hour. If, however, $s_{ih} = T_j$ then that means that class C_i has a lesson with teacher T_j that hour.

Now consider a bipartite graph G, with bipartition (V_1, V_2) where $V_1 = \{x_1, \ldots, x_n\}$ and $V_2 = \{y_1, \ldots, y_m\}$, which has biadjacency matrix **B**. Then there is a one-to-one correspondence between proper t-colourings of G and timetables of length t which respect the requirement matrix **B**. This correspondence is that the column numbers of **S** form the colour set $\mathcal{C}$ and $s_{ih} = T_j$ if and only if one of the edges with ends x_i and y_j is coloured with colour h. In other words we can obtain a solution to this simple timetabling problem directly from König's Colouring Theorem.

Corollary 8.1.2   *There is a timetable of length t corresponding to the matrix* **B** *if and only if*

$$\sum_{i=1}^{n} b_{ij} \leq t \quad \text{for } j = 1, \ldots, m$$

$$\text{and} \sum_{j=1}^{m} b_{ij} \leq t \quad \text{for } i = 1, \ldots, n. \qquad \square$$

We now have both a solution to this timetabling problem and, since the problem reduces to König's Theorem, an algorithm to find such a timetable when it exists. However, in practice there are many other considerations, and restrictions which might govern a desirable timetable. Throughout this chapter we shall consider various types of restricted edge colourings many of which can be thought of as ways to pander to various staff, or student requirements. The first of these is to take into consideration the size of the classes to be taught, and resources which teachers may wish to call upon. Suppose that the classes $C_1, \ldots, C_n$ are grouped together into subsets $V_{11}, V_{12}, \ldots, V_{1p}$ of classes of similar size, and that there are $\alpha_i \leq |V_{1i}|$ classrooms in which the classes in V_{1i} can be taught. Similarly, suppose that the teachers are also grouped together into subsets $V_{21}, V_{22}, \ldots, V_{2q}$, so that each teacher in a group requires some piece of equipment for their lesson; there are $\beta_j \leq |V_{2j}|$ pieces of equipment for the teachers in group V_{2j}. We might ask how short a timetable can be arranged to teach each class in a suitable classroom, with its required equipment.

In terms of an edge colouring problem, we begin with the bipartite graph G with bipartition (V_1, V_2), as before. Once again we shall colour the edges of G with the hour numbers. Let $V_{si}(H)$ be the subset of vertices in V_{si} which are incident with an edge in a matching H. Then we see that the restrictions of classrooms and equipment amount to the conditions of the following theorem.

Theorem 8.1.3 (de Werra (1970)) *The smallest integer t^* such that the graph G has a t^*-colouring f satisfying*

$$\left.\begin{array}{ll} |V_{1i}(M_l)| \leq \alpha_i, & i = 1, \ldots, p, \\ |V_{2j}(M_l)| \leq \beta_j, & j = 1, \ldots, q, \end{array}\right\} \quad l = 1, \ldots, t^*$$

(where $M_l = M(f, l)$), is given by $t^* = \max\{\Delta(G), a, b\}$ where

$$a = \max_{1 \leq i \leq p} \left\lceil \frac{1}{\alpha_i} \sum_{x \in V_{1i}} d(x) \right\rceil, \quad \text{and} \quad b = \max_{1 \leq j \leq q} \left\lceil \frac{1}{\beta_j} \sum_{x \in V_{2j}} d(x) \right\rceil.$$

Proof. First notice that t^* certainly can be no smaller than claimed, otherwise either $t^* < \Delta(G)$, or $t^* \alpha_i < \sum_{x \in V_{1i}} d(x)$ or $t^* \beta_j < \sum_{x \in V_{2j}} d(x)$ for some i or j, none of which can be the case.

Let $t^* = \max\{\Delta, a, b\}$. We shall find the t^*-colouring we need by constructing a t^*-regular bipartite graph G' from G.

With each subset V_{1i} we associate a set X_i of $|V_{1i}| - \alpha_i$ new vertices added to V_2. Since $\sum_{x \in V_{1i}} (t^* - d(x)) = t^* |V_{1i}| - \sum_{x \in V_{1i}} d(x) \geq t^* (|V_{1i}| - \alpha_i)$ we can add new edges between V_{1i} and X_i, so that all the new vertices and some of the vertices of V_{1i} have degree t^*. Similarly we add new sets of vertices Y_j to V_1 so that each Y_j has $|V_{2j}| - \beta_j$ vertices, and add new edges so that all of these vertices and some of the vertices of V_{2j} also have degree t^*. If some vertices of the original V_1 and V_2 still have degree less than t^* then we add some additional vertices A_1 to V_1 and A_2 to V_2 and enough edges from V_1 to A_2 and A_1 to V_2, and finally from A_1 to A_2 so that every vertex in V_1 and V_2 has degree precisely t^*. This final graph we call G' (see figure 8.1.2).

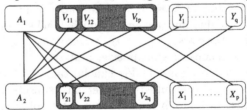

Figure 8.1.2 The construction of G'

This t^*-regular graph G' has a t^*-colouring, by König's Colouring Theorem. We need only check that this colouring, when restricted to G, satisfies our constraints. To do this we notice that there are $|V_{1i}|$ edges of colour h incident with the vertices of V_{1i} in G' (for each $i = 1, \ldots, p$), but at least $|V_{1i}| - \alpha_i$ of these edges have an end which is an added vertex in V_2'. Thus there are at most α_i edges of each colour incident with V_{1i} which are also edges of G, as we require. We may argue similarly to show that the colouring also satisfies the conditions for each V_{2j}. The result follows. $\qquad\square$

Finally for this section we shall consider *equitable* k-colourings of a bipartite graph. A k-colouring of G is called equitable if at each vertex the numbers of incident edges, in any pair of colours, differ by at most one.

Theorem 8.1.4 (de Werra (1971a)) *For any* $k \geq 2$, *a bipartite graph* G *has an equitable* k-*colouring.*

Proof. If $k \geq \Delta(G)$ then an equitable k-colouring is simply a proper k-colouring, and so its existence is clear from König's Colouring Theorem. If $k < \Delta(G)$, we build a new graph G' by splitting each vertex v into $\lfloor d(v)/k \rfloor$ vertices of degree k and possibly one extra vertex of degree $d(v) - k\lfloor d(v)/k \rfloor$. The partitioning of the edges in this splitting is arbitrary, other than ensuring that each vertex receives the correct degree.

Observe that G' is still a bipartite graph, but has maximum degree k. Thus König's Colouring Theorem provides a proper k-colouring of the edges of G'. Now collapse G' back to G, and consider the induced colouring of the edges. Since we had a colouring of G' every colour appeared at each vertex in G' at most once. Thus each vertex v in G can be adjacent to at most $\lceil d(v)/k \rceil$ edges of each colour. On the other hand, corresponding to each vertex v there were $\lfloor d(v)/k \rfloor$ vertices in G' of degree k. Thus there must be at least $\lfloor d(v)/k \rfloor$ edges of each colour adjacent to v. $\qquad\square$

Actually Theorem 8.1.4 also has an interpretation in terms of timetables. It is natural to require a timetable in which the lessons for each class and each teacher are spread throughout the week as evenly as possible. To construct such a timetable we could set k equal to the number of teaching days in a week, and construct an equitable k-colouring of G. Each F_i, $i = 1, \ldots, k$, could then be properly coloured in $k_i = \Delta(G[F_i])$ colours, to obtain a timetable for the ith day. Indeed a stronger theorem than Theorem 8.1.4 is also true.

Theorem 8.1.5 (de Werra (1971c)) *For any* $k \geq 2$ *a bipartite graph* G *has an equitable* k-*colouring so that* $|q_i(x,y) - q_j(x,y)| \leq 1$ *for every* $1 \leq i < j \leq k$, *and every pair of adjacent vertices* x *and* y *(where* $q_i(x,y)$ *is the number of edges joining* x *to* y *which are coloured* i). $\qquad\square$

Exercises

8.1.1 △ Prove König's Colouring Theorem using regularisation and the existence of a perfect matching in a regular bipartite graph.

8.1.2   Show that $K_{t,t}$ has two t-colourings f and g such that $M(f,i) \neq M(g,j)$ for each pair i,j, $1 \leq i,j \leq t$.

8.1.3 △ Show that if G and H are disjoint simple bipartite graphs then $\chi'(G \times H) = \Delta(G \times H)$.

8.1.4 ▽ Show that if $n \geq 4$, $n \neq 5$ then there exists an n-colouring of an n-cube such that the edges in any cycle of length 4 are coloured with distinct colours. (Faudree et al. (1993))

8.1.5 ▽ A matching $M = \{e_1, \ldots, e_n\}$ of a graph G is called an *induced matching* if each end of e_i is not adjacent to each end of e_j for every $1 \leq i < j \leq n$. Prove that a simple cubic bipartite graph has a proper 9-colouring f such that $M(f, i)$ is an induced matching for each $i = 1, \ldots, 9$. (Steger, Yu (1993))

8.2 Interval edge colourings

In this section we shall consider special types of edge colourings introduced and developed by Asratian and Kamalian (1987). An *interval t-colouring* of a graph G is a proper t-colouring such that at least one edge is coloured by each of the colours and the edges incident with each vertex of G are coloured by $d(x)$ consecutive colours; in other words, the set $\{f(e) : e \text{ incident with } x\}$ is an interval of integers for each $x \in V(G)$. If G is a bipartite graph, then such a colouring has a natural interpretation as a timetable in which the lessons for each class and each teacher are without interruption. Let $\mathfrak{U} = \bigcup_{t \geq 1} \mathfrak{U}_t$ where $\mathfrak{U}_t$ is the set of graphs which have an interval t-colouring. We begin with a necessary condition for membership of $\mathfrak{U}$.

Proposition 8.2.1 *If a graph G is in $\mathfrak{U}$ then $\chi'(G) = \Delta(G)$.*

Proof. Consider an interval t-colouring f of G. Let

$$E_j = \{e \in E(G) : f(e) \equiv j \,(\text{mod } \Delta(G))\}, \quad j = 1, \ldots, \Delta(G).$$

It is not difficult to see that each E_j is a matching in G. Therefore colouring the edges of E_j by colour j for $j = 1, \cdots, \Delta(G)$ gives a $\Delta(G)$-colouring of G and thus $\chi'(G) = \Delta(G)$. The proof is complete. □

Of course, every bipartite graph satisfies the condition $\chi'(G) = \Delta(G)$, but not every bipartite graph is a member of $\mathfrak{U}$.

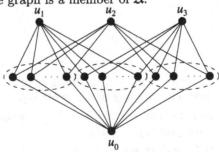

Figure 8.2.1

Proposition 8.2.2 *For any $n \geq 5$ there exists a bipartite graph on $3n + 4$ vertices which is not a member of $\mathfrak{U}$.*

Proof. Let G_n be the simple bipartite graph with bipartition (V_1, V_2), where $V_1 = \{u_0, u_1, u_2, u_3\}$ and $V_2 = \{x_1, x_2, \ldots, x_{3n}\}$, in which u_0 is adjacent to all vertices in V_2, u_1 is adjacent to $x_1, \ldots, x_{2n}$, u_2 is adjacent to $x_1, \ldots, x_n, x_{2n+1}, \ldots, x_{3n}$ and u_3 is adjacent to $x_{n+1}, \ldots, x_{3n}$ (see figure 8.2.1).

Clearly, $\Delta(G) = 3n$. Suppose that G_n has an interval t-colouring for some $t \geq 3n$. Then there are vertices x_i and x_j so that $f(u_0 x_i) = k$ and $f(u_0 x_j) = k + 3n - 1$, for some $k \geq 1$. By construction, at least one of the vertices u_1, u_2 and u_3 must be adjacent to both x_i and x_j; call such a vertex u. Then $f(u x_i) \leq k+2$, since $d(x_i) = 3$, and so $f(u x_j) \leq (k+2)+2n-1 = 2n+k+1$. Hence, since $d(x_j) = 3$, $f(u_0 x_j) \leq 2n + k + 3$, which contradicts the choice of x_j, provided $n \geq 5$.   □

Sevast'janov (1990) showed that the problem of deciding whether a given bipartite graph is a member of $\mathfrak{U}$ is NP-complete. Nevertheless some important classes of graphs can be dealt with.

Proposition 8.2.3 *The set $\mathfrak{U}$ contains all regular bipartite graphs, trees and complete bipartite graphs.*

Proof. If G is a k-regular bipartite graph, then it has a k-colouring, which is itself, therefore, also an interval k-colouring.

It is simple to see, by applying induction on the number of vertices in G, that, if G is a tree, it has an interval $\Delta(G)$-colouring.

Finally, if $G = K_{m,n}$ with colour classes $V_1 = \{x_1, \ldots, x_m\}$ and $V_2 = \{y_1, \ldots, y_n\}$, then we can define a colouring f by

$$f(x_i y_j) = i + j - 1, \quad i = 1, \ldots, m, \quad j = 1, \ldots, n.$$

It is easy to check that this defines an interval $(m + n - 1)$-colouring of G.   □

Let $G \in \mathfrak{U}$. Then we denote by $\chi'_{int}(G)$ the least t for which there exists an interval t-colouring of G. When G is bipartite with biadjacency matrix $\mathbf{B}$ the number $\chi'_{int}(G)$ can be interpreted as the minimum length of a timetable without interruptions, corresponding to the matrix $\mathbf{B}$. The next theorem gives an upper bound for $\chi'_{int}(G)$ when G is bipartite, and indeed for every t for which there is an interval t-colouring of G.

Theorem 8.2.4 (Asratian, Kamalian (1994)) *Let G be a connected bipartite graph with diameter $d(G)$ and bipartition (V_1, V_2), and let Δ_1 and Δ_2 be the maximum degrees of vertices in V_1 and V_2, respectively. If G has an interval t-colouring and $\Delta_1 \geq \Delta_2$ then*

$$t \leq \begin{cases} \frac{1}{2}d(G)(\Delta_1 + \Delta_2) - d(G) + 1 & \text{if } d(G) \text{ is even,} \\ \frac{1}{2}(d(G)+1)\Delta_1 + \frac{1}{2}(d(G)-1)\Delta_2 - d(G) + 1 & \text{if } d(G) \text{ is odd.} \end{cases}$$

Proof. Consider f, an interval t-colouring of G. If $t = \Delta_1$ then the theorem is trivially satisfied. Otherwise there are distinct edges e and e' so that $f(e) = 1$ and $f(e') = t$. Let e and e' have ends u_1, u_2 and v_1, v_2 respectively. Furthermore, let P_{ij}, for $i, j = 1, 2$, be a shortest path joining u_i to v_j, and let P be a shortest path amongst these four paths. Then without loss of generality we can suppose that P joins u_1 to v_1.

Consider the path P_{21}. If P_{21} has equal length to that of P, then G has an odd cycle, which is impossible. Thus P_{21} is strictly longer than P, and P must have length at most $d(G) - 1$.

Let $P = x_1 e_1 x_2 e_2 \ldots x_k e_k x_{k+1}$, where $x_1 = u_1$ and $x_{k+1} = v_1$. Then

$$f(e_1) \leq d_G(x_1),$$
$$f(e_{i+1}) \leq f(e_i) + d_G(x_i) - 1, \quad \text{for } i = 1, \ldots, k-1,$$
$$\text{and } t = f(e') \leq f(e_k) + d_G(x_{k+1}) - 1.$$

From these inequalities we obtain $t \leq \sum_{i=1}^{k+1} d_G(x_i) - k = 1 + \sum_{i=1}^{k+1}(d_G(x_i) - 1)$.

Put $\alpha_i = \begin{cases} \Delta_1 & \text{if } i \text{ is odd,} \\ \Delta_2 & \text{if } i \text{ is even,} \end{cases}$ for $i = 1, \ldots, d(G)$. Clearly, if $x \in V_1$ then

$$\sum_{i=1}^{k+1}(d_G(x_i) - 1) \leq \sum_{i=1}^{d(G)}(\alpha_i - 1). \tag{8.2.1}$$

Since $\Delta_1 \geq \Delta_2$, (8.2.1) is true also for the case when $x_1 \in V_2$. But

$$\sum_{i=1}^{d(G)} \alpha_i \leq \begin{cases} \frac{1}{2}d(G)(\Delta_1 + \Delta_2) + 1 & \text{if } d(G) \text{ is even,} \\ \frac{1}{2}(d(G)+1)\Delta_1 + \frac{1}{2}(d(G)-1)\Delta_2 + 1 & \text{if } d(G) \text{ is odd.} \end{cases}$$

This completes the proof. $\qquad\square$

Corollary 8.2.5   *If a connected bipartite graph G has an interval t-colouring then $t \le d(G) \cdot (\Delta(G) - 1) + 1$.*

Theorem 8.2.4 implies that $\chi'_{int}(K_{n,n+1}) \le 2n$. Actually it is not difficult to see that $\chi'_{int}(K_{n,n+1}) = 2n$; so the bound in the theorem is sharp. However, for simple bipartite graphs we have a more explicit upper bound. The result and proof are identical for bipartite and triangle-free graphs, and so we shall formulate the theorem in its more general form. To prove the theorem we need the following lemma.

Lemma 8.2.6   *Let f be a proper t-colouring of a graph G and let H be a connected subgraph of G, so that the edges of H incident with each vertex $x \in V(H)$ are coloured with consecutive colours. If S is the set of colours used in H, then H has an interval $|S|$-colouring. Moreover H has an interval t-colouring if $\{1, t\} \subseteq S$.*

Proof. Let $t_1 = \min\{f(e) : e \in E(H)\}$ and $t_2 = \max\{f(e) : e \in E(H)\}$. Since H is connected, there exists a path $P = v_0 e_1 v_1 \ldots v_{k-1} e_k v_k$ in H where $f(e_1) = t_2, f(e_k) = t_1$. If $f(e_i) \neq s$, $i = 1, \ldots, k$, for some colour $s \in S$ then $t_1 < s < t_2$ and there exists an integer $i_0, 1 \le i_0 < k$, such that $f(e_{i_0}) > s$ and $f(e_{1+i_0}) < s$. Therefore, there is an edge of H coloured s which is incident with x_{i_0}. So for any s, $t_1 \le s \le t_2$, there is an edge in H coloured s. Therefore $|S| = t_2 - t_1 + 1$.

Now consider the $|S|$-colouring f_1 of H where $f_1(e) = f(e) - t_1 + 1$ for each edge $e \in E(H)$. Clearly, f_1 is an interval $|S|$-colouring of H. If $t_1 = 1$ and $t_2 = t$ then $|S| = t$ and f_1 is an interval t-colouring of H. The proof is complete. $\square$

Theorem 8.2.7 (Asratian, Kamalian (1987, 1994)) *If a simple triangle-free graph G has an interval t-colouring then $t \le |V(G)| - 1$.*

Proof. Suppose that the theorem is false. Let t be the least integer for which the set $\mathfrak{U}_t$ contains a simple triangle-free graph on at least t vertices, and let G be such a graph with a minimum number of edges. Without loss of generality we can assume that G is a connected graph with $|E(G)| > 1$.

Consider an interval t-colouring f of G. We denote by $\mathcal{M}$ the set of those paths which start with an edge coloured t and finish with an edge coloured 1. To each path $P \in \mathcal{M}$ with a sequence of edges $e_1, \ldots, e_k$ ($k \ge 2$) corresponds the sequence $f(P)$ of the colours of its edges, where $f(P) = (f(e_1), \ldots, f(e_k))$. We shall show that there exists a P_0 in $\mathcal{M}$ for which $f(P_0)$ is decreasing.

Let $f(e') = t$, $e' = x_0 x_1$ and $d_G(x_1) \ge d_G(x_0)$. Since $|E(G)| > 1$ then $d_G(x_1) \ge 2$. We shall construct a sequence X of vertices of G in the following way:

We start with $X = \{x_0, x_1\}$. Suppose that after i steps we have obtained $X = \{x_0, x_1, ..., x_i\}$ where x_i is the last constructed vertex. If $N(x_i) \setminus X = \emptyset$ or $f(x_iy) > f(x_{i-1}x_i)$ for each $y \in N(x_i) \setminus X$ then the construction of X is completed. Otherwise we find the vertex x_{i+1} in $N(x_i) \setminus X$ for which $f(x_ix_{i+1}) = \min_{y \in N(x_i) \setminus X} f(x_iy)$. We add the vertex x_{i+1} to X and consider the sequence $X = \{x_0, x_1, ..., x_i, x_{i+1}\}$ in the next step.

Suppose that X has been already constructed and $X = \{x_0, x_1, ..., x_k\}$. Obviously X defines a path $P_0 = x_0e_1x_1...x_{k-1}e_kx_k$ where $f(e_k) < t$. We define the graph H as $H = G - x_k$, if $d_G(x_k) = 1$, and otherwise $H = G - e_k$.

Let us show that H is connected. Assume that it is not connected. Then $H = G - e_k$. We denote by F_1 and F_2 the connected components of H, where $x_{k-1} \in V(F_1), x_k \in V(F_2)$. Let H_1 and H_2 be the subgraphs of G induced by the sets $V(F_1) \cup \{x_k\}$ and $V(F_2) \cup \{x_{k-1}\}$ respectively. Denote by S_i the subset of colours used in $H_i, i = 1, 2$. The edges of H_i incident with each vertex $x \in V(H_i)$ are coloured by consecutive colours. Therefore, by Lemma 8.2.6, H_i has an interval $|S_i|$-colouring, $i = 1, 2$. Since $|E(H_i)| < |E(G)|$ we have $|S_i| \leq |V(H_i)| - 1, i = 1, 2$. Finally,

$$t = |S_1 \cup S_2| = |S_1| + |S_2| - |S_1 \cap S_2|$$
$$\leq |S_1| + |S_2| - 1 \leq |V(H_1)| + |V(H_2)| - 3 = |V(G)| - 1.$$

This inequality contradicts the choice of G; consequently H is connected.

Let us show that $f(e_k) = 1$. Assume $f(e_k) > 1$. The edges of H incident with each vertex $x \in V(H)$ are coloured by consecutive colours. Since $1 < f(e_k) < t$ there are edges in H coloured by 1 and t. Therefore, by Lemma 8.2.6, H has an interval t-colouring. We have $H \in \mathfrak{U}_t$ and $t \geq |V(G)| \geq |V(H)|$. This inequality contradicts the choice of G because H is triangle-free and $|E(H)| < |E(G)|$. Therefore $f(e_k) = 1$.

So we have constructed a path $P_0 \in \mathcal{M}$ for which the sequence $f(P_0)$ is decreasing. Let θ be the set of all the shortest paths P in $\mathcal{M}$ with decreasing $f(P)$. We denote by k the length of paths of θ. Now we define subsets $\theta_1, ..., \theta_k$ by letting $\theta_1 = \theta$ and θ_i be the subset of paths from θ_{i-1} with the greatest colour on the ith edge, $i = 2, ..., k$.

Let $P_1 = x_0e_1x_1...x_{k-1}e_kx_k$ be a path from θ_k, and for each $i = 1, ..., k-1$, let $\mathcal{A}(i) = \{y \in N(x_i) : f(e_{i+1}) < f(x_iy) < f(e_i)\}$. Evidently, $|\mathcal{A}(i)| = f(e_i) - f(e_{i+1}) - 1, i = 1, ..., k-1$.

We shall show that $\mathcal{A}(i) \cap \{x_0, ..., x_k\} = \emptyset, i = 1, ..., k-1$. Assume that there exist i_0 and j_0 for which $x_{i_0} \in \mathcal{A}(j_0)$ or $x_{j_0} \in \mathcal{A}(i_0)$. Let us define a path P'. If $i_0 \neq 0, j_0 \neq k$ then $P' = x_0x_1...x_{i_0}x_{j_0}...x_k$. If $i_0 = 0$ then $P' = x_1x_0x_{j_0}...x_k$. If $j_0 = k$ then $P' = x_0x_1...x_{i_0}x_kx_{k-1}$. In all cases $f(P')$ is decreasing and P' is shorter than P_1 which contradicts the choice of P_1.

Indeed, $\mathcal{A}(i) \cap \mathcal{A}(j) = \emptyset, 1 \leq i < j \leq k - 1$; for suppose that there exist $i_0, j_0, 1 \leq i_0 < j_0 \leq k - 1$, for which $\mathcal{A}(i_0) \cap \mathcal{A}(j_0) \neq \emptyset$. Since G is triangle-free, $j_0 - i_0 \geq 2$. Let $v \in \mathcal{A}(i_0) \cap \mathcal{A}(j_0)$. Consider a new path $P'' = x_0 x_1 \ldots x_{i_0} v x_{j_0} \ldots x_{k-1} x_k$. Clearly $f(P'')$ is decreasing. If $j_0 - i_0 \geq 3$ then P'' is shorter than P_1. If $j_0 - i_0 = 2$ we have $f(x_{i_0} v) > f(e_{1+i_0})$. So in both cases we have a contradiction to the choice of P_1.

Finally we can conclude that

$$|V(G)| \geq k + 1 + \sum_{i=1}^{k-1} |\mathcal{A}(i)| = k + 1 + \sum_{i=1}^{k-1} (f(e_i) - f(e_{i+1}) - 1)$$

$$= k + 1 + t - 1 - (k - 1) = 1 + t.$$

This contradicts the choice of G. The proof is complete. □

Thus $\chi'_{int}(G) \leq |V(G)| - 1$ for a simple bipartite graph, and indeed $K_{n,n+1}$ shows that this bound is also sharp. In fact the bound is sharp for $K_{m,n}$ whenever $\gcd(m, n) = 1$ (see Exercise 8.2.9).

It is also interesting to consider t-colourings of bipartite graphs which have the interval property for only one colour class. Let G be a bipartite graph with bipartition (V_1, V_2). A t-colouring is called a V_1-*interval* t-*colouring* if edges incident with each vertex of V_1 are coloured with consecutive colours. Such a colouring, of course, could correspond to a timetable for which each class has its lessons without interruption. The existence of such a colouring is no longer a problem, since G always has a V_1-interval $|E(G)|$-colouring; simply colour the edges incident with each vertex of V_1 with the next new set of consecutive colours. We denote by $\chi'_{int}(G, V_1)$ the smallest t for which G admits a V_1-interval t-colouring.

Proposition 8.2.8 (Asratian, Kamalian (1987, 1994)) *Let G be a bipartite graph with bipartition (V_1, V_2). Then there is a V_1-interval t-colouring for every t satisfying $\chi'_{int}(G, V_1) \leq t \leq |E(G)|$.*

Proof. The proof is by induction on the size of V_1. When $|V_1| = 1$ the result is clear. Thus suppose that the theorem holds for all graphs with $|V_1| \leq p$ and suppose that $|V_1| = p + 1$. Assume that there exists a V_1-interval t-colouring for some $t < |E(G)|$, from this colouring we shall construct a V_1-interval $(t + 1)$-colouring. Consider those vertices in V_1 incident with edges coloured t; let x_1 be such a vertex with the smallest degree. Evidently there exists an edge e_1 coloured by colour $t + 1 - d_G(x_1)$ which is also incident with x_1. The proof now breaks into cases.

Case A. There is an edge coloured by colour $t + 1 - d_G(x_1)$ other than e_1

In this case we can just recolour e_1 with colour $t + 1$ to obtain a V_1-interval $(t + 1)$-colouring.

Case B. e_1 is the only edge coloured with $t + 1 - d_G(x_1)$

Let s be the largest numbered colour with which more than one edge is coloured. Then $1 \leq s < t$.

Case B(i). $t + 1 - d_G(x_1) < s$

We recolour the edges coloured with colour i, for $i = t + 1 - d_G(x + 1), \ldots, s$ by colour $i + t - s$. Similarly, we recolour the edges coloured with colour i, for $i = s + 1, \ldots, t$, by colour $i - s + t - d_G(x_1)$. It is evident that this new colouring is also a V_1-interval t-colouring. Also since edges coloured s become coloured t the set of ends, in V_1, of edges coloured t contains more than one vertex. Choose x_2 from amongst these vertices so that x_2 has the smallest degree. Now we recolour the edge incident with x_2 coloured with the smallest colour, with the colour $t + 1$, to give a V_1-interval $(t + 1)$-colouring of G.

Case B(ii). $s < t + 1 - d_G(x_1)$

We can construct our new colouring by appealing to the induction hypothesis. Let $G' = G - x_1$, and so $V_1 := V_1 \backslash \{x\}$. Then G' certainly has a V_1-interval $(t - d_G(x_1))$-colouring, since the edges incident with x_1 were coloured with colours which occurred only once. Also $t - d_G(x_1) < |E(G')|$, thus by the induction hypothesis there exists a V_1-interval $(t + 1 - d_G(x_1))$-colouring of G'. Finally, if we replace x_1 and colour the edges incident with it with the colours $t + 2 - d_G(x_1), \ldots, t + 1$, we have a colouring of G as required. The result follows by induction. $\qquad\qquad$ □

It is certainly clear, for any bipartite graph, that $\chi'_{int}(G, V_1) \geq \Delta(G)$, but unfortunately, the general problem of deciding whether $\chi'_{int}(G, V_1) \leq k$ for a given k is NP-complete (see Appendix). Nevertheless, for many graphs G which correspond to situations which arise in practice we have $\chi'_{int}(V_1, G) = \Delta(G)$. Moreover, such graphs have a proper $\Delta(G)$-colouring such that the edges incident with each vertex $x \in V_1$ are coloured with precisely the colours $1, 2, \ldots, d_G(x)$. Such a colouring we call a V_1-*sequential colouring*. It can be thought of as a timetable in which all the classes have lessons without interruption, and all begin at the same time.

It is an obvious question to ask why practice is so different from theory. Probably the main reason is (see Asratian (1980)) that a common property of timetables in practice is that the number of lessons for each class is no less than the number of lessons taught by each teacher teaching that class. We shall give a formal proof of all of this in the following proposition.

Proposition 8.2.9 *Let G be a bipartite graph with bipartition (V_1, V_2). If $d(x) \geq d(y)$ for each pair of adjacent vertices $x \in V_1$ and $y \in V_2$, then G has a V_1-sequential colouring.*

Proof. The proof is by induction on $\Delta(G)$. The result is trivial for $\Delta(G) = 1$. Suppose then that the hypothesis holds whenever $\Delta(G) < k$ and consider a graph G with $\Delta(G) = k$. By Corollary 6.1.7 G has a matching M which covers all the vertices of degree $\Delta(G)$. Delete from M any edge which is not incident with a vertex of maximum degree, and colour the edges of the remaining matching M_0 with colour $\Delta(G)$. Then $\Delta(G - M_0) = k - 1$ and by induction $G - M_0$ admits a V_1-sequential colouring, which together with the coloured edges of M_0 provides a V_1-sequential colouring of G. $\square$

As a final remark we mention that the problem of deciding whether a general bipartite graph G has a V_1-sequential colouring is NP-complete even for the case $\Delta(G) = 3$ (see Asratian and Kamalian (1987)).

Exercises

8.2.1 △ Show that if f is an interval t-colouring of a graph G and $g(e) = t + 1 - f(e)$ for each edge $e \in E(G)$, then g is also an interval t-colouring.

8.2.2 △ Show that the graphs G_2, G_3 and G_4 defined as in the proof of Proposition 8.2.2 are in $\mathfrak{U}$.

8.2.3   Show that for any positive integer r there exists a bipartite graph G with $\chi'_{int}(G, V_1) \geq \Delta(G) + r$.

8.2.4   Show that a simple graph G has an interval $|E(G)|$-colouring if and only if G is a caterpillar.

8.2.5   Show that any bipartite graph G with $\Delta(G) \leq 3$ is a member of $\mathfrak{U}$.

8.2.6   Show that all doubly convex graphs are in $\mathfrak{U}$.

8.2.7 ▽ Show that $bip(G)$ is in $\mathfrak{U}$ for any n-regular graph G. (Hanson, Loten (1996))

8.2.8   Let T be a tree, and for each pair of pendant vertices x, y, denote by $\tau(x, y)$ the number of edges e such that the unique (x, y)-path in T contains at least one end of e. Prove that T has an interval t-colouring if and only if $\Delta(T) \leq t \leq \max_{x,y} \tau(x, y)$. (Kamalian (1989))

8.2.9 ▽ Show that the graph $K_{m,n}$ has an interval t-colouring if and only if $m + n - \gcd(m, n) \leq t \leq m + n - 1$. (Kamalian (1989))

8.2.10 ▽ Show that there are a bipartite graph, and positive integers $t_1 < t_2 < t_3$, such that $G \in \mathfrak{U}_{t_1}, G \in \mathfrak{U}_{t_3}$, but $G \notin \mathfrak{U}_{t_2}$. (Sevast'janov (1990))

8.2.11 ▽ An *interval cyclic t-colouring* of a graph G is a proper t-colouring such that the edges incident with each vertex x are coloured by $d_G(x)$ colours from the set $\{1, 2, \ldots, t\}$ which are consecutive modulo t. Prove that G has an interval cyclic t-colouring for any $t \geq \Delta(G)$ if it can be represented in the form $G' = K_2 \cup P^1 \cup \ldots \cup P^k$ where each P^i is a path of odd length which joins two adjacent vertices of $K_2 \cup P^1 \cup \ldots \cup P^{i-1}$, but has no other vertices in common. (de Werra, Solot (1991))

8.3 List colourings

In this section we shall consider some problems about edge colourings which respect lists of colours assigned to the vertices or edges. These types of restrictions have many uses in timetabling, and a variety of other guises some of which we shall see a little later in the book. As usual, we shall assume that the sets of colours are subsets of positive integers.

Problem 1. Let G be a graph and for each edge $a \in E(G)$ let $L_e(a)$ be a set (or *list*) of colours assigned to a. Can G be given a proper edge colouring in which each edge a receives a colour from its list $L_e(a)$?

Such a colouring when it exists is called an L_e-*list colouring*. It was conjectured by Vizing in 1975 (at a conference in Odessa, as was communicated to us by Kostochka), that provided each edge receives a list of $\chi'(G)$ colours, a list colouring of G must exist. Independently of Vizing, Dinitz posed the same question in 1979, but for the complete bipartite graph $K_{n,n}$ (see Erdős, Rubin and Taylor (1979)). Despite attempts by various authors it was not until the early 1990's that any serious headway was made. Janssen (1993) relying on new methods by Alon and Tarsi (1992) gave a solution to Vizing's conjecture for $K_{n,n-1}$ (Janssen's solution is purely an existence result). Then in 1994 Galvin gave a full solution to Vizing's conjecture for all bipartite graphs which provides a polynomial algorithm to find the required colouring. A more detailed history of this problem can be found in the paper of Chetwynd and Häggkvist (1992) and the book of Jensen and Toft (1995) (section 12.20). Note that deciding whether even a bipartite graph has an L_e-list colouring is NP-complete in general.

Below we shall give a version of Galvin's proof, but first we need to define a little notation. Let G be a bipartite graph with bipartition (V_1, V_2), and f be a proper edge colouring of G. Then the colouring $f : E(G) \to \mathbb{N}$ defines a partial order on the edges of G as follows: let e_1 and e_2 be two edges incident with some vertex v; if $v \in V_1$ and $f(e_1) > f(e_2)$, or if $v \in V_2$ and $f(e_1) < f(e_2)$, then we shall say that e_1 is *preferred* to e_2, and write $e_2 \prec_f e_1$.

Theorem 8.3.1 *Let G be a bipartite graph with lists on its edges defined by L_e. Let $f : E(G) \to \mathbb{N}$ be a proper edge colouring of G such that*

$$|\{e' : e \prec_f e'\}| < |L_e(e)| \quad \text{for each } e \in E(G).$$

Then G has an L_e-list colouring.

Proof. Consider the subgraph of G induced by the set of edges W which contain some element, which for convenience we shall call 1, in their lists. For each $v \in V(G[W])$ we may use the ordering of the edges to define a linear ordering of the vertices in $N(v)$. Let $u_1, u_2 \in N(v)$, and let e_1 and e_2 be the most preferable edges joining v to u_1 and u_2 respectively. We say that $u_1 \prec u_2$ if and only if $e_1 \prec_f e_2$. In this way each vertex of $G[W]$ is equipped with a linear ordering of its neighbours.

Now by Theorem 5.4.2 there must be a stable matching M with respect to these linear orderings (rankings) in $G[W]$ which is also a maximal matching. We colour the edges of M with the colour 1, and delete the colour 1 from the lists of the edges of $W \backslash M$. We shall proceed to colour all the edges of G by repeating this process: choosing a colour i; finding a maximal stable matching in the graph spanned by the uncoloured edges which have i in their lists; colouring those edges in the matching i and removing i from the other edges' lists. For this process to succeed we must ensure that no list ever becomes empty prematurely.

Claim: Whenever an element i is deleted from the list of an edge, at least one uncoloured edge preferred to that edge is coloured i.

Suppose that W is the set of edges whose lists contain colour i, and that M is a stable maximal matching in $G[W]$ and e is an edge of $W \backslash M$. Then since M is maximal there must be some edge $e' \in M$ which is incident with e, and furthermore the stability of M ensures that $e \prec_f e'$. This proves the claim.

This claim suffices to prove the result. For initially the list of each edge a has more elements than there are edges which are preferred to a. Thus the claim ensures that when all the edges preferable to a have been coloured, at least one colour j must remain in the list $L_e(a)$. When the stable matching for the colour j is constructed, a will be a member and will be coloured j. Thus the process succeeds in colouring all the edges of the graph. $\qquad\square$

In light of Theorem 8.3.1 we define $\mathrm{pref}_f(e)$ to be the number of edges which are preferable to an edge $e \in E(G)$ with respect to the edge colouring f.

Theorem 8.3.2 (Galvin (1995)) *Let G be a bipartite graph. Then provided each edge receives a list of $\Delta(G)$ colours there is a proper edge colouring which respects the lists.*

Proof. Observe that if every edge has a list of $\Delta(G)$ colours then any edge colouring f fulfills the conditions of Theorem 8.3.1, and so the result follows. $\qquad\square$

Indeed we can say more by choosing our edge colouring f rather more carefully, but first for this we need the following lemma due to Borodin, Kostochka and Woodall (1996).

Lemma 8.3.3 *Let G be a bipartite graph with bipartition (V_1, V_2) which has no isolated vertices. If $|V_1| \leq |V_2|$ then G contains a non-empty matching M such that whenever an edge $e \in E(G)$ has ends $x \in V_1$ and $y \in V_2$ and y is an end of an edge of M, then x is also an end of some edge of M.*

Proof. Clearly $|N(V_2)| \leq |V_1| \leq |V_2|$. Let $X \subseteq V_2$ be a minimal non-empty subset of vertices such that $|N(X)| \leq |X|$. Then, clearly $|N(X)| = |X|$. If $X = \{y\}$ then, since G has no isolated vertices, $|N(X)| = \{x\}$ and an edge with ends x and y provides the matching M we need. Otherwise $|N(X)| \geq 2$ and, since $|N(Y)| > |Y|$ for every non-empty subset $Y \subset X$, Hall's Theorem ensures a perfect matching M between X and $N(X)$ which has the required properties. $\square$

In fact, in general there is no need for all of the lists to contain $\Delta(G)$ elements whilst still guaranteeing the existence of an L_e-list colouring. Instead we have the following two-sided strengthening of Galvin's result.

Theorem 8.3.4 (Borodin, Kostochka, Woodall (1996)) *Let G be a bipartite graph, and for each edge $e \in E(G)$ let $|L_e(e)| = \max\{d(x), d(y)\}$ where x and y are the ends of e. Then G has an L_e-list colouring.*

Proof. We prove the result by induction on the number of edges in G. We shall prove that there is a proper edge colouring $f : E(G) \to \mathbb{N}$ such that $\text{pref}_f(e) < |L_e(e)|$ for each $e \in E(G)$; the result will then follow from Theorem 8.3.1.

Let G have bipartition V_1 and V_2, and let V_1' and V_2' be obtained from V_1 and V_2 by deleting any isolated vertices.

If $|V_1'| \leq |V_2'|$, let M be the matching whose existence was proved in Lemma 8.1.3. By the induction hypothesis, $G - M$ has a proper edge colouring $f' : E(G - M) \to \mathbb{N}$ such that $\text{pref}_{f'}(e) < |L_e'(e)|$ for each $e \in E(G - M)$, where $|L_e'(e)| = \max\{d'(x), d'(y)\}$, d' denotes degree in $G - M$, and x and y are the ends of the edge e. Let $f(e) = f'(e) + 1$ if $e \in E(G - M)$, and let $f(e) = 1$ for each edge of M.

Now, consider an edge e with ends $x \in V_1'$ and $y \in V_2'$. Then, $\text{pref}_f(e) \leq \text{pref}_{f'}(e) + 1$ if $e \notin M$. Furthermore, if $\text{pref}_f(e) = \text{pref}_{f'}(e) + 1$ then y is an end of an edge of M, which by Lemma 8.3.3 implies that x is also an end of an edge of M and so $|L_e(e)| = |L_e'(e)| + 1$. If $e \in M$ then $\text{pref}_f(e) = d(x) - 1 < d(x)$. Thus $\text{pref}_f(e) < |L_e(e)|$ for each edge e, as required.

If $|V_1'| > |V_2'|$ then Lemma 8.3.3 implies the existence of a non-empty matching M such that whenever $e \in E(G)$ with ends $x \in V_1$ and $y \in V_2$ and x is an end of an edge of M, then y is also. We now define f' as above, but define $f = f'(e)$ if $e \in E(G - M)$ and let $f(e)$ be greater than any colour used in colouring $G - M$ for $e \in M$. It is simple to check that as before, $\mathrm{pref}_f(e) < |L_e(e)|$ for each edge e, as required. $\qquad \square$

In Section 12.1 we shall use Theorem 8.3.4 to give another result on list edge colourings, but this time for general graphs.

Rather than placing lists on the edges, what if we place lists of allowed colours on the vertices of either one or both colour classes? Consider the following problem.

Let G be a bipartite graph with bipartition (V_1, V_2) and for each vertex $u \in V(G)$ let $L(u)$ be a set of colours assigned to u. Can G be given a proper edge colouring in which each edge e, with ends x and y, receives a colour which lies in $L(x) \cap L(y)$?

If we interpret such a colouring as a timetable in our usual way, with V_1 as the set of classes and V_2 as the set of teachers, then such a colouring corresponds to a timetable in which the list $L(u)$ corresponds to the set of hours in which the class or teacher is available. In general this problem is also NP-complete (see Even, Itai and Shamir (1976)). However, Theorem 8.3.4 gives the following result.

Theorem 8.3.5 *If $|L(x) \cap L(y)| \geq \max\{d(x), d(y)\}$ for every pair of adjacent vertices x and y then the required colouring exists.* $\qquad \square$

It might seem at first sight that placing lists on the vertices is a little less restrictive than its edge counterpart, but this is not actually the case. There is a surprising interplay between these two problems, and in fact, any edge list colouring problem can be posed as an edge colouring problem with lists on the vertices of an auxiliary graph. Furthermore, by a further reduction we may assume that the restrictive lists lie only on the vertices of one colour class. We shall give the first reduction below, but the second we leave as an exercise.

Let G be a bipartite graph with edge lists L_e. From G construct a new graph G' by replacing each edge e, with ends x and y, by the path $x x_e y_e y$ and then replace each edge $x_e y_e$ by $|L_e(e)| - 1$ parallel edges. Now let $L(x_e) = L_{(}y_e) = L_e(e)$ for each $e \in E(G)$ and for each $u \in V(G)$ let $L(u)$ to be the set of all possible colours. Then it can easily be seen that G' has an edge L-colouring if and only if G has a list L_e-colouring.

Problem 2. Let G be a bipartite graph with bipartition (V_1, V_2). For each vertex $u \in V_1$, let $L(u)$ be a set of at least $d_G(u)$ colours assigned to u.

Can G be given a proper edge colouring in which each edge incident with a vertex $u \in V_1$ receives a colour from the set $L(u)$?

Such a colouring when it exists is called an *edge L-colouring* of G, and the function L is called a V_1-*scheme* of G. When $|L(u)| = d_G(u)$ for every vertex $u \in V_1$ we call L a *tight V_1-scheme*. We begin with a series of results about tight V_1-schemes.

Theorem 8.3.6 (Häggkvist (1983b, 1989)) *Let G be a bipartite graph with bipartition (V_1, V_2). Let s and t be two colours, and L_1 and L_2 be two tight V_1-schemes for G fulfilling*

$$L_1(v) = \begin{cases} L_2(v)\backslash\{s\} \cup \{t\} & \text{if } s \in L_2(v) \text{ and } t \notin L_2(v), \\ L_2(v) & \text{otherwise.} \end{cases}$$

Then G has an edge L_2-colouring if G has an edge L_1-colouring.

Proof. Consider the subgraph H of G induced by those edges coloured s or t in an edge L_1-colouring of G. Let C be a component of H which contains a vertex v_0 for which $L_1(v_0) \neq L_2(v_0)$. Then, by assumption, since $L_1(v_0)$ and $L_2(v_0)$ differ, $s \notin L_1(v_0)$, $t \in L_1(v_0)$, and v_0 has degree 1 in C. Thus C must be a path $C = v_0 v_1 \ldots v_m$ whose edges alternate in colour, which begins with an edge coloured t. What is more, since if $L_1(v)$ ever contains s it must also contain t, the path also ends with an edge of colour t, and so must be of odd length. Hence we may interchange the colours along this path, retaining a proper edge colouring, and changing only the set of colours which appear at v_0. It follows that by carrying out this process in all components of H which contain vertices at which L_1 and L_2 differ we will produce an edge L_2-colouring. $\square$

We shall call the operation of producing L_1 from L_2 a *compression*, and based on this idea of compression we can impose a partial ordering on the set of V_1-schemes of G. For two different tight V_1-schemes of a graph G, L_1 and L_2, we say that $L_1 \prec L_2$ if there is a finite sequence of compressions which transforms L_2 into L_1. With this ordering Theorem 8.3.6 immediately gives the following corollary.

Corollary 8.3.7 *Let L_1 and L_2 be two tight V_1-schemes for G with $L_1 \prec L_2$. Then if G has an edge L_1-colouring then G also has an edge L_2-colouring.*

Given the nature of this ordering of V_1-schemes, a particular type of scheme becomes of interest. We shall denote by L_{seq} the tight V_1-scheme of G, where $L_{seq}(v) = \{1, 2, \ldots, d_G(v)\}$ for each vertex $v \in V_1$. So a proper colouring corresponding to L_{seq} is a V_1-sequential colouring, which we introduced in the previous section. Clearly $L_{seq} \prec L$ for each tight V_1-scheme L of G, hence Corollary 8.3.7 implies the following theorem.

Theorem 8.3.8 (Häggkvist (1983b)) *Let G be a bipartite graph, with bipartition (V_1, V_2). If G has a V_1-sequential colouring then it has an edge L-colouring for any V_1-scheme L.*   □

The next three results, due to Häggkvist(1983b) have several applications later, in Sections 11.4 and Section 12.2. These results were originally proved by using Theorem 8.3.8 but they can also be deduced directly from Theorem 8.3.4.

Proposition 8.3.9 *Let G be a bipartite graph with bipartition (V_1, V_2), and L a V_1-scheme of G. Then G has an edge L-colouring if $|L(x)| \geq d_G(y)$ for every pair of adjacent vertices $x \in V_1$ and $y \in V_2$.*   □

Proof. From G construct a graph G' by adding, for each vertex $x \in V_1$, a new vertex z_x and $|L(x)| - d_G(x)$ edges joining x to z_x. Now $|L(x)| = d_{G'}(x) \geq d_{G'}(y)$ and so by Proposition 8.2.9 G' has a V_1-sequential colouring, which with Theorem 8.3.8 completes the proof.   □

Corollary 8.3.10   *Let G be a bipartite graph with bipartition (V_1, V_2), and let $|L(v)| \geq \Delta(G)$ for each $v \in V_1$. Then G has an edge L-colouring.*
                                                                          □

Theorem 8.3.11 *Let G be a bipartite graph with bipartition (V_1, V_2) in which $d_G(x) \geq d_G(y)$ for every pair of adjacent vertices $x \in V_1$ and $y \in V_2$. For each vertex $u \in V_1$ let $L(u)$ be a set of colours which are the set of allowed colours at u, and for each $v \in V_2$ let $D(v)$ be a set of disallowed colours. If $|L(u)| \geq d_G(u) + k$ for each $u \in V_1$ and $|D(v)| = k$ for each $v \in V_2$, then G has a proper edge colouring in which the colour of each edge receives a colour which is allowed at both its ends.*   □

Proof. Let $V_1 = \{u_1, \ldots, u_m\}$, $V_2 = \{v_1, \ldots, v_n\}$, $X = \{x_1, \ldots, x_n\}$ and $Y = \{y_1, \ldots, y_{kn}\}$. We define a new bipartite graph G' with bipartition $(V_1 \cup X, V_2 \cup Y)$ as follows: let $G'[V_1 \cup V_2] = G$; then join x_i to v_i by $|D(v_i)|$ parallel edges, for $i = 1, \ldots, n$, and u_i to each of $y_i, y_{i+r}, \ldots, y_{i+(k-1)n}$ by single edges, for $i = 1, \ldots, m$.

By Proposition 8.2.9 G has a V_1-sequential colouring f. We can use this colouring to construct a $(V_1 \cup X)$-sequential colouring of G'. For each vertex of $x \in X \cup V_1$ we colour the new edges incident with x with the colours $1, 2 \ldots, k$, and for each edge in G we colour its copy in G' with $f(e) + k$. Thus, G' has a $(V_1 \cup X)$-sequential colouring, and so Theorem 8.3.8 implies that it has an edge L'-colouring for any $(V_1 \cup X)$-scheme. In particular G' has an edge L'-colouring for the scheme

$$L'(u_i) = L(u_i) \quad \text{for } i = 1, \ldots, m,$$
$$L'(x_i) = D(v_i) \quad \text{for } i = 1, \ldots, n,$$

and since any edge L'-colouring of G' induces an edge colouring of G with the required properties, the theorem is proved. $\square$

Exercises

8.3.1   Let G be a bipartite graph and $L_e(e) = \{1, 2, \ldots, r(e)\}$, for each $e \in E(G)$ where $r(e) = \max\{d(x_e), d(y_e)\}$ and x_e, y_e are the ends of e. Show without using Theorem 8.3.4 that G has an L_e-list colouring. (Geller, Hilton (1974))

8.3.2 ▽ Let a bipartite graph G have a proper edge colouring in which the edges incident with each vertex $x \in V(G)$ are coloured with the colours $\{1, 2, \ldots, d(x)\}$. Show without using Theorem 8.3.4 that G has a L_e-list colouring if for each edge $e \in E(G)$, $|L_e(e)| \geq \max\{d(x), d(y)\}$ where x and y are the ends of e.

8.4 Colour-feasible sequences

Let G be a bipartite graph with q edges, and let $n_1, \ldots, n_t$ be a sequence of positive integers. The subject of this section will be to determine if there exists a proper t-colouring of G in which precisely n_i edges receive colour i, for each $i = 1, \ldots, t$. This problem was first posed and investigated by Folkman and Fulkerson (1969). If such a colouring exists then the sequence $(n_1, \ldots, n_t)$ is called *colour-feasible* for G. For example, the sequences $(5, 2, 2)$ and $(4, 4, 1)$ are colour-feasible for the graph shown in figure 8.4.1, but the sequence $(5, 3, 1)$ is not.

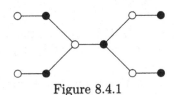

Figure 8.4.1

In terms of the timetabling problem (see Section 8.1), a proper t-colouring, corresponding to the sequence $(n_1, \ldots, n_t)$, can be interpreted, for example, as a timetable of length t with the added restriction that only n_i classrooms are available for teaching in each hour i, $1 \leq i \leq t$. In general the decision problem for the colour-feasibility of a sequence $N = (n_1, \ldots, n_t)$ for G is NP-complete. This follows immediately from the NP-completeness of the existence problem for V_1-sequential colourings and the following property.

Property 8.4.1 *Let (V_1, V_2) be a bipartition of G, and p_k be the number of vertices in V_1 with degree at least k, $k = 1, \ldots, \Delta(G)$. Then G has a V_1-sequential colouring if and only if the sequence $P = (p_1, \ldots, p_{\Delta(G)})$ is colour-feasible for G*   □

However, given a little more notation, we can formulate a simple necessary condition (due to de Werra (1971b)) for the colour-feasibility of a sequence. Indeed later we shall show that in some cases this condition is also sufficient.

An edge subset $F \subseteq E(G)$ is called a *k-matching* of G if each vertex of G is incident with at most k edges of F. Then, of course, a 1-matching is simply a matching. A k-matching with the maximum number of edges is called a *maximum k-matching* of G and its cardinality we shall denote by $q_k(G)$.

Property 8.4.2 *If a sequence $N = (n_1, \ldots, n_t)$ is colour-feasible for G, then*

$$\left.\begin{array}{l} t \geq \Delta(G), \quad |E(G)| = \sum_{i=1}^{t} n_i \\[2mm] \text{and} \\[1mm] q_k(G) \geq \sum_{i=1}^{k} n_i, \quad \text{for } k = 1, \ldots, \Delta(G) - 1. \end{array}\right\} \quad (8.4.1)$$

Proof. Let f be a proper t-colouring corresponding to N. We need only concern ourselves with the last inequality, and for that need only observe that each set $\bigcup_{i=1}^{k} M(f, i)$ is a k-matching. The result follows.   □

It is evident that we need only consider non-increasing sequences of positive integers, but what other properties are satisfied by the set $C(G)$ of colour-feasible sequences for G?

Let $\mathcal{D}_q = \bigcup_{t=1}^{q} \mathcal{D}_q^t$, where $\mathcal{D}_q^t$ denotes the set of non-increasing sequences of t positive integers which sum to q. For two sequences $P = (p_1, \ldots, p_m)$ and $N = (n_1, \ldots, n_t)$ from $\mathcal{D}_q$, the sequence P is said to *majorise* N, written $P \succeq N$, if $m \leq t$ and $\sum_{i=1}^{l} p_i \geq \sum_{i=1}^{l} n_i$, for each $l = 1, \ldots, m - 1$. Clearly the majorisation relation can be viewed as a partial order on the set $\mathcal{D}_q$. We shall show that the majorisation relation is also a partial order on the set $C(G)$.

Theorem 8.4.3 (Folkman, Fulkerson (1969)) *If a sequence $P \in \mathcal{D}_q$ is colour-feasible for G, then every sequence $N \in \mathcal{D}_q$ for which $P \succ N$ is also colour-feasible for G.*

Proof. Let $P = (p_1, \ldots, p_m)$, $N = (n_1, \ldots, n_t)$ and $P \succ N$. Then, clearly $m \leq t$. Consider a proper m-colouring, f, corresponding to the sequence P. The proof breaks into two cases.

Case A. $m = t$

Let l be the last integer in the interval $1 \leq l \leq t - 1$ for which $\sum_{i=1}^{l} p_i > \sum_{i=1}^{l} n_i$. Then $p_{l+1} < n_{l+1}$, and there are integers in the interval $1 \leq i \leq l$ for which $p_i > n_i$; let k be the last such. Then $p_k > n_k \geq n_{l+1} > p_{l+1}$. Let $\bar{P} = (\bar{p}_1, \ldots, \bar{p}_t)$ be obtained from P,

$$\bar{p}_i = \left\{ \begin{array}{ll} p_i & \text{if } i \neq k, l+1, \\ p_i - 1 & \text{if } i = k, \\ p_i + 1 & \text{if } i = l+1, \end{array} \right\} \quad 1 \leq i \leq t.$$

Consider a subgraph of G induced by the set $M(f, k) \cup M(f, l+1)$. Every component of the subgraph is either a path or a cycle. Moreover, since $p_k > p_{l+1}$, there is a path-component which starts and ends with edges of colour k. By interchanging the colours k and $l+1$ along this path, we obtain a proper t-colouring corresponding to $\bar{P}$. If $\bar{P} \neq N$ then we can iterate this procedure on $\bar{P}$, and so on. After a finite number of steps we will obtain a proper t-colouring corresponding to N, and so N is colour-feasible for G.

Case B. $m < t$

Let l be maximal with $p_l > 1$ and $\sum_{i=1}^{l} p_i > \sum_{i=1}^{l} n_i$. By recolouring an edge of colour l with the new colour $m + 1$ we obtain a proper $(m + 1)$-colouring corresponding to the sequence $P' = (p'_1, \ldots, p'_{m+1})$ where

$$p' = \left\{ \begin{array}{ll} p_i & \text{if } i \neq l, m+1, \\ p_i - 1 & \text{if } i = l, \\ 1 & \text{if } i = m+1, \end{array} \right\} \quad 1 \leq i \leq m+1.$$

Clearly $P \succ P' \succeq N$. If $m + 1 < t$ we repeat this process for P', and so on. After $t - m$ steps we obtain a colour-feasible sequence P^* which contains t numbers such that $P \succ P^* \succeq N$. The result then follows from the argument in case A. $\qquad \square$

Using Theorem 8.4.3 we can characterise the colour-feasible sequences for a regular bipartite graph.

Corollary 8.4.4 *A non-increasing sequence $N = (n_1, \ldots, n_t)$ is colour-feasible for an r-regular bipartite graph G on $2m$ vertices if and only if*

$$\left. \begin{array}{l} t \geq r, \quad mr = \displaystyle\sum_{i=1}^{t} n_i \\ \text{and} \\ mk \geq \displaystyle\sum_{i=1}^{k} n_i \quad \text{for } k = 1, \ldots, r - 1. \end{array} \right\} \tag{8.4.2}$$

Proof. By König's Colouring Theorem, G has a proper r-colouring. In other words the sequence $P = (m, \ldots, m)$ (of length r) is colour-feasible for

G. Then the condition (8.4.2) precisely amounts to saying that $P \succeq N$, and so by Theorem 8.4.3 N is colour-feasible for G. Conversely, if N is colour-feasible for G, then by Property 8.4.2 the condition (8.4.1) holds. But (8.4.1) and (8.4.2) are equivalent since $q_k(G) = km$. $\qquad\square$

The next result actually also follows from Theorem 8.4.3, but we shall give a direct proof, which shows clearly how to generate a feasible colouring.

Theorem 8.4.5 (de Werra (1971b)) *A sequence $N = (n_1, \ldots, n_t)$ is colour-feasible for a bipartite graph G if $t \geq \Delta(G)$, $\sum_{i=1}^{t} n_i = |E(G)|$ and $n_1 \geq n_2 \geq \ldots \geq n_t \geq n_1 - 1$.*

Proof. By König's Colouring Theorem there is a proper t-colouring f of G. If there are two colours k and l such that $|M(f,k)| - |M(f,l)| = d \geq 2$ then there exists a maximal (k,l)-coloured path which starts and ends with edges coloured k. By interchanging the colours along this path we obtain a proper t-colouring f' with $|M(f',k)| - |M(f',l)| = d - 2$, and $M(f',i) = M(f,i)$ for every $i \neq k, l$. By repeatedly applying this procedure we will obtain a proper t-colouring corresponding to the sequence N. $\qquad\square$

This result has a nice application in terms of timetabling. Suppose that we want to know the minimum length t_{min} of a timetable corresponding to a requirement matrix B, but we are limited to using only d classrooms. In terms of edge colourings this means that we would like to find the minimum number of colours with which the graph G can be properly coloured so that each colour appears on at most d edges. If G has q edges and $t = \max\{\Delta(G), \lceil q/d \rceil\}$, then clearly $t_{min} \geq t$. But, on the other hand, Theorem 8.4.5 shows that G has a proper t-colouring in which each colour appears at most $\lceil q/t \rceil$ times. Thus, since $t \geq q/d$ implies that $\lceil q/t \rceil \leq d$, we must have $t_{min} = \max\{\Delta(G), \lceil q/d \rceil\}$.

With each non-increasing sequence $N = (n_1, \ldots, n_t)$ we can associate a new sequence $s(N) = (s(0), s(1), \ldots, s(l))$, where $s(0) = 0$ and $s(i+1) = \max\{j : 1 + n_j \geq n_{s(i)+1}\}$, for $i = 0, 1, \ldots, l - 1$. For example if $N = (9, 8, 8, 7, 4, 3)$ then $s(N) = (0, 3, 4, 6)$. In fact the integer l in the definition of $s(N)$ is the minimum number of disjoint, balanced subsequences of N, i.e. the number of subsequences, in each of which any two members differ by at most 1. With this definition the sequences N considered in Theorem 8.4.5 have an $s(N)$ which contains only two entries, $s(N) = (s(0), s(1)) = (0, t)$. The next result provides a sufficient condition for a sequence N, with at least three entries in its sequence $s(N)$, to be colour-feasible for a graph G.

Theorem 8.4.6 (Asratian (1980)) *Let $N = (n_1, \ldots, n_t)$ be a non-increasing sequence with $s(N) = (s(0), s(1), \ldots, s(l))$, $l \geq 2$. Then N is colour-feasible for a bipartite graph G if every pair of vertices of degree more than $s(2)$ are non-adjacent in G and the condition (8.4.1) holds.*

Proof. Let X_1 be a maximum $s(1)$-matching of G. We shall construct edge subsets $X_2, \ldots, X_l$ in the following way: suppose that $X_1, \ldots, X_{i-1}$ are already constructed $(i \leq l)$. If $s(i) \geq \Delta(G)$, put $X_i = E(G)$. Otherwise, at each vertex u with $d_G(u) > s(i)$ delete precisely $d_G(u) - s(i)$ edges from $E(G) \backslash X_{i-1}$. The remaining edges of $E(G) \backslash X_{i-1}$ together with X_{i-1} form the next edge subset X_i. It is not difficult to check that every alternating path $P = a_0 a_1 \ldots a_{2r+1}$ relative to X_i with end edges in $E(G) \backslash X_i$ has the property that if a_0 is incident with less than $s(i)$ edges of X_i then each of the vertices $a_1, a_3, \ldots, a_{2r+1}$ is incident with exactly $s(i)$ edges of X_i, and each of the vertices $a_2, a_4, \ldots a_{2r}$ is incident with less than $s(i)$ edges of X_i. This property implies that X_i is a maximum $s(i)$-matching of G (see exercise 7.1.3). Clearly $X_1 \subseteq X_2 \subseteq \ldots \subseteq X_l$. From X_1 choose a subset of edges F_1 of size $\sum_{i=1}^{s(1)} n_i$, and from X_j choose a subset F_j of size $\sum_{i=1}^{s(j)} n_i$ containing F_{j-1}, for each $j = 2, \ldots, l$. This is possible since condition (8.4.1) holds. Now we will prove that the edges in F_j can be properly coloured with colours $1, \ldots, s(j)$ such that precisely n_i edges are coloured with colour i, for $i = 1, \ldots, s(j)$.

For the case $j = 1$ this follows from Theorem 8.4.5. Suppose that the required colouring is already constructed for F_j, $j < l$. We will colour edges from $F_{j+1} \backslash F_j$ with colours from $\mathcal{C} = \{1, 2, \ldots, s(j+1)\}$.

Let e be an edge which has so far not been coloured, and u and v be the ends of e. If there are a colour $\alpha \in \mathcal{C}$ for which there is no edge coloured α adjacent to e, then use colour α to colour e. Otherwise, since F_{j+1} is an $s(j+1)$-matching, and we have $s(j+1)$ colours, there are a colour $t_v \in \mathcal{C}$ which is not used to colour an edge incident with u, and a colour $t_u \in \mathcal{C}$, $t_u \neq t_v$, which is not used to colour an edge incident with v. The maximal (t_u, t_v)-coloured path P which originates at u cannot pass through v, otherwise $E(P) \cup \{e\}$ forms an odd cycle in G, which contradicts G being bipartite. Thus if we interchange the two colours t_u and t_v along P, the colour t_u will no longer be used on an edge adjacent to either vertex, and we can colour e with t_u.

Suppose that the proper colouring f of the edges of F_{j+1} produced by this procedure is such that n'_i edges are coloured with colour i, for each $i = 1, \ldots, s(j+1)$. We may assume (possibly after permuting the colours) that $n'_1 \geq n'_2 \geq \ldots \geq n'_{s(j+1)}$. It is not difficult to see that the above procedure of colouring certainly guarantees that $n'_i \geq n_i$, for each $i = 1, \ldots, s(j)$. Let $k(j+1)$ denote the maximal i with $n'_i > 0$. Then $s(j) < k(j+1) \leq s(j+1)$ and $(n'_1, \ldots, n'_{k(j+1)}) \succeq (n_1, \ldots, n_{s(j+1)})$ because $n_{1+s(j)} - n_{s(j+1)} \leq 1$.

If $(n'_1, \ldots, n'_{k(j+1)}) \neq (n_1, \ldots, n_{s(j+1)})$ then by using the algorithm suggested in the proof of Theorem 8.4.3 we can polynomially transform the colouring f to a proper $s(j+1)$-colouring of the edges of F_{j+1} corresponding to the sequence $(n_1, n_2, \ldots, n_{s(j+1)})$. $\qquad \square$

By using Theorems 8.4.5 and 8.4.6 we can now identify three cases when the necessary condition 8.4.1 is also sufficient.

Corollary 8.4.7 *Let G be a bipartite graph in which every pair of vertices of degree at least 3 are non-adjacent. Then a sequence $N = (n_1, \ldots, n_t)$ is colour-feasible for G if and only if the condition (8.4.1) holds.* □

Corollary 8.4.8 (de Werra (1971b)) *Let G be a bipartite graph. The non-increasing sequence $N = (n_1, \ldots, n_t)$ with $s(N) = (s(0), s(1), s(2))$ is colour-feasible for G if and only if the condition (8.4.1) holds.*

Proof. Clearly, $s(2) = t$. Suppose that condition (8.4.1) holds. Since $t \geq \Delta(G)$, the conditions of Theorem 8.4.6 are trivially satisfied. □

Corollary 8.4.9 (Folkman, Fulkerson (1969)) *Let $N = (n_1, \ldots, n_t)$ be a non-increasing sequence such that $n_1 = \ldots = n_k > n_{k+1} = \ldots = n_t$. Then N is colour-feasible for G if and only if the condition (8.4.1) holds.*

Proof. If $n_1 - n_t \geq 2$ then $s(N) = (0, k, t)$ and the assertion follows from Corollary 8.4.8. If $n_1 - n_t = 1$ then the assertion follows from Theorem 8.4.5. □

The last result has a useful interpretation in terms of timetabling. It may happen that at some hours a limited number of classrooms are available in one building, and at other hours some additional rooms are available in another building. Corollary 8.4.9 guarantees that this extra resource can be used to its full extent.

Exercises

8.4.1 △ Let G be a bipartite graph with bipartition (V_1, V_2), where $d(x) \geq d(y)$ for each pair of adjacent vertices $x \in V_1$ and $y \in V_2$. Show that a non-increasing sequence $N = (n_1, \ldots, n_t)$ is colour-feasible for G if and only if the condition (8.4.1) holds.

8.4.2 Prove that for a bipartite graph G, every maximal element of the poset $(C(G), \succeq)$ of colour-feasible sequences contains $\Delta(G)$ members. (Folkman, Fulkerson (1969))

8.4.3 ▽ Let G be a bipartite graph with q edges and maximum degree 3. Prove that the set $C(G)$ has at most $\lfloor (q + 15)/12 \rfloor$ maximal elements. (de Werra (1971b))

8.4.4 Let G be a bipartite graph with bipartition (V_1, V_2) and let $R = (r_1, \ldots, r_n)$ and $S = (s_1, \ldots, s_m)$ be the degree sequences of the vertices in V_1 and V_2, respectively. Show that if a sequence N is colour-feasible for G then the pairs (R, N) and (N, S) are graphical. (Talanov, Ševčenko (1972))

8.4.5 Let G be a graph (not necessarily bipartite) in which every two vertices of degree more than 2 are non-adjacent. Show that the set $C(G)$ of colour-feasible sequences has a unique maximal element. (Asratian (1978))

8.4.6 ▽ Let G be a bipartite graph and $H(G) = (h_1, \ldots, h_{\Delta(G)})$ be a sequence, with $h_1 = q_1(G)$ and $h_{i+1} = q_{i+1}(G) - q_i(G)$, for $i = 1, \ldots, \Delta(G) - 1$. Show that

(a) $h_1 \geq h_2 \geq \ldots \geq h_{\Delta(G)}$,

(b) the set $C(G)$ has a unique maximal element if and only if $H(G)$ is colour-feasible for G. (de Werra (1971b))

8.4.7 Show that in a bipartite graph G the sequence $H(G)$ is colour-feasible if $|E(G)| \leq \Delta(G) + 5$.

8.4.8 Show that, for a bipartite graph G, the sequence $H(G)$ is the least upper bound of the set $C(G)$.

8.5 Transformations of proper colourings

We have already seen that by using edge colourings we can model many of the restrictions which arise in timetabling problems. Unfortunately, we have also seen that for the most part the addition of these restrictions makes construction problems NP-hard. One possibility is to solve these problems approximately. We can begin by finding any timetable corresponding to the requirement matrix, and then use some transformations to make our solution a better and better approximation to the optimal solution. This section is devoted to formalising this approach in terms of transformations of edge colourings. We shall see that any proper t-colouring of a t-regular bipartite graph can be obtained from any other, by using only two simple types of transformation. Solving this problem allows us immediately to solve the more general question of transforming between edge-colourings of an arbitrary bipartite graph. Let us begin by describing the two transformations.

Let G be a t-regular bipartite graph, f a proper t-colouring of G, and let $C = v_0 e_1 v_1 e_2 \ldots e_{2k-1} v_{2k-1} e_{2k} v_0$ be an even cycle of G, in which the colour of all the even numbered edges is α. If the colour of all the odd numbered edges is β, then we could carry out the first of our transformations, a 2-*transformation* of G (along C), by exchanging the colours of the edges along the cycle.

Suppose instead that the odd numbered edges are each coloured with one of two colours β and γ. Then the cycle C is called a 3-*colour* (α, β, γ)-*cycle* or sometimes a 3-coloured cycle. In this case we can carry out our second type of transformation: divide the set

$$\left(M(f,\beta) \cup M(f,\gamma) \cup \{e_2, e_4, \ldots, e_{2k}\} \right) \setminus \{e_1, , e_3 \ldots, e_{2k-1}\}$$

into two matchings P_1 and P_2, and define a new proper t-colouring g by

$$g(e) = \begin{cases} \alpha & \text{if } e \in (M(f,\alpha) \setminus \{e_2, e_4, \ldots, e_{2k}\}) \cup \{e_1, e_3, \ldots, e_{2k-1}\}, \\ \beta & \text{if } e \in P_1, \\ \gamma & \text{if } e \in P_2, \\ f(e) & \text{if } f(e) \notin \{\alpha, \beta, \gamma\}. \end{cases}$$

This transformation we shall call a 3-*transformation* of G (along C). Such a 3-transformation also only changes the original colouring locally: if E_0 denotes the set $M(f,\alpha) \cup M(f,\beta) \cup M(f,\gamma)$, the transformation first changes the matching $M(f,\alpha)$ along the cycle C, and then colours the rest of E_0 with the remaining two colours β and γ.

The effects of this pair of transformations can perhaps be better understood from the 3-colourings f and g of $K_{3,3}$, shown in figure 8.5.1. It is easy to check that g cannot be obtained from f by a sequence of 2-transformations alone, but g is the result of a 3-transformation of f along the 3-coloured cycle C.

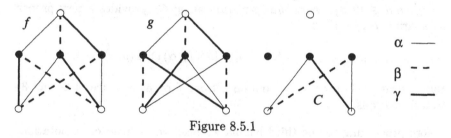

Figure 8.5.1

As we have already mentioned, this section revolves around transformations of t-colourings. The following theorem solves this problem completely for regular bipartite graphs.

Theorem 8.5.1 (Asratian, Mirumian (1991)) *Let $t \geq 3$ and G be a t-regular bipartite graph. Then every proper t-colouring of G can be obtained from any other by a sequence of 2- and 3-transformations.* □

We will give a simplified proof of this theorem (see Asratian (1996)). For this we first need a little more notation and three preliminary lemmas.

Let f and g be two distinct proper t-colourings of G. We shall say that f and g *differ* by an m-coloured subgraph if there is a set of colours S, of size m, so that $M(f,j) \neq M(g,j)$ for each $j \in S$, but $M(f,j) = M(g,j)$ for each $j \notin S$. We denote by $G(f,g,j)$ the coloured subgraph induced by the edge subset $M(f,j) \triangle M(g,j)$, where each edge $e \in M(f,j) \triangle M(g,j)$ has colour $f(e)$. Since G is regular, all the components of $G(f,g,j)$ are even cycles, and the following lemma is evident.

Lemma 8.5.2 *If f and g differ by a 2-coloured subgraph then g can be obtained from f by a sequence of 2-transformations.* $\square$

A little less evident is the next lemma.

Lemma 8.5.3 *If f and g differ by a 3-coloured subgraph then g can be obtained from f by a sequence of 2- and 3-transformations.*

Proof. Suppose that we have colours α, β and γ so that $M(f,j) \neq M(g,j)$ if and only if $j \in \{\alpha, \beta, \gamma\}$. We construct a sequence $\{f_k\}, k \geq 0$, of proper t-colourings of G as follows.

Put $f_0 = f$. Suppose that we have already constructed a sequence $f_0, ..., f_k$ of proper t-colourings of G where $M(f_k, j) = M(g,j)$ for $j \notin \{\alpha, \beta, \gamma\}$. If the colouring f_k differs from g by a 2-coloured subgraph then, by Lemma 8.5.2, g can be obtained from f_k by a sequence of 2-transformations. If the colouring f_k differs from g by a 3-coloured subgraph then $M(f_k, j) \neq M(g,j)$ for each $j \in \{\alpha, \beta, \gamma\}$ and the subgraph $G(f_k, g, \alpha)$ contains a d-coloured cycle C_k for some $d \in \{2,3\}$. A d-transformation along C_k provides a new proper t-colouring f_{k+1} such that

$$|M(f_{k+1}, \alpha) \cap M(g, \alpha)| > |M(f_k, \alpha) \cap M(g, \alpha)|.$$

By repeating this process, we obtain g from f by a sequence of 2- and 3-transformations. $\square$

Before proceeding to the third lemma we need yet a little more notation. Let f and g be two distinct proper t-colourings of G with $M(f,t) \neq M(g,t)$. Let $F(f,g,j,t)$ denote the set of all (t,j)-coloured paths of even length in the graph $G(f,g,t)$ and let

$$F(f,g,t) = \bigcup_{j=1}^{t-1} F(f,g,j,t).$$

From each $P \in F(f,g,j,t)$ we can produce a new t-colouring of G, which is not proper, by interchanging the colours t and j on P; this colouring we shall denote by $f(P)$. Then $M(f(P), t)$ is the set of edges of colour t under the colouring $f(P)$. Finally we let $h(P; f, g, t)$ be the number of edges in $M(f(P), t) \cap M(g, t)$ and

$$h(f, g, t) = \max_{P \in F(f,g,t)} h(P; f, g, t).$$

Lemma 8.5.4 Let $t \geq 4$, and f and g be two distinct proper t-colourings of G. If $M(f, t) \neq M(g, t)$ then there exists a sequence of proper t-colourings $f = f_0, \ldots, f_k$ such that

$$|M(f_k, t) \cap M(g, t)| > |M(f, t) \cap M(g, t)| \qquad (8.5.1)$$

with f_i and f_{i+1} differing by a 2- or 3-coloured subgraph, for each $i = 0, \ldots, k-1$.

Proof. If the subgraph $G(f, g, t)$ contains a 2- or 3-coloured cycle C, then we can take $k = 1$, produce f_1 from f by applying an appropriate 2- or 3-transformation along C, and we are done. If this is not the case we must work a little harder.

Put $f_0 = f$. Suppose that $f_0, f_1, \ldots f_i$ are already defined, and $G(f_i, g, t)$ does not contain a 2- or 3-coloured cycle. Then we construct a new proper t-colouring f_{i+1} of G with $h(f_{i+1}, g, t) > h(f_i, g, t)$. Consider a path P of maximum even length from the set $F(f_i, g, t)$. Clearly P is a (t, s)-coloured path for some $s \in \{1, \ldots, t-1\}$ and $h(P; f_i, g, t) = h(f_i, g, t)$. Let C be the component of $G(f_i, g, t)$ which contains P, and let

$$P = v_0 e_1 v_1 e_2 \ldots v_{2n-1} e_{2n} v_{2n},$$
$$C = v_0 e_1 v_1 e_2 \ldots v_{2m-1} e_{2m} v_0, \quad m \geq n+2$$

where $f_i(e_{2j-1}) = t$ and $f_i(e_{2j}) = s$ for each $j = 1, \ldots, n$. Then $f_i(e_{2n+1}) = t$ and, by the maximality of P, $f_i(e_{2n+2}) = l$ for some $l \notin \{s, t\}$.

Consider the improper colouring f_{i1} obtained from f_i by interchanging the colours s and t along the edges of P, and interchanging the colours of e_{2n+1} and e_{2n+2}. Then f_{i1} has the following properties:

(1a) There is no edge incident with v_0 of colour t, but there are two edges of colour s, and one of each colour $j \neq s$, $1 \leq j \leq t-1$.

(1b) There is no edge incident with v_{2n} of colour s, but there are two of colour l, and one of each colour $j \neq s, l$, $1 \leq j \leq t$.

(1c) There are no edges incident with v_{2n+2} of colour l, but there are two of colour t, and one of each colour $j \neq l$, $1 \leq j \leq t-1$.

(1d) At each vertex other than v_0, v_{2n} and v_{2n+2} each colour appears on precisely one edge.

Now consider the subgraph H induced by the sets $M(f_{i1}, s) \cup M(f_{i1}, l)$. By *(1a)–(1d)*, H contains an even (s, l)-coloured path P_1 of even length, with end vertices v_0 and v_{2n}. Let us interchange the colours along the path P_1 to obtain a new t-colouring f_{i2}. Then f_{i2} is not proper, and has the following properties:

(2a) There is no edge incident with v_0 of colour t, but there are two edges of colour l, and one of each colour $j \neq l$, $1 \leq j \leq t - 1$.

(2b) There are no edges incident with v_{2n+2} of colour l, but there are two of colour t, and one of each colour $j \neq l$, $1 \leq j \leq t - 1$.

(2c) At each vertex other than v_0 and v_{2n+2} each colour appears on precisely one edge.

It follows from $(2a) - (2c)$ that there is an (t, l)-coloured path P_2 of even length with end vertices v_0 and v_{2n+2}. This path P_2 finally allows the definition of a proper t-colouring f_{i+1} from f_{i2} by interchanging the colours t and l along P_2. It is clear that $M(f_{i+1}, j) = M(f_i, j)$ for each $j \notin \{l, s, t\}$, that $f_{i+1}(P_2) = f_{i2}$ and that

$$M(f_{i+1}(P_2), t) \cap M(g, t) = (M(f_i, t) \cap M(g, t)) \cup \{e_{2i} : i = 1, 2, ..., n + 1\}.$$

Furthermore,

$$M(f_i(P), t) \cap M(g, t) = (M(f_i, t) \cap M(g, t)) \cup \{e_{2i} : i = 1, 2, ..., n\}.$$

Therefore, $|M(f_{i+1}(P_2), t) \cap M(g, t)| > |M(f_i(P), t) \cap M(g, t)|$. This implies that $h(f_{i+1}, g, t) > h(f_i, g, t)$ because

$$h(f_i, g, t) = h(P; f_i, g, t) = |M(f_i(P), t) \cap M(g, t)|$$

by the choice of P, and

$$h(f_{i+1}, g, t) \geq h(P_2; f_{i+1}, g, t) = |M(f_{i+1}(P_2), t) \cap M(g, t)|.$$

Thus, if $G(f_i, g, t)$ does not contain a 2- or 3-coloured cycle then there is a proper t-colouring f_{i+1} of G with $h(f_{i+1}, g, t) > h(f_i, g, t)$. Since $h(f_i, g, t) < h(f_{i+1}, g, t) \leq |V(G)|/2$ for each $i \geq 0$, there exists an integer j such that the subgraph $G(f_j, g, t)$ will contain a d-coloured cycle C' for some $d \in \{2, 3\}$. Then put $k = j + 1$ and let f_k be obtained from f_j by a d-transformation along C'. Then the sequence $f_0, \ldots, f_k$ satisfies the requirements of the lemma. $\qquad \square$

Proof of Theorem 8.5.1. We shall prove the theorem by induction on t. For $t = 3$ the result follows from Lemmas 8.5.2 and 8.5.3. Let us turn then to the induction step and suppose that G is a t-regular bipartite graph and that the induction hypothesis holds for $(t - 1)$-regular graphs, $t \geq 4$.

Let ϕ and ψ be two distinct t-colourings of G. The proof breaks into two cases.

Case A. $M(\phi, t) = M(\psi, t)$

Then the graph $G' = G - M(\phi, t)$ is $(t-1)$-regular. Let ϕ' and ψ' be the two distinct proper $(t-1)$-colourings of G' induced by ϕ and ψ, respectively. Then, since by the inductive hypothesis ϕ' can be obtained from ψ' by a sequence of 2- and 3-transformations, the same must be true of ϕ and ψ.

Case B. $M(\phi, t) \neq M(\psi, t)$

Then, by Lemma 8.5.4, we can obtain a sequence of proper t-colourings $\phi = \phi_0, \phi_1, \ldots, \phi_k$ so that $M(\phi_k, t) = M(\psi, t)$ and ϕ_{i+1} differ from ϕ_i by a 2- or 3-coloured subgraph, for each $i = 0, 1, \ldots, k-1$. Then Lemmas 8.5.3 and 8.5.4 imply that ϕ_k can be obtained from ϕ by a sequence of 2- and 3-transformations. Finally, then, as in case A, ψ can be obtained from ϕ_k by the induction hypothesis, and the proof of the theorem is complete. $\square$

It is not difficult to see that this proof provides a polynomial algorithm for transforming one proper t-colouring of a regular bipartite graph into another. Of course if transformations exist for regular graphs it is natural to wonder about general bipartite graphs. Conveniently we can use Theorem 8.5.1 to provide such a transformation.

Consider an arbitrary bipartite graph H, a proper m-colouring f and a proper n-colouring g. Take two disjoint copies of H, H' and H'', with $V(H') = \{x' : x \in V(H)\}$ and $V(H'') = \{x'' : x \in V(H)\}$ and let $t = \max\{m, n\}$. Then we can define a t-regular bipartite graph G obtained from H' and H'' by joining x' and x'' with $t - d_H(x)$ parallel edges, for each vertex $x \in V(H)$. The graph G in figure 8.5.2 is the result of this construction applied to the graph H in the same diagram, when $t = 3$.

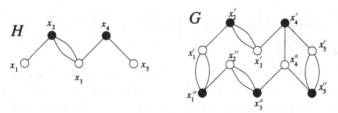

Figure 8.5.2

The colouring f of H induces a proper t-colouring ϕ of G in the following way: we colour the copies H' and H'' in the same way as H, and then colour the set of parallel edges joining x' to x'' with those colours from $\{1, \ldots, t\}$ which are not used to colour an edge incident with x in H. Similarly, the colouring g induces a proper colouring ψ on G. Thus, since Theorem 8.5.1 ensures that ϕ can be transformed into ψ by a sequence of

2- and 3-transformations, it is clear that these transformations also define a sequence of proper colourings of H, beginning with f and ending with g.

The above procedure provides a proof of the following result also obtained by Asratian and Mirumian (1991).

Theorem 8.5.5 *Let H be an arbitrary bipartite graph. If f is a proper m-colouring and g is a proper n-colouring of H, then f can be transformed into g such that each intermediate colouring is proper and differs from the previous colouring by a 2- or 3-coloured subgraph.* □

It is worth mentioning that the above theorems do not hold for non-bipartite graphs (see Exercise 8.5.4).

Exercises

8.5.1   Prove that any two proper edge-colourings of a tree can be obtained, one from the other, by a sequence of colour interchanges along maximal 2-coloured paths.

8.5.2 △   Is it true that for $n > 3$, any proper n-colouring of $K_{n,n}$ can be obtained from any other by a sequence of 2-transformations?

8.5.3 ▽   Prove that any proper n-colouring of $K_{n,n}$ can be obtained from any other by a sequence of 3-transformations if and only if n is not a power of 2. (Asratian, Mirumian (1995))

8.5.4   Show that the graph K_6 has two proper 5-colourings which cannot be obtained from one another by using transformations on only 2- and 3-coloured subgraphs. (Asratian, Mirumian (1992))

8.6 Uniquely colourable bipartite graphs

A simple graph G is called *uniquely edge colourable* if any two proper $\chi'(G)$-colourings of G induce the same partition of the edge set $E(G)$, i.e. they differ only in the naming of the colours. For example, the graph $K_{1,n}$ is uniquely edge colourable, but the graph $K_{2,3}$ is not: two proper 3-colourings of $K_{2,3}$ which cannot be obtained from one another by renaming the colours are shown in figure 8.6.1.

It is not difficult to observe that every uniquely edge colourable, simple bipartite graph G with edges coloured with the colours $1, \ldots, \Delta(G)$ has the property that each 2-coloured subgraph is connected, i.e. is a path or an even cycle. Using this property it is not difficult to characterise uniquely colourable, simple bipartite graphs.

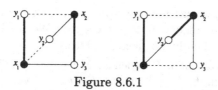

Figure 8.6.1

Proposition 8.6.1 *A simple bipartite graph is uniquely edge colourable if and only if it is either a path, an even cycle or a star.*[†]   □

But what of bipartite graphs with multiple edges? Of course, a similar definition to that for simple graphs can also be made for graphs in general, and this case is dealt with in Exercise 8.6.2. On the other hand, in applications (see for example the interconnection between timetables and edge colourings in Section 8.1), it is more natural not to distinguish between proper edge colourings which differ only by the colours on some parallel edges. With this in mind, we will say that two proper $\Delta(G)$-colourings of a bipartite graph G are *equivalent* if, possibly after interchanging the colours on some parallel edges, they induce the same partition of the edges into matchings. Figure 8.6.2 shows two proper 3-colourings of a bipartite graph which are equivalent in this way. To see this equivalence, we need only interchange the colours 1 and 2 on the two left-most parallel edges in figure 8.6.2 (a). It is then easy to see that we obtain a 3-colouring which induces the same partition into matchings as the colouring in figure 8.6.2 (b).

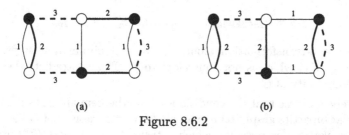

(a)   (b)

Figure 8.6.2

We call a bipartite graph G *uniquely edge colourable* if any two proper $\Delta(G)$-colourings of G are equivalent. It is not difficult to check that the graph shown in figure 8.6.2 is one such graph. The following property follows easily from Theorem 8.5.5.

Property 8.6.2 *Let a bipartite graph G be properly edge coloured with $\Delta(G)$ colours. Then G is uniquely edge colourable if and only if each 3-coloured subgraph of G is also uniquely edge colourable.*   □

[†]Indeed, Thomason (1978) proved that amongst all simple graphs G (not necessarily bipartite) with $\chi'(G) \neq 3$, only a path, even cycle and star are uniquely edge colourable.

Using this property, we can provide a polynomial characterisation of unique edge colourability for regular bipartite graphs. We shall use this result in the next chapter on doubly stochastic matrices.

Theorem 8.6.3 (Asratian, Kostochka, Mirumian (1987)) *Let G be a k-regular bipartite graph, $k \geq 3$, with bipartition (V_1, V_2) and let f be a proper k-colouring of G. Then G is uniquely edge colourable if and only if the following conditions hold:*

(1) there is at most one cycle of length at least 4 in each 2-coloured subgraph;

(2) for each 3-coloured subgraph H, there is at most one vertex from V_1 adjacent to three distinct vertices in H.

Proof. It follows from Property 8.6.2 that we need only consider the case when G is 3-regular. Suppose then that G is uniquely edge colourable and is 3-regular. Then *(1)* is evident. Suppose, however, that there are two vertices x_1 and x_2 from V_1 that are each adjacent to three distinct vertices in V_2. This implies that there are a $(1, 2)$-coloured cycle C_1, and a $(1, 3)$-coloured cycle C_2, each of length 4 or more, such that both contain x_1 and x_2. Without loss of generality, we shall assume that these cycles are $C_1 = u_1 e_1 u_2 e_2 \ldots u_{2n} e_{2n} u_1$ and $C_2 = v_1 e_1' v_2 e_2' \ldots v_{2l} e_{2l}' v_1$, where $u_1 = x_1$, $f(e_1) = 2$, $v_1 = x_2 = u_{2s+1}$ and $f(e_1') = 3$.
Let m be the minimum number for which $v_{2m} \in \{u_2, u_4, \ldots, u_{2n}\}$, and let $v_{2m} = u_{2k}$. Then we have a 3-colour $(1, 2, 3)$-cycle C_3,

$$C_3 = u_{2k} e_{2k} u_{2k+1} \ldots u_{2s+1} e_1' v_2 e_2' \ldots e_{2m}' v_{2m},$$

and from a 3-transformation along C_3 we will obtain a new proper 3-colouring g of G which is not equivalent to f. This contradicts the unique edge colourability of G.

Conversely, suppose that the conditions of the theorem are not sufficient for a 3-regular bipartite graph to be uniquely edge colourable. Let G be a graph, on as few vertices as possible, which satisfies both *(1)* and *(2)* but is not uniquely edge colourable. Then G is connected, and *(2)* ensures that there is a vertex $x \in V_1$ which is adjacent to only two vertices in $y_1, y_2 \in V_2$. If we delete the vertex x and collapse the two vertices y_1 and y_2 into one, then it is not hard to check that the graph we obtain is also 3-regular, bipartite, satisfies the conditions of the theorem, and is not uniquely edge colourable. This contradicts the choice of G. $\square$

It is not difficult to see that for connected bipartite graphs the first condition in Theorem 8.6.3 can be omitted, giving the following corollary.

Corollary 8.6.4 *Let G be a connected, k-regular bipartite graph, $k \geq 3$, with bipartition (V_1, V_2) and let f be a proper k-colouring of G. Then G*

is uniquely edge colourable if and only if for each 3-coloured subgraph H, there is at most one vertex from V_1 adjacent to three distinct vertices in H. □

It is clear that the condition of Corollary 8.6.4 can be checked in polynomial time, but how might we construct graphs which satisfy this condition? To answer this question we require an operation of *extension* for uniquely edge colourable graphs.

Let G' be a connected, uniquely edge colourable, k-regular bipartite graph with a proper k-colouring f'. Choose two adjacent vertices u and v and an edge e joining them. We form a new graph G and a proper k-colouring f of G in the following way: delete the edge e, add two new vertices u' and v', and join u to v' and v to u' with edges of colour $f'(e)$; finally join u' and v' with $k-1$ parallel edges, properly coloured with the remaining $k-1$ colours. If G satisfies the condition of Corollary 8.6.4 then we say that G is obtained from G' by the operation of extension. Examples of this operation, for $k = 3$, are shown in figure 8.6.3.

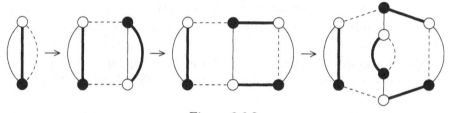

Figure 8.6.3

It is now clear that by using the operation of extension we can inductively construct uniquely edge colourable, k-regular bipartite graphs from the properly coloured graph $G_2(k)$ (the graph of two vertices joined by k parallel edges). Indeed a stronger result is true (see Asratian, Kostochka and Mirumian (1987)).

Proposition 8.6.5 *Any connected, uniquely edge colourable, k-regular bipartite graph on $2n$ vertices can be obtained from the properly coloured graph $G_2(k)$ by a sequence of $n-1$ extensions.* □

Finally, we shall formulate a criterion for a general bipartite graph to be uniquely edge colourable.

Theorem 8.6.6 (Asratian, Kostochka, Mirumian (1987)) *Let G be a bipartite graph, $\Delta(G) \geq 3$, with bipartition (V_1, V_2), properly coloured by $\Delta(G)$ colours. Then G is uniquely edge colourable if and only if the following conditions hold:*

(1) at most one of the non-trivial components of each 2-coloured subgraph is not a cycle of length two;

(2) for each 3-coloured subgraph H, the set V_i contains at most one vertex x_i with $2 \leq d_H(x_i) \leq 3$ which has $d_H(x_i)$ distinct neighbours, for $i = 1, 2$.

$\square$

Exercises

8.6.1   Deduce Proposition 8.6.1 from Theorem 8.6.6.

8.6.2   Let G be a bipartite graph having at least two parallel edges. Show that any two $\Delta(G)$-colourings of G induce the same partition of $E(G)$ if and only if G is either obtained from a star by multiplying some edges or isomorphic to a graph on four vertices, v_1, v_2, v_3 and v_4, in which v_1 is joined to v_2, v_4 to v_3, and v_2 is joined to v_3 by at least two parallel edges.

8.6.3   Let G be a connected, uniquely edge colourable, k-regular bipartite graph, on at least three vertices. Show that

(a)   there is a vertex v_0 in G which is adjacent to only two vertices, v_1 and v_2,

(b)   the graph G' obtained from $G - v_0$ by collapsing v_1 and v_2 into one vertex is also a uniquely edge colourable, k-regular bipartite graph.

Application

8.7 Rearrangeable telephone networks

In each of the previous sections of this chapter we have given interpretations of edge colouring results in timetabling. Later, we shall see some more combinatorial applications, but here we shall apply the fact that any k-regular bipartite graph has a proper edge k-colouring to construct a minimum rearrangeable telephone network. This network models the situation of a telephone network in which we already have some existing calls and would like to add some additional connection. In a rearrangeable network a new call can always be added by rearranging the routings of the existing calls.

Let us now formulate this problem in terms of graphs. Let G be a graph, with n distinguished *input vertices* $\{x_1, \ldots, x_n\} \subset V(G)$ and n distinguished *output vertices* $\{y_1, \ldots, y_n\} \subset V(G)$, these play the parts of n incoming telephone lines and n possible recipients for the calls. We call G a *rearrangeable network* if for any permutation $\pi \in S_n$ there are some n vertex-disjoint paths, which respectively join x_1 to $y_{\pi(1)}$, x_2 to $y_{\pi(2)}, \ldots$, and x_n to $y_{\pi(n)}$.

Imagine that those edges of the network which are used in the connecting paths are switched *on*, and the others are switched *off*. In this way each set of connecting paths can be thought of as a binary vector of length m, where m is the number of edges in G. It is then clear that if the network can represent all $n!$ possible permutations 2^m must be at least $n!$. Therefore G must have at least $\log_2(n!) = O(n \log_2 n)$ edges. We shall show that rearrangeable networks with $O(n \log_2 n)$ edges actually exist.

The networks we shall consider are based on so called *Clos networks*; the Clos (m, k)-network is constructed as follows:

Let the $n = mk$ input vertices be partitioned into m subsets of k vertices, $I_1, \ldots, I_m$, where each $I_i = \{x_{(i-1)k+l} : l = 1, \ldots, k\}$. Similarly, let the output vertices be partitioned into subsets $O_1, \ldots, O_m$, where $O_i = \{y_{(i-1)k+l} : l = 1, \ldots, k\}$ for $i = 1, \ldots, m$. As 'internal' vertices of the graph there are subsets of vertices

$$
\begin{aligned}
I_i' &= \{x_{(i-1)k+l}' : l = 1, \ldots, k\}, &&\text{for } i = 1, \ldots, m, \\
O_i' &= \{y_{(i-1)k+l}' : l = 1, \ldots, k\}, &&\text{for } i = 1, \ldots, m, \\
L_j &= \{z_{(j-1)m+l} : l = 1, \ldots, m\}, &&\text{for } j = 1, \ldots, k, \\
R_j &= \{z_{(j-1)m+l}' : l = 1, \ldots, m\}, &&\text{for } j = 1, \ldots, k.
\end{aligned}
$$

To construct the Clos (m, k)-network, place a complete bipartite graph between each I_i and I_i', for $i = 1, \ldots, m$, between each O_i and O_i', for $i = 1, \ldots, m$, and between each L_j and R_j, for $j = 1, \ldots, k$. Finally join the jth vertex of I_i' to the ith vertex of L_j and the jth vertex of O_i' to the ith vertex of R_j, for each appropriate i and j. The structure of this graph is, perhaps more clearly, shown by the example in figure 8.7.1. The figure also shows one configuration of the network, corresponding to the permutation $\pi = \begin{pmatrix} 1 & 2 & 3 & 4 & 5 & 6 \\ 4 & 2 & 5 & 6 & 3 & 1 \end{pmatrix}$, depicting edges which are 'on' with solid lines, and edges which are 'off' with dotted lines.

Proposition 8.7.1 (Duguid (1959) and Slepian (1952)) *Every Clos (m, k)-network with $m \geq k$ is rearrangeable.*

Proof. Let $n = mk$ and let $\pi \in S_n$. We must show that there is a set of vertex-disjoint paths which 'mimic' the permutation π.

For each $i = 1, \ldots, m$ we define the set of vertices $K_i = \{y_{\pi(j)} : x_j \in I_i\}$. Then we define a bipartite graph H with bipartition (V_1, V_2) where $V_1 = \{u_1, \ldots, u_m\}$, $V_2 = \{v_1, \ldots, v_m\}$ and u_i is joined to v_j by $|O_i \cap K_j|$ parallel

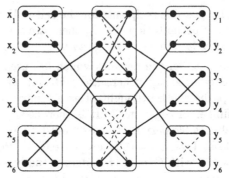

Figure 8.7.1 A Clos $(3, 2)$-network

edges each labelled with a distinct element of $O_i \cap K_j$. Then H is a k-regular bipartite graph, and so has a proper edge k-colouring, f.

This colouring is all we require to establish the paths from x_i to $y_{\pi(i)}$. Consider the matching $M(f, r)$ in H. Then for each edge $e \in M(f, r)$ with ends u_i and v_{j_i} we can associate the input subset I_i with the output subset O_{j_i}. In this way each input set corresponds to a unique output set in such a way that some input $x_i^* \in I_i$ is mapped by π to the output vertex $y_{j_i}^* \in O_{j_i}$ which labels the edge joining u_i to v_{j_i} in H which is coloured r. We now switch 'on' the edges which join x_i^* to the rth vertex in I_i', the edges which join $y_{j_i}^*$ to the rth vertex of O_{j_i}' and those edges which complete the path connecting x_i^* to $y_{j_i}^*$ through the bipartite graph on L_r and R_r (for each $i = 1, \ldots, m$). So using $M(f, r)$ we produce m vertex-disjoint paths. If we carry out this process for each $r = 1, \ldots k$, we produce all mk vertex-disjoint paths which mimic the permutation π. □

Theorem 8.7.2 *There is a rearrangeable network with n inputs and n outputs, and $O(n \log_2 n)$ edges.*

Proof. The construction consists of combining Clos networks together. We shall suppose that $n = 2^p$, otherwise fictitious input and output vertices can be added. A Clos $(2^{p-1}, 2)$-network has two complete 2^{p-1} by 2^{p-1} bipartite graphs at its centre. We could replace these two complete bipartite graphs by two Clos $(2^{p-2}, 2)$-networks, and the total network would behave in precisely the same way as the original. If we do this and continue recursively, replacing each complete bipartite graph by an appropriate Clos network, we finally obtain a network consisting only of copies of $K_{2,2}$ joined together by various 'connecting' edges. The construction ensures that this final graph is a rearrangeable network, and it can easily be seen that its edges number at most

$$(2p - 1)(n/2) \cdot 4 + 2(p - 1)n = n(6p - 4) = n(6(\log_2 n) - 4) = O(n \log_2 n)$$

as required. □

Chapter 9

Doubly stochastic matrices and bipartite graphs

9.1 Convex representations of doubly stochastic matrices

$\mathbf{A}$ *doubly stochastic matrix* of order n is a non-negative $n \times n$ matrix $\mathbf{A}$ in which all rows and column sums are equal to 1. A *permutation matrix* $\mathbf{T}$ is a square matrix with exactly one 1 in each row and column, the rest of the entries being zero. Clearly, then, a permutation matrix is also doubly stochastic. A set $S = \{a_{1j_1}, a_{2j_2}, \ldots, a_{nj_n}\}$ of elements of $\mathbf{A}$ is called a *diagonal* of $\mathbf{A}$ if $(j_1, j_2, \ldots, j_n)$ is a permutation of the set $\{1, \ldots, n\}$. A diagonal is *positive* if all its constituent elements are positive.

Lemma 9.1.1 (König (1916)) *Every doubly stochastic matrix has a positive diagonal.*

Proof. Let $\mathbf{A} = [a_{ij}]$ be a doubly stochastic matrix of order n. Using $\mathbf{A}$ we may define a simple bipartite graph G with bipartition (V_1, V_2) where $V_1 = \{r_1, \ldots, r_n\}$, the set of rows of $\mathbf{A}$, and $V_2 = \{c_1, c_2, \ldots, c_n\}$, the set of columns. Then $r_i c_j \in E(G)$ if and only if $a_{ij} > 0$. With this definition, proving the lemma amounts to showing that G has a perfect matching.

Suppose that G does not have a perfect matching. Then by Hall's Theorem there must be some subset $X \subset V_1$ with $|N(X)| < |X|$. In other words there is some set of l rows which have non-zero entries in some $k < l$ columns. However, summing the entries row by row we have l, but the summing column by column does not exceed k, implying that $l \leq k$, a clear contradiction. So G has a perfect matching and $\mathbf{A}$ a positive diagonal. $\qquad\square$

The following theorem was proved by Birkhoff (1946) and independently by von Neumann (1953).

Theorem 9.1.2 (Birkhoff–von Neumann Theorem) *Each doubly stochastic matrix* $\mathbf{A}$ *can be represented as a convex combination of permutation matrices, that is,*

$$\mathbf{A} = c_1 \mathbf{T}_1 + c_2 \mathbf{T}_2 + \ldots + c_k \mathbf{T}_k \qquad (9.1.1)$$

where the c_i *are positive real numbers with* $\sum_{i=1}^{k} c_i = 1$ *and* $\mathbf{T}_1, \mathbf{T}_2, \ldots, \mathbf{T}_k$ *are distinct permutation matrices.*

Proof. Let $\nu(\mathbf{A})$ denote the number of positive elements in $\mathbf{A}$; clearly, $\nu(\mathbf{A}) \geq n$. We shall prove the theorem by induction on $\nu(\mathbf{A})$. If $\nu(\mathbf{A}) = n$ then $\mathbf{A}$ is a permutation matrix and the assertion is clear. Suppose now that the assertion holds whenever $\nu(\mathbf{A}) < r$ and consider a doubly stochastic matrix $\mathbf{A}$ with $\nu(\mathbf{A}) = r$. By Lemma 9.1.1 $\mathbf{A}$ has a positive diagonal $S = \{a_{1j_1}, a_{2j_2}, \ldots, a_{nj_n}\}$. Let $\mathbf{T} = [t_{ij}]$ be the permutation matrix with $t_{ij} = 1$ if and only if $j = j_i$ for $1 \leq i \leq n$.

Let $\lambda = \min_{1 \leq i \leq n} a_{ij_i}$. Then $\lambda < 1$ and we can form a new doubly stochastic matrix $\mathbf{A}' = (\mathbf{A} - \lambda\mathbf{T})/(1 - \lambda)$ which must have $\nu(\mathbf{A}') < \nu(\mathbf{A}) = r$. Hence, by induction, there are positive numbers $\lambda_1, \lambda_2, \ldots, \lambda_k$ and permutation matrices $\mathbf{T}_1, \ldots, \mathbf{T}_k$ such that $\frac{1}{1 - \lambda}(\mathbf{A} - \lambda\mathbf{T}) = \sum_{i=1}^{k} \lambda_i \mathbf{T}_i$. Then setting $c_{k+1} = \lambda$, $\mathbf{T}_{k+1} = \mathbf{T}$ and $c_i = (1 - \lambda)\lambda_i$ we have $\sum_{i=1}^{k+1} c_i = 1$ and $\mathbf{A} = \sum_{i=1}^{k+1} c_i \mathbf{T}_i$. $\qquad \square$

Corollary 9.1.3 *If* $\mathbf{A}$ *is a non-negative matrix with column sums and row sums all equal to some value* σ, *then* $\mathbf{A}$ *can be represented in the form* $\mathbf{A} = c_1 \mathbf{T}_1 + c_2 \mathbf{T}_2 + \ldots + c_k \mathbf{T}_k$ *where the* c_i *are positive real numbers with* $\sum_{i=1}^{k} c_i = \sigma$ *and* $\mathbf{T}_1, \mathbf{T}_2, \ldots, \mathbf{T}_k$ *are distinct permutation matrices.* $\qquad \square$

Corollary 9.1.4 *If* $\mathbf{A}$ *is a non-negative integral matrix with equal row and column sums, then* $\mathbf{A}$ *can be written as a sum of permutation matrices.*

We shall call a representation such as that in (9.1.1) a *convex representation* of a doubly stochastic matrix. In general such a representation is not even unique up to reordering the terms. For example the following doubly stochastic matrix has two representations.

$$A = \begin{pmatrix} 1/3 & 1/3 & 1/3 \\ 1/3 & 1/3 & 1/3 \\ 1/3 & 1/3 & 1/3 \end{pmatrix}$$

$$= \frac{1}{3}\begin{pmatrix} 1 & 0 & 0 \\ 0 & 1 & 0 \\ 0 & 0 & 1 \end{pmatrix} + \frac{1}{3}\begin{pmatrix} 0 & 1 & 0 \\ 0 & 0 & 1 \\ 1 & 0 & 0 \end{pmatrix} + \frac{1}{3}\begin{pmatrix} 0 & 0 & 1 \\ 1 & 0 & 0 \\ 0 & 1 & 0 \end{pmatrix}$$

$$= \frac{1}{3}\begin{pmatrix} 1 & 0 & 0 \\ 0 & 0 & 1 \\ 0 & 1 & 0 \end{pmatrix} + \frac{1}{3}\begin{pmatrix} 0 & 1 & 0 \\ 1 & 0 & 0 \\ 0 & 0 & 1 \end{pmatrix} + \frac{1}{3}\begin{pmatrix} 0 & 0 & 1 \\ 0 & 1 & 0 \\ 1 & 0 & 1 \end{pmatrix}$$

A criterion for unique convex representation will be discussed in the following section.

Exercises

9.1.1 △ Prove König's Colouring Theorem by using Corollary 9.1.4.

9.1.2   Let $\mathbf{A} = (a_{ij})$ be a doubly stochastic matrix of order $n \geq 2$ with a convex representation $\mathbf{A} = \sum_{i=1}^{k} c_i \mathbf{T}_i$. Furthermore, let $f(\mathbf{A}) = \sum_{i=1}^{n} \sum_{j=1}^{n} d_{ij} a_{ij}$, where $d_{ij} \geq 0$ for each i, j, $1 \leq i \leq n$, $1 \leq j \leq n$. Prove that $f(\mathbf{A}) = \sum_{i=1}^{n} c_i f(\mathbf{T}_i)$.

9.1.3   Suppose that a society consists of n men and n women. Let d_{ij} be a measure of the 'happiness' derived by man i and woman j living together, and let a_{ij} denote the fraction of time which man i spends living with woman j (i.e. the matrix $\mathbf{A} = (a_{ij})$ is doubly stochastic).

(a)   Show that the total happiness in the society $\sum_{i=1}^{n} \sum_{j=1}^{n} d_{ij} a_{ij}$ is maximised when $\mathbf{A}$ is a permutation matrix – in other words when marriage is monogamous.

(b)   Using a similar argument, show that the total happiness is minimised by another arrangement of monogamous marriages(!).

9.2 Matrices with a unique convex representation

$\mathbf{L}$et $\mathbf{A} = [a_{ij}]$ be a doubly stochastic matrix of order n with a convex representation

$$\mathbf{A} = c_1\mathbf{T}_1 + c_2\mathbf{T}_2 + \ldots + c_k\mathbf{T}_k \qquad (9.2.1)$$

where $\mathbf{T}_r = [t_{ij}^{(r)}]$ is a permutation matrix, $1 \le r \le k$. We shall give a criterion for when (9.2.1) is the unique convex representation of $\mathbf{A}$ (up to reordering the terms). We may assume that $k \ge 2$ in (9.2.1) or $\mathbf{A}$ is a permutation matrix, and the representation is clearly unique.

Let $\mathbf{T} = [t_{ij}] = \sum_{r=1}^{k} \mathbf{T}_r$ and consider the bipartite graph G with bipartition (V_1, V_2) where $V_1 = \{x_1, x_2, \ldots, x_n\}$, $V_2 = \{y_1, y_2, \ldots, y_n\}$ and vertices x_i and y_j are joined by t_{ij} edges ($1 \le i, j \le n$). Then G is a k-regular bipartite graph. Denote by $L(G, k)$ the set of all proper edge colourings of G with the colours $1, 2, \ldots, k$. Then each k-colouring $f \in L(G, k)$ induces a collection $\mathbf{R}(f, 1), \mathbf{R}(f, 2), \ldots, \mathbf{R}(f, k)$ of $n \times n$ permutation matrices where $\mathbf{R}(f, r) = [h_{ij}^{(r)}]$ and

$$h_{ij}^{(r)} = \left\{ \begin{array}{ll} 1 & \text{if there is an edge joining } x_i \text{ and } y_j \text{ coloured } r, \\ 0 & \text{otherwise,} \end{array} \right\} \ 1 \le i, j \le n.$$

Furthermore, the collection of matrices $\mathbf{T}_1, \mathbf{T}_2, \ldots, \mathbf{T}_k$ from (9.2.1) induces a proper k-colouring of G. Associate with each edge e joining x_i and y_j a number $f_0(e)$ from the set $\{r : t_{ij}^{(r)} = 1\}$; each edge receiving a different number. Clearly, then, $f_0 : E(G) \longrightarrow \{1, 2, \ldots, k\}$ is a proper k-colouring of G. These observations lead us to the following lemma.

Lemma 9.2.1 *The convex representation (9.2.1) of the doubly stochastic matrix $\mathbf{A}$ is unique if and only if the graph G is uniquely edge colourable.*

Proof. Suppose, first, that there is a proper k-colouring $f_1 \in L(G, k)$ which is not isomorphic to the proper k-colouring f_0 induced by $\mathbf{T}_1, \mathbf{T}_2, \ldots, \mathbf{T}_k$. Then $\sum_{r=1}^{k} \mathbf{T}_r = \sum_{r=1}^{k} \mathbf{R}(f_1, r)$. Without loss of generality we shall assume that $c_1 = \min_{1 \le r \le k} c_r$. This gives that

$$\mathbf{A} = c_1 \sum_{r=1}^{k} \mathbf{T}_r + \sum_{r=1}^{k} (c_r - c_1)\mathbf{T}_r,$$

and

$$\mathbf{A} = c_1 \sum_{r=1}^{k} \mathbf{R}(f_1, r) + \sum_{r=1}^{k} (c_r - c_1)\mathbf{T}_r. \qquad (9.2.2)$$

Since f_0 and f_1 are non-isomorphic the two collections of permutation matrices $\mathbf{T}_1, \mathbf{T}_2, \ldots, \mathbf{T}_k$ and $\mathbf{R}(f_1, 1), \mathbf{R}(f_1, 2), \ldots, \mathbf{R}(f_1, k)$ are not equal. Therefore, after simplifying (9.2.2) we obtain a convex representation different from that in (9.2.1).

Conversely, suppose that there is a convex representation of $\mathbf{A}$ which is different from (9.2.1):

$$\mathbf{A} = b_1 \mathbf{D}_1 + b_2 \mathbf{D}_2 + \ldots + b_l \mathbf{D}_l.$$

The proof now breaks into two cases.

Case A. $\{\mathbf{D}_1, \mathbf{D}_2, \ldots, \mathbf{D}_l\} \not\subseteq \{\mathbf{T}_1, \mathbf{T}_2, \ldots, \mathbf{T}_k\}$

Without loss of generality assume that $\mathbf{D}_1 \notin \{\mathbf{T}_1, \mathbf{T}_2, \ldots, \mathbf{T}_k\}$. Let $\mathbf{D}_1 = [d_{ij}]$ and the positive elements of $\mathbf{D}_1$ be $d_{1j_1}, \ldots, d_{nj_n}$. Then $a_{ij_i} > 0$ and in G there exists an edge e_i with ends x_i and y_{j_i} for each $i = 1, \ldots, n$. The collection $M = \{e_1, e_2, \ldots, e_n\}$ then form a perfect matching in G. The graph $G - M$ is $(k - 1)$-regular, and so has a proper $(k - 1)$-colouring. This implies that G has a proper k-colouring f_2 so that $f_2(e_i) = k$ for each $1 \leq i \leq n$. Then f_2 is not isomorphic to f_0 since $\mathbf{D}_1 \notin \{\mathbf{T}_1, \mathbf{T}_2, \ldots, \mathbf{T}_k\}$. That is G is not uniquely edge colourable.

Case B. $\{\mathbf{D}_1, \mathbf{D}_2, \ldots, \mathbf{D}_l\} \subseteq \{\mathbf{T}_1, \mathbf{T}_2, \ldots, \mathbf{T}_k\}$

Then certainly $k \geq l \geq 2$. We assume without loss of generality that $\mathbf{D}_r = \mathbf{T}_r$ for $r = 1, \ldots, l$. There is some $1 \leq r \leq l$ such that $b_r > c_r$. Possibly by renumbering, we assume that $b_1 > c_1$. Then

$$\sum_{r=2}^{k} c_r \mathbf{T}_r = (b_1 - c_1)\mathbf{T}_1 + \sum_{r=2}^{l} b_r \mathbf{T}_r$$

and the doubly stochastic matrix $\mathbf{A}' = \sum_{r=2}^{k} c_r \mathbf{T}_r / (1 - c_1)$ has two different convex representations:

$$\mathbf{A}' = \frac{b_1 - c_1}{1 - c_1} \mathbf{T}_1 + \sum_{r=2}^{l} \frac{b_r}{1 - c_1} \mathbf{T}_r \qquad (9.2.3)$$

$$\text{and} \quad \mathbf{A}' = \frac{c_2}{1 - c_1} \mathbf{T}_2 + \frac{c_3}{1 - c_1} \mathbf{T}_3 + \ldots + \frac{c_k}{1 - c_1} \mathbf{T}_k. \qquad (9.2.4)$$

Consider the proper k-colouring f_0. Observe that the matching $M(f_0, r)$ is induced by the matrix $\mathbf{T}_r$. The $(k - 1)$-regular graph $G - M(f_0, 1)$ has a proper $(k - 1)$-colouring f_0' induced by the matrices $\mathbf{T}_2, \ldots, \mathbf{T}_k$. Recall, now, that $\mathbf{A}'$ has the two representations (9.2.3) and (9.2.4), and $\mathbf{T}_1$ is not present in the second. Then Case A ensures that $G - M(f_0, 1)$ has a proper edge colouring f_3 with colours $2, \ldots, k$, not isomorphic to f'_0. Putting $f_3(e) = 1$ for each edge $e \in M(f_0, 1)$ we obtain a proper k-colouring of G not isomorphic to f_0, as required. $\qquad \square$

Associated with each permutation matrix $\mathbf{T}_r = [t_{ij}^{(r)}]$ there is the corresponding permutation π_r where $t_{i\pi_r(i)}$ is the unique element in row i of $\mathbf{T}_r$ which is 1. Then from Lemma 9.2.1 and Theorem 8.6.3 we have the following theorem.

Theorem 9.2.2 (Asratian, Kostochka, Mirumian (1987)) *Let* $\mathbf{A}$ *be a doubly stochastic matrix of order* n *which has a convex representation (9.2.1) where* $k \geq 2$. *Then (9.2.1) is a unique convex representation of* $\mathbf{A}$ *if and only if the following two conditions hold:*

(1) There is at most one cycle of length more than 2 in the permutation $\begin{pmatrix} \pi_a(1) & \pi_a(2) & \cdots & \pi_a(n) \\ \pi_b(1) & \pi_b(2) & \cdots & \pi_b(n) \end{pmatrix}$ *for each* $1 \leq a < b \leq k$.

(2) If $k \geq 3$ *then for any* $1 \leq a < b < c \leq k$ *the three numbers* $\pi_a(i), \pi_b(i)$, *and* $\pi_c(i)$ *are distinct for at most one* i, $1 \leq i \leq n$. $\qquad \square$

For another criterion for $\mathbf{A}$ to have a unique convex representation see Brualdi and Gibson (1977).

Exercises

9.2.1 △ Prove that the product of two doubly stochastic matrices is also doubly stochastic.

9.2.2 Construct two doubly stochastic matrices $\mathbf{A}_1$ and $\mathbf{A}_2$ with unique convex representations so that $\mathbf{A}_1 \cdot \mathbf{A}_2$ has at least two convex representations.

9.3 Permanents and perfect matchings

In this section we will prove the bound on the number of perfect matchings in a regular bipartite graph which we mentioned earlier. This we shall do using doubly stochastic matrices. But first we shall introduce a new definition.

Let $\mathbf{A} = [a_{ij}]$ be an $n \times n$ matrix, then the *permanent* of $\mathbf{A}$, per $\mathbf{A}$, is defined by the formula

$$\text{per } \mathbf{A} = \sum_{\pi \in S_n} a_{1\pi(1)} \cdot a_{2\pi(2)} \cdot \ldots \cdot a_{n\pi(n)}$$

where S_n denotes the set of all permutations of the set $\{1, 2, \ldots, n\}$.

Proposition 9.3.1 *The number of perfect matchings $\phi(G)$ of an n by n bipartite graph G is equal to the permanent of its biadjacency matrix $\mathbf{B}(G)$. If G is k-regular then $\phi(G) = k^n \mathrm{per}\,\mathbf{A}$ where $\mathbf{A} = \frac{1}{k}\mathbf{B}(G)$ is a doubly stochastic matrix.*

Proof. Let G have bipartition (V_1, V_2), where $V_1 = \{x_1, \ldots, x_n\}$ and $V_2 = \{y_1, \ldots, y_n\}$. Consider a perfect matching of G, $M = \{e_1, e_2, \ldots, e_n\}$ where e_i has ends x_i and y_{j_i}, $1 \leq i \leq n$. Then, since M is a matching, $(j_1, j_2, \ldots, j_n)$ is a permutation in S_n. We will say that a matching $M' = \{e_1', e_2', \ldots, e_n'\}$ is *parallel* to M if e_i and e_i' have the same ends, for each $1 \leq i \leq n$. Then the set of all matchings which are parallel to M corresponds to the positive diagonal $S = \{b_{1j_1}, b_{2j_2}, \ldots, b_{nj_n}\}$ in $\mathbf{B}(G)$. Conversely, the positive diagonal S induces $b_{1j_1} \cdot b_{2j_2} \cdot \ldots \cdot b_{nj_n}$ distinct perfect matching of G which are parallel to M. Thus $\phi(G) = \mathrm{per}\,\mathbf{B}(G)$.

If G is k-regular then all the row and column sums of $\mathbf{B}(G)$ are equal to k. Hence, $\mathbf{A} = \frac{1}{k}\mathbf{B}(G)$ is a doubly stochastic matrix and $\phi(G) = k^n \mathrm{per}\,\mathbf{A}$. $\square$

In 1927 Van der Waerden conjectured that if $\mathbf{A}$ is a doubly stochastic matrix of order n then $\mathrm{per}\,\mathbf{A} \geq n!/n^n$ and that equality is achieved only when $\mathbf{A} = \frac{1}{n}\mathbf{J}_n$ where $\mathbf{J}_n$ is the $n \times n$ matrix in which every element is 1. This conjecture was proved by Egorychev (1980) and independently by Falikman (1981). We shall give a proof of this conjecture. For this we shall follow the outline of Schrijver (1983). The ingredients are two results, the first one being a special case of an inequality for 'mixed volumes' of convex bodies due to Fenchel (1936) and Alexandroff (1938).

Theorem 9.3.2 *Let $\mathbf{B}$ be a non-negative $n \times (n-2)$ matrix, and $\mathbf{x}$ and $\mathbf{y}$ be column vectors of length n with $\mathbf{x} \geq \mathbf{0}$. Then*

$$\mathrm{per}^2(\mathbf{B}, \mathbf{x}, \mathbf{y}) \geq \mathrm{per}(\mathbf{B}, \mathbf{x}, \mathbf{x}) \cdot \mathrm{per}(\mathbf{B}, \mathbf{y}, \mathbf{y}).$$

Furthermore, if $\mathbf{B}$ and $\mathbf{x}$ are strictly positive, equality holds if and only if $\mathbf{y} = \lambda\mathbf{x}$. (Here $(\mathbf{B}, \mathbf{x}, \mathbf{x})$ is the $n \times n$ matrix formed by adding two extra columns to $\mathbf{B}$ which are each the column vector $\mathbf{x}$.)

Proof. The proof is by induction on n. The case $n = 2$ is trivial, and so we shall suppose that $n = k$ and the theorem is true for all smaller values of n. By continuity we may assume that the entries of $\mathbf{B}$ and $\mathbf{x}$ are strictly positive.

Define the matrix $\mathbf{Q} = [q_{ij}]$ by $q_{ij} = \mathrm{per}(\mathbf{B}, \mathbf{e}_i, \mathbf{e}_j)$ where $\mathbf{e}_i$ and $\mathbf{e}_j$ denote the ith and jth standard column basis vectors. Then $\mathrm{per}(\mathbf{B}, \mathbf{x}, \mathbf{y}) = \mathbf{x}^T \mathbf{Q}\mathbf{y} = \mathbf{y}^T \mathbf{Q}\mathbf{x}$.

We begin by showing that $\mathbf{Q}$ is non-singular. Suppose the contrary, that there is some non-zero vector $\mathbf{z}$ so that $\mathbf{Q}\mathbf{z} = \mathbf{0}$. Then $\mathrm{per}(\mathbf{B}, \mathbf{z}, \mathbf{e}_j) = 0$ for each $1 \leq j \leq n$. Observe, however, that each of these permanents is simply the permanent of the $(n-1) \times (n-1)$ matrix, $(\mathbf{B}, \mathbf{z})$, with the jth row removed. Let $\mathbf{c}$ be the last column of $\mathbf{B}$ and $\mathbf{B} = (\mathbf{C}, \mathbf{c})$. Then by induction we have

$$\operatorname{per}^2(\mathbf{C}, \mathbf{c}, \mathbf{z}, \mathbf{e}_j) \geq \operatorname{per}(\mathbf{C}, \mathbf{c}, \mathbf{c}, \mathbf{e}_j) \cdot \operatorname{per}(\mathbf{C}, \mathbf{z}, \mathbf{z}, \mathbf{e}_j), \quad 1 \leq j \leq n. \quad (9.3.1)$$

By assumption the entries of $\mathbf{C}$ and $\mathbf{c}$ are positive, and so we also have $\operatorname{per}(\mathbf{C}, \mathbf{c}, \mathbf{c}, \mathbf{e}_j) > 0$. Combining this with (9.3.1) we have $\operatorname{per}(\mathbf{C}, \mathbf{z}, \mathbf{z}, \mathbf{e}_j) \leq 0$ for every $j = 1, \ldots, n$ and so

$$0 = \operatorname{per}(\mathbf{C}, \mathbf{z}, \mathbf{z}, \mathbf{c}) = \sum_{j=1}^{n} c_j \operatorname{per}(\mathbf{C}, \mathbf{z}, \mathbf{z}, \mathbf{e}_j) \leq 0.$$

Hence (9.3.1) holds with equality, and by induction $\mathbf{z} = \lambda \mathbf{c}$ for some λ. If $\lambda \neq 0$ then $0 = \operatorname{per}(\mathbf{B}, \mathbf{z}, \mathbf{e}_j) = \lambda \operatorname{per}(\mathbf{B}, \mathbf{c}, \mathbf{e}_j)$ which is a contradiction. Thus $Ssince \lambda = 0$, $\mathbf{z}$ is the zero vector, and $\mathbf{Q}$ must be non-singular.

In fact we can say more about $\mathbf{Q}$. Let $\mathbf{Q}_\varepsilon = [q'_{ij}]$ where

$$q'_{ij} = \operatorname{per}(\varepsilon \mathbf{B} + (1 - \varepsilon)\mathbf{J}, \mathbf{e}_i, \mathbf{e}_j),$$

$\mathbf{J}$ is the $n \times (n-2)$ matrix of all 1's, and $0 \leq \varepsilon \leq 1$. Then since the matrix $\varepsilon \mathbf{B} + (1 - \varepsilon)\mathbf{J}$ has positive entries, the above shows that $\mathbf{Q}_\varepsilon$ is also non-singular. The matrix $\mathbf{Q}_\varepsilon$ when $\varepsilon = 0$ has a zero diagonal and all other entries as $(n-2)!$; a matrix with only one positive and $n-1$ negative eigenvalues. Therefore, by continuity, the matrix when $\varepsilon = 1$, $\mathbf{Q}$, must also have only one positive eigenvalue and $n-1$ negative eigenvalues.

The definition of $\mathbf{Q}$ allows us to reformulate the assertion of the theorem as

$$\left(\mathbf{x}^T \mathbf{Q} \mathbf{y}\right)^2 \geq \left(\mathbf{x}^T \mathbf{Q} \mathbf{x}\right) \cdot \left(\mathbf{y}^T \mathbf{Q} \mathbf{y}\right). \quad (9.3.2)$$

This inequality holds trivially if $\mathbf{x}$ and $\mathbf{y}$ are linearly dependent. If $\mathbf{x}$ and $\mathbf{y}$ are linearly independent then, since $\mathbf{Q}$ has $n-1$ negative eigenvalues, the 2-dimensional linear hull of $\mathbf{x}$ and $\mathbf{y}$ intersects the $(n-1)$-dimensional linear hull of the eigenvectors of $\mathbf{Q}$, corresponding to these negative eigenvalues, in some non-zero vector $\alpha \mathbf{x} + \beta \mathbf{y}$ for some α and β not both zero. Thus

$$(\alpha \mathbf{x} + \beta \mathbf{y})^T \mathbf{Q} (\alpha \mathbf{x} + \beta \mathbf{y}) < 0. \quad (9.3.3)$$

Since $\mathbf{x}$ is positive per $(\mathbf{B}, \mathbf{x}, \mathbf{x})$ is also greater than zero, and so $\beta \neq 0$. Thus, by rescaling we can take $\beta = 1$ and the left hand side of (9.3.3) becomes a quadratic polynomial in α:

$$\alpha^2 \left(\mathbf{x}^T \mathbf{Q} \mathbf{x}\right) + 2\alpha \left(\mathbf{x}^T \mathbf{Q} \mathbf{y}\right) + \mathbf{y}^T \mathbf{Q} \mathbf{y}.$$

What is more, (9.3.3) implies that the discriminant of this polynomial has to be positive, or in other words that (9.3.2) must hold with strict inequality.

$\square$

Clearly the set of all doubly stochastic matrices of order n is a closed bounded subset of $\mathbb{R}^{n^2}$. Hence it is also compact, and so since the permanent is a continuous function there are doubly stochastic matrices minimising the permanent function.

The second ingredient of the proof of Van der Waerden's conjecture considers such a minimising element. The following result is due to Marcus and Newman (1959), and London (1971).

Theorem 9.3.3 *If* $\mathbf{A} = [a_{ij}]$ *is a doubly stochastic matrix minimising the permanent, then* per $\mathbf{A}_{ij} \geq$ per $\mathbf{A}$ *for each minor* $\mathbf{A}_{ij}$ *of* $\mathbf{A}$.

Proof. Consider the directed graph $\vec{G}$ whose underlying graph G is bipartite with bipartition (V_1, V_2), where $V_1 = \{u_1, u_2, \ldots, u_n\}$ and $V_2 = \{v_1, v_2, \ldots, v_n\}$, and edge set defined by

$$\vec{u_i v_j} \in E(\vec{G}) \text{ if and only if per } \mathbf{A}_{ij} \leq \text{per } \mathbf{A},$$

$$\vec{v_j u_i} \in E(\vec{G}) \text{ if and only if per } \mathbf{A}_{ij} \geq \text{per } \mathbf{A} \text{ and } a_{ij} > 0.$$

Suppose that for some $1 \leq i \leq j \leq n$ we have per $\mathbf{A}_{ij} <$ per $\mathbf{A}$. Without loss of generality we may assume that $i = j = 1$. Consider the directed graph $\vec{G}$. We shall first show that $\vec{u_1 v_1}$ is not contained in a directed cycle in $\vec{G}$. Suppose that C is a directed cycle of $\vec{G}$ containing $\vec{u_1 v_1}$. Then given $\varepsilon > 0$ we may form a new doubly stochastic matrix $\mathbf{A}_\varepsilon$ by replacing a_{ij} by $a_{ij} + \varepsilon$ if $\vec{u_i v_j} \in E(C)$ and by $a_{ij} - \varepsilon$ if $\vec{v_j u_i} \in E(C)$. The permanent of this matrix is a polynomial in ε:

$$\text{per } \mathbf{A}_\varepsilon = \text{per } \mathbf{A} + \varepsilon \left(\sum_{\vec{u_i v_j} \in E(C)} \text{per } \mathbf{A}_{ij} - \sum_{\vec{v_j u_i} \in E(C)} \text{per } \mathbf{A}_{ij} \right) + O(\varepsilon^2). \quad (9.3.4)$$

Then by the definition of $\vec{G}$, the coefficient of ε in (9.3.4) is negative, and if ε is small enough $\mathbf{A}_\varepsilon$ is a doubly stochastic matrix with per $\mathbf{A}_\varepsilon <$ per $\mathbf{A}$; a clear contradiction.

Consider now those vertices which can be reached from u_1 along a directed path, $v_1, v_2, \ldots, v_k, u_{l+1}, \ldots, u_n$, say. So $k, l \geq 1$ and G has no edges $\vec{u_i v_j}$ with $i \geq l+1$ and $j \geq k+1$, nor edges $\vec{v_j u_i}$ with $i \leq l$ and $j \leq k$. In other words

$$\left. \begin{array}{l} \text{per } \mathbf{A}_{ij} > \text{per } \mathbf{A} \text{ whenever } i > l \text{ and } j > k, \\ \text{per } \mathbf{A}_{ij} < \text{per } \mathbf{A} \text{ or } a_{ij} = 0 \text{ whenever } i \leq l \text{ and } j \leq k. \end{array} \right\} \quad (9.3.5)$$

This together with the identities $\sum_{j=1}^{n} a_{ij} = 1$ and $\sum_{j=1}^{n} a_{ij} \operatorname{per} \mathbf{A}_{ij} = \operatorname{per} \mathbf{A}$ gives rise to the following sequence of inequalities:

$$(n-k-l)\operatorname{per} \mathbf{A} = \sum_{i>l}\sum_{j=1}^{n} a_{ij}\operatorname{per} \mathbf{A}_{ij} - \sum_{j\leq k}\sum_{i=1}^{n} a_{ij}\operatorname{per} \mathbf{A}_{ij}$$

$$= \left(\sum_{i>l}\sum_{j>k} a_{ij}\operatorname{per} \mathbf{A}_{ij} - \sum_{i\leq l}\sum_{j\leq k} a_{ij}\operatorname{per} \mathbf{A}_{ij} \right)$$

$$\geq \left(\sum_{i>l}\sum_{j>k} a_{ij} - \sum_{i\leq l}\sum_{j\leq k} a_{ij} \right) \operatorname{per} \mathbf{A}$$

$$= \left(\sum_{i>l}\sum_{j=1}^{n} a_{ij} - \sum_{i=1}^{n}\sum_{j\leq k} a_{ij} \right) \operatorname{per} \mathbf{A}$$

$$= (n-k-l)\operatorname{per} \mathbf{A}. \tag{9.3.6}$$

Clearly we must have equality all through (9.3.6), but due to (9.3.5) this can only be true if all terms in (9.3.6) are zero. Thus $n = k + l$.

Since $k, l \geq 1$ it follows that both k and l are at most $n-1$. Hence, from (9.3.5), $\operatorname{per} \mathbf{A}_{nn} > \operatorname{per} \mathbf{A} > 0$. So there is a permutation $\pi \in S_{n-1}$ with $a_{i\pi(i)} > 0$ for $1 \leq i \leq n-1$. As $k > (n-l)-1$ this implies that $a_{ij} > 0$ for at least one pair of $i \leq l$ and $j \leq k$, contradicting the equality in (9.3.6). Hence $\operatorname{per} \mathbf{A}_{11} \geq \operatorname{per} \mathbf{A}$ as required. $\qquad\square$

Theorem 9.3.4 (Egorychev (1980), Falikman (1981)) *If $\mathbf{A}$ is a doubly stochastic matrix of order n then $\operatorname{per} \mathbf{A} \geq n!/n^n$.*

Proof. Let $\mathbf{A} = (\mathbf{B}, \mathbf{x}, \mathbf{y})$ be a doubly stochastic matrix which minimises the permanent. By appealing to compactness once again, we may also suppose that $\mathbf{A}$ is chosen so that $\sum_{i,j=1}^{n} a_{ij}^2$ is also minimal. Then by Theorem 9.3.3

$$\operatorname{per} (\mathbf{B}, \mathbf{x}, \mathbf{x}) = \sum_{i=1}^{n} x_i \operatorname{per} (\mathbf{B}, \mathbf{x}, \mathbf{e}_i) \geq \operatorname{per} (\mathbf{B}, \mathbf{x}, \mathbf{y}) \sum_{i=1}^{n} x_i = \operatorname{per} (\mathbf{B}, \mathbf{x}, \mathbf{y}).$$

Similarly, $\operatorname{per} (\mathbf{B}, \mathbf{y}, \mathbf{y}) \geq \operatorname{per} (\mathbf{B}, \mathbf{x}, \mathbf{y})$. However, by Theorem 9.3.2,

$$\operatorname{per}^2(\mathbf{B}, \mathbf{x}, \mathbf{y}) \geq \operatorname{per} (\mathbf{B}, \mathbf{x}, \mathbf{x}) \operatorname{per} (\mathbf{B}, \mathbf{y}, \mathbf{y}).$$

So, since $\operatorname{per} (\mathbf{B}, \mathbf{x}, \mathbf{y}) > 0$ it follows that $\operatorname{per} (\mathbf{B}, \mathbf{x}, \mathbf{x}) = \operatorname{per} (\mathbf{B}, \mathbf{x}, \mathbf{y}) = \operatorname{per} (\mathbf{B}, \mathbf{y}, \mathbf{y})$ which then implies that

$$\text{per}\left(\mathbf{B}, \frac{1}{2}\mathbf{x} + \frac{1}{2}\mathbf{y}, \frac{1}{2}\mathbf{x} + \frac{1}{2}\mathbf{y}\right) = \frac{1}{4}\text{per}\left(\mathbf{B}, \mathbf{x}, \mathbf{x}\right) + \frac{1}{2}\text{per}\left(\mathbf{B}, \mathbf{x}, \mathbf{y}\right)$$
$$+ \frac{1}{4}\text{per}\left(\mathbf{B}, \mathbf{y}, \mathbf{y}\right)$$
$$= \text{per}\left(\mathbf{B}, \mathbf{x}, \mathbf{y}\right).$$

Since per $\left(\mathbf{B}, \frac{1}{2}\mathbf{x} + \frac{1}{2}\mathbf{y}, \frac{1}{2}\mathbf{x} + \frac{1}{2}\mathbf{y}\right)$ is also doubly stochastic, it also is a doubly stochastic matrix minimising the permanent.

Let $\mathbf{x}$ and $\mathbf{y}$ be the last two columns of $\mathbf{A}$. Then if $\mathbf{A} \neq \frac{1}{n}\mathbf{J}_n$ we may assume without loss of generality that $\mathbf{A} = (\mathbf{B}, \mathbf{x}, \mathbf{y})$ with $\mathbf{x} \neq \mathbf{y}$. But then the matrix $(\mathbf{B}, \frac{1}{2}\mathbf{x} + \frac{1}{2}\mathbf{y}, \frac{1}{2}\mathbf{x} + \frac{1}{2}\mathbf{y})$ has a smaller square sum, contradicting our choice of $\mathbf{A}$. Thus $\mathbf{A} = \frac{1}{n}\mathbf{J}_n$ and per $\mathbf{A} = n!/n^n$.   □

From Proposition 9.3.1 and Theorem 9.3.4 we have the following.

Proposition 9.3.5 *Let G be a k-regular, n by n bipartite graph. Then the number of perfect matchings of G is at least $n!(k/n)^n$.*   □

In fact a result of Brègman (1973) shows that for a k-regular n by n bipartite graph $\phi(G) \leq (k!)^n/k$. Applying Stirling's formula to the upper and lower bounds for $\phi(G)$ then gives that

$$\lim_{k,n \to \infty} \frac{1}{k} \left(\phi(G)\right)^{1/n} = \frac{1}{e}.$$

This shows, rather surprisingly, that the number of perfect matchings is hardly affected by the structure of a k-regular bipartite graph.

Exercise

9.3.1 △ Prove that a bipartite graph in which every vertex has even degree has an even number of perfect matchings. (Little (1972))

Chapter 10

Coverings

10.1 Some examples of covering problems

Many problems in graph theory can be formulated as an instance of the following, somewhat general, covering problem:

> We are given two sets X and Y, and with each element $x \in X$ there is an associated subset $K(x)$ of elements of Y (which are covered in some sense by the element x). We know that $\bigcup_{x \in X} K(x) = Y$; our task is to find a subset $X_0 \subseteq X$ of minimum cardinality such that $\bigcup_{x \in X_0} K(x) = Y$.

The set X is called the *covering set* and the set Y is the *covered set*. Every subset $X' \subseteq X$ satisfying $\bigcup_{x \in X'} K(x) = Y$ is called a *covering* of Y by X. Our problem is to find a covering consisting of as few elements as possible, a *minimum covering* of Y by X. As an example we might take X to be the set of all matchings in a bipartite graph G, Y to be the set of edges $E(G)$, and $K(M) = M$ for each matching $M \in X$. Then a proper $\Delta(G)$-colouring of G induces a minimum covering of Y by X. There are many other possible such examples, several are given in the exercises for this section. Later, in Section 12.2, we will consider covering the edges of a non-bipartite graph with bipartite subgraphs with prescribed properties. First, however, there are several special forms of coverings for bipartite graphs which merit particular attention. So let G be a bipartite graph with bipartition (V_1, V_2) without isolated vertices.

Problem 1 (A minimum covering of V_2 by V_1)

Here the covering set X is the vertex set V_1, the covered set is V_2, and $K(x) = N_G(x)$ for each $x \in V_1$. It is necessary to find a subset $X_0 \subseteq V_1$ of minimum cardinality such that $N_G(X_0) = V_2$.

Example

A museum wants to employ a number of interpreters to give tours in a number of different languages. How can the museum employ as few of the job applicants as possible, whilst still being able to offer tours in all the languages?

This problem is easily reformulated as an instance of Problem 1. We simply construct a bipartite graph, with V_1 as the set of applicants for the job, and V_2 as the set of languages which the museum would like to have. A minimum covering of V_2 by V_1 now exactly corresponds to the fewest interpreters needed to deal with the group.

Example (The set covering problem)

Given a set $X = \{x_1, \ldots, x_m\}$ and a family $\mathcal{S} = \{S_1, \ldots, S_n\}$ of subsets of X such that $\bigcup_{i=1}^{n} S_i = X$, it is necessary to find the subfamily $\mathcal{R}$ consisting of the fewest subsets such that $\bigcup_{S \in \mathcal{R}} S$ also equals X.

The set covering problem can be reformulated as an instance of Problem 1. Construct a bipartite graph G with bipartition (V_1, V_2) where $V_1 = \mathcal{S}$ and $V_2 = X$ and an edge joins $S_i \in \mathcal{S}$ to $a \in X$ whenever $a \in S_i$. Then a solution of the set covering problem is a minimum covering of V_2 by V_1 in G.

Problem 1 in general is NP-hard, but there are various algorithms which provide approximate solutions. The simplest is the so-called *greedy covering algorithm*, which we give below.

Greedy covering algorithm

- **Begin** with the bipartite graph G with colour classes V_1 and V_2.

- **Let** $C = \emptyset$.

- **Repeat while** $V_2 \neq \emptyset$

  - Choose a vertex v of maximum degree in V_1.

  - Delete v and its neighbourhood from G.

  - Let $G = G - \{v\} - N_G(v)$, and add v to the set C.

- **End while**

The following result was obtained independently by several authors, see for example Sapozhenko (1972).

Proposition 10.1.1 *Let G be a bipartite graph with bipartition (V_1, V_2) where $|V_1| = n$ and $|V_2| = m$. If every vertex in V_2 has degree at least s then the greedy covering algorithm constructs a covering C of cardinality at most*

$$1 + \frac{n}{s}\left(1 + \log\frac{sm}{n}\right). \tag{10.1.1}$$

Proof. Let U_k be the subset of vertices remaining in V_2 after the kth iteration of the algorithm; then $U_0 = V_2$. It suffices to show that

$$|U_k| \leq m \prod_{i=0}^{k-1} \left(1 - \frac{s}{n-i}\right) \qquad (10.1.2)$$

since then for any $k \geq 0$ the number of remaining iterations necessary to complete the algorithm is bounded by $|U_k|$. So the total number of iterations, i.e. the number of elements in C, is bounded by

$$|C| \leq k + |U_k| \leq k + m \left(1 - \frac{s}{n}\right)^k \leq k + me^{-sk/n}$$

and taking $k = \lceil (n/s) \log(sm/n) \rceil$ we have $|C| \leq 1 + \frac{n}{s} \left(1 + \log \frac{sm}{n}\right)$.

It remains to show inequality (10.1.2). To see this consider the number of edges remaining in the graph after iteration k. Clearly there must be at least $s|U_k|$ such edges, and also there must be precisely $n - k$ vertices remaining in V_1. Thus there is a vertex $S \in V_1$ of degree at least $|U_k|s/(n-k)$, showing that $|U_{k+1}| \leq |U_k| - \frac{s}{n-k}|U_k|$. Inequality (10.1.2) and so the result now easily follow. $\qquad\square$

Clearly this greedy algorithm only provides a rough estimate for the true minimum covering, in general. Nevertheless, Kuzjurin (1980) proved that if every vertex in V_2 has degree s then, under natural restrictions, the size of a minimum covering in almost all such bipartite graphs is asymptotically (10.1.1), and so for almost all cases (in this sense) the greedy algorithm provides asymptotically good results.

Problem 2 (A minimum vertex dominating set)

A *vertex dominating set* in a graph G is defined to be a subset of vertices $D \subseteq V(G)$ such that each vertex $x \in V(G) \backslash D$ is adjacent to some vertex of D. The problem is to find a vertex dominating set D_0 consisting of as few vertices as possible.

In terms of coverings D_0 is the minimum covering under the assumption that the covering and covered sets X and Y are both $V(G)$ and each vertex x covers itself and those vertices to which it is adjacent, i.e. $K(x) = \{x\} \cup N_G(x)$.

Example

Imagine that an army wishes to control some square region of territory. This region is divided into equally sized smaller square regions. If an army post controls one of these small squares it also controls each of the squares with which it has a common border. The problem is to find the smallest number and locations of army posts which could control the whole territory.

To reformulate this problem we construct the graph G, with vertex set the set of squares and two vertices joined by an edge if their corresponding squares have a common border. Now a minimum vertex dominating set of this graph is exactly an optimal positioning of the army posts. In fact it is not difficult to check that this graph G is also bipartite.

Damaschke, Müller and Kratsch (1990) suggested a polynomial algorithm for finding a minimum dominating set in convex bipartite graphs. However, Problem 2 is NP-hard, even for chordal bipartite graphs (see Müller and Brandstädt (1987)).

Problem 3 (A minimum edge dominating set)

A subset R of edges is called an *edge dominating set* if every edge in $E(G)\backslash R$ is adjacent to an element of R. The problem is to find an edge dominating set of minimum cardinality. This is a covering problem where the covering and covered sets X and Y are both $E(G)$ and each edge covers itself and all adjacent edges.

Example

Imagine that we have a telephone switching network, built to route phone calls from incoming lines to outgoing connections – where we shall assume that only one call can pass through any connection at a given time. The problem is to assess the worst case scenario, and see how few telephone calls could 'block' the network.

This problem is equivalent to finding a matching which is also a minimum edge dominating set in the following simple bipartite graph. The graph has colour classes consisting of the set of incoming lines and the set of outgoing connections, with a line and connection joined by an edge if that line can be switched to that connection. It is clear then that a matching which is also a minimum edge dominating set exactly corresponds to a configuration of the fewest telephone calls which would block the network. But in fact this problem is equivalent to Problem 3 since, if we have any minimum edge dominating set R, we can transform it to an equally sized matching which is also an edge dominating set. This can be achieved as follows: consider two adjacent edges in R, uv and vw say, and let D be the set of edges, other than vw, which are incident with w. Observe that $D \neq \emptyset$, and $R \cap D = \emptyset$, since otherwise we could remove vw from R and retain an edge dominating set, which contradicts the minimality of R. Thus there must be an edge $wz \in D$ which is covered only by vw. Then $R \cup \{wz\}\backslash\{vw\}$ is also a minimum edge dominating set, but has fewer pairs of adjacent edges than R. Continuing in this way we obtain the required matching.

Once again, Problem 3 is NP-hard even for bipartite graphs with maximum degree 3 (see Yannakakis and Gavril (1980)).

Problem 4 (A minimum edge covering)

An *edge covering* of a graph G is defined to be a set $F \subseteq E(G)$ such that each vertex is the end of some edge of F. The problem is to find an edge covering consisting of as few edges as possible – a *minimum edge covering*. Here the covering set is $E(G)$, the covered set is $V(G)$ and each edge covers precisely its ends.

In contrast with the previous three problems, this problem can be solved polynomially for any graph G, using the following algorithm of Norman and Rabin (1959):

> First find a maximum matching M_0 of G, and for each vertex x uncovered by M_0 choose an edge e_x incident with x. Then the union of M_0 and the set M_1 of chosen edges is a minimum edge covering of G.

That $C_0 = M_0 \cup M_1$ is an edge covering, is clear. This covering contains $p - |M_0|$ edges, where $p = |V(G)|$; we shall show that any minimum edge covering C contains precisely this many edges. Clearly, C can contain no edge both of whose ends are incident with edges from C. This implies that C is a union of stars (considered as sets of edges). If one edge from each star is selected then the obtained set M is a matching. Clearly, $|M| + |C| = p$ and $|M| \leq |M_0|$, since M_0 is a maximum matching. Thus, $|C| \geq p - |M_0| = |C_0|$, which completes the proof.

Indeed we have proved the following result.

Theorem 10.1.2 *In any non-trivial, connected graph G, a minimum edge covering C and a maximum matching M satisfy $|M| + |C| = |V(G)|$.* $\square$

To close this section we note the folowing result due to Gupta (1967).

Theorem 10.1.3 *Let G be a bipartite graph with minimum degree $k \geq 2$. Then $E(G)$ is the union of k disjoint edge coverings.*

Proof. By Theorem 8.1.4 G has an equitable k-colouring. For each vertex there is an edge of each colour incident with x, since the minimum degree is k. Thus the set of edges of each colour is an edge covering of G. $\square$

Exercises

10.1.1 △ Show that Problem 1 can be reduced to the set covering problem.

10.1.2   Show that every vertex dominating set in a k-regular graph on p vertices contains at least $p/(k+1)$ vertices.

10.1.3 ▽ Show that an n-cube Q_n has a vertex dominating set D such that $d_{Q_n}(x, y) \geq 3$ for every pair of vertices $x, y \in D$ if and only if $n = 2^r - 1$ for some $r \geq 2$. (Hamming (1950))

10.1.4 A graph G is called *domination perfect* if each of its induced sub-graphs has a minimum vertex dominating set in which each pair of vertices is non-adjacent. Prove that a bipartite graph is domination perfect if and only if it does not contain one of the graphs shown in figure 10.1.1 as an induced subgraph. (Zverovich (1990))

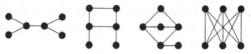

Figure 10.1.1

10.1.5 Show that $K_{m,n}$ can be represented as a union of edge-disjoint stars $K_{1,t}$ if and only if

(a) $m \equiv 0 (\bmod t)$ if $n < t$,

(b) $n \equiv 0 (\bmod t)$ if $m < t$,

(c) $nm \equiv 0 (\bmod t)$ if $n \geq t$ and $m \geq t$. (Yamamoto et al. (1975))

10.1.6 ▽ A graph G is called a *star forest* if each of its components is a star. Show that if $n \geq 8$ the minimum number of edge-disjoint star forests whose union is $K_{n,n}$ is $\lceil (n+1)/2 \rceil + 2$. (Egawa et al. (1986))

10.1.7 ▽ Prove that the graph $K_{m,n}$ can be represented as a union of edge-disjoint cycles of length k if and only if

(a) $m = n \equiv 0 (\bmod 2)$,

(b) $k \geq 4$ and $k \equiv 0 (\bmod 2)$,

(c) $2n \equiv 0 (\bmod 2k)$ and it is false that $m = n = k = 6$.

(Enomoto, Miyamoto, Ushio (1988))

10.1.8 Show that the graph $K_{n,n+1}$ can be represented as an edge-disjoint union of paths of distinct even length.

10.1.9 △ Let T be a tree with n pendant vertices. Show that the minimum number of paths whose union is T equals $\lceil n/2 \rceil$.

10.2 Vertex coverings and independent sets

In the previous section we outlined a general covering problem. Here we shall consider another variation of this problem in a graph G, when the covering set is $V(G)$, the covered set is $E(G)$, and each vertex covers all edges with which it is incident.

A *vertex covering* of a graph G is defined to be a subset of vertices $S \subseteq V(G)$ such that each edge $e \in E(G)$ is incident with some vertex in S. A vertex covering consisting of the fewest vertices is called a *minimum vertex covering*.

As is well known, the problem of finding the cardinality of a minimum vertex covering is NP-hard for an arbitrary graph (see Karp (1972)). However, for bipartite graphs the next theorem, due to König (1931, 1933), shows that computing the cardinality of these sets is relatively easy; indeed as easy as finding a maximum matching.

Theorem 10.2.1 (König's Minimax Theorem) *For any bipartite graph G, the cardinality of a minimum vertex covering equals the cardinality of a maximum matching.*

Proof. Let G have bipartition (V_1, V_2) with $|V_1| \leq |V_2|$, and, without loss of generality, assume that G has no isolated vertices. Let M be a maximum matching of G and $V_1(M)$ be the set of ends of edges of M which lie in V_1. If $V_1(M) = V_1$ then V_1 is certainly a vertex covering, and since any vertex covering must contain at least one end of each edge in a matching, V_1 is also a minimum vertex covering of G.

Consider, then, when $V_1(M) \subset V_1$, and define a set $S = S(M)$ in the following way: a vertex $x \in V(G)$ belongs to S if and only if some free vertex $y \in V_1 \backslash V_1(M)$ is connected to x by an alternating path, relative to M. First, we shall show that the set $C = (V_1(M) \backslash S) \cup (V_2 \cap S)$ is a vertex covering of G.

Suppose the contrary, and that there is an edge $e \in E(G)$ with ends $z_1 \in V_1 \backslash C$ and $z_2 \in V_2 \backslash C$. Then $z_2 \in V_2 \backslash S$. The vertex z_1 cannot be in $V_1 \cap S$ since then z_2 must be in S. So z_1 is free. But this also implies that z_2 must be in S; a contradiction. Thus C is a vertex covering. Then it must contain at least one end of each edge in M, i.e. $|C| \geq |M|$. On the other hand no edge $e \in M$ can have both its ends u and v in C. For then $v \in V_2 \cap S$ and we could extend the alternating path from some free vertex to v to include u and so u must also be in S, and so $u \notin C$. This implies that $|C| \leq |M|$ since $C \subseteq V(M)$. Combining these two inequalities gives the result as required. $\square$

Actually the proof of Theorem 10.2.1 provides an algorithm for constructing a minimum vertex covering in polynomial time. We need only notice that the set $S(M)$ can be constructed using a similar technique to that used in the algorithm of Hopcroft and Karp (see Section 5.2). Thus the problem of finding a minimum vertex covering, reduces to a simple two stage process:

- Find a maximum matching M of G.
- **If** $V_1(M) = V_2$ **then** V_1 is a minimum vertex covering;
- **Else** construct the set $S(M)$.
- $C = (V_1(M) \backslash S) \cup (V_2 \cap S)$ is a minimum vertex covering.

The link with the algorithm of Hopcroft and Karp in the construction of S is no coincidence. In fact we can deduce Hall's Theorem from König's

Minimax Theorem, but this we shall see later (see Section 11.1). König originally proved Theorem 10.2.1 not in terms of graphs, but in terms of matrices. In a 0-1 matrix, two 1's are called *independent* if they are not in the same row or column, and a set of rows and columns are said to *cover* all the 1's if they collectively contain every 1. Then Theorem 10.2.1 is a reformulation of saying that the maximum number of independent 1's in a 0-1 matrix is equal to the minimum number of rows and columns in a cover of the 1's.

Rather pleasingly, there is a duality between vertex coverings and *independent* sets of vertices (or simply *independent sets*). An independent set is a set of vertices in which no pair of vertices is adjacent; a *maximum independent set* of G is an independent set of largest cardinality. Our first result about this duality is the following result of Gallai (1959).

Proposition 10.2.2 *In any graph G, on $p \geq 2$ vertices, a maximum independent set I and a minimum vertex covering C satisfy $|C| + |I| = p$.*

Proof. Since no two vertices in the set $V(G) \backslash C$ can be joined by an edge, $V(G) \backslash C$ is independent and so $|I| \geq p - |C|$. Conversely $V(G) \backslash I$ forms a vertex covering, since it must span every edge. Thus $|C| \leq p - |I|$ and the result follows.   □

Corollary 10.2.3 *A vertex covering C of a graph G is minimum if and only if $I = V(G) \backslash C$ is a maximum independent set of G.*   □

Corollary 10.2.3 implies that, although for general graphs to find a maximum independent set of vertices is NP-hard, in a bipartite graph this can be done in polynomial time. To do this we need only apply the algorithm above to find a minimum vertex covering C, and then $I = V(G) \backslash C$ is a maximum independent set.

It is tempting to imagine that a minimum vertex covering in a bipartite graph is always itself an independent set, but this is not always the case, for instance consider the graph obtained from two disjoint copies of $K_{1,3}$ by joining the two vertices of degree 3. However, the next result gives a criterion for exactly when this does occur.

Proposition 10.2.4 *Let G be a bipartite graph and n the number of edges in a maximum matching. Then G has an independent minimum vertex covering if and only if for some vertex $x \in V(G)$ the set $\{x\} \cup \bigcup_{i \geq 1} N_{2i}(x)$ contains n vertices.*

Proof. First notice that, by Theorem 10.2.1, the cardinality of a minimum vertex covering is n, and furthermore that for each vertex $y \in V(G)$ the set

$C(y) = \{y\} \cup \bigcup_{i \geq 1} N_{2i}(y)$ is a vertex covering. Hence if for some vertex y the set $C(y)$ contains n vertices it is a minimum vertex covering.

Conversely, suppose that G has a minimum vertex covering C which is an independent set. Consider a vertex $x \in C$. Then, clearly, $N_1(x) \cap C = \emptyset$ and so to cover the edges in $E(N_1(x), N_2(x))$ we must have $N_2(x) \subseteq C$. Continuing in this way we see that $N_{2i}(x) \subseteq C$ and $N_{2i-1}(x) \cap C = \emptyset$ for $i \geq 1$. The result follows. $\qquad \square$

We could go further and ask for graphs which have a unique minimum vertex covering which is also independent. Corollary 10.2.3 implies that this is equivalent to a graph having a unique maximum independent set with a complement which is also independent. Such a graph is called a *strong unique independence graph*. The next two results provide a characterisation of such graphs.

Proposition 10.2.5 (Hopkins, Staton (1985)) *A tree T is a strong unique independence tree if and only if the distance between any two pendant vertices is even.*

Proof. First suppose that T is a unique strong independence tree. Then the vertices of the maximum independent set I and the set $C = V(G) \backslash I$ constitute the colour classes of a bipartition of T. Thus, since any path must pass through vertices of alternate colour classes, the distance between any two vertices of I must be even. However, the uniqueness of I guarantees that every pendant vertex is in I, for otherwise if v is a pendant vertex with neighbour $w \in I$ we have another maximum independent set $I \backslash \{w\} \cup \{v\}$.

To prove the converse we proceed by induction on p, the number of vertices in T. The result is trivial for $p = 1$. Suppose then that the induction hypothesis holds for some $p \geq 1$ and that T has the property that the distance between every pair of pendant vertices is even, but T has $p + 1$ vertices. If T is a path, then it must have an odd number of vertices, and so a unique maximum independent set. Otherwise let v be a pendant vertex of T, and let w be the vertex of degree at least 3 which lies closest to v. Since T is a tree there is a unique path $vv_1v_2 \ldots v_nw$ joining v and w. Then $T - \{v, v_1, \ldots, v_n\}$ is a tree T' on at most p vertices, which also has the property that the distance between any two pendant vertices is even. Thus by induction T' is a strong unique independence tree. The proof breaks into two cases.

First suppose that in T' w does not lie in the unique maximum independent set I'. Then since every pendant vertex must lie in I' the distance from w to any pendant vertex must be odd. Thus n must be even, or v would contradict the property of T. Then I' can be extended to a maximum independent set in T in a unique way, and T is a strong unique independence tree.

Suppose then that w does lie in I'. Then n is odd and so once again I' can be extended to a maximum independent set in T in a unique way.   □

Theorem 10.2.6 (Hopkins, Staton (1985)) *A connected graph is a strong unique independence graph if and only if it is bipartite and has a spanning tree which is a strong unique independence tree.*

Proof. Suppose that G is bipartite and has a spanning strong independence tree T. Let I be the maximum independent set in T, then $V(G)\backslash I$ is also independent in T. Then each edge not in T joins a vertex in I to a vertex not in I; for the addition of each such edge forms a cycle, and if an edge joined two vertices of I, the cycle would be of odd length, contradicting the assumption that G is bipartite. It follows that I and also $V(G)\backslash I$ are independent in G, and so that G is a strong unique independence graph.

Conversely, suppose that G is a strong unique independence graph, with maximum independent set I. Then since $C = V(G)\backslash I$ is also independent it is clear that G is bipartite with (C, I) forming a bipartition. Consider a subset $\emptyset \neq A \subseteq C$. Then $|N(A)| > |A|$ since otherwise $(I\backslash N(A)) \cup A$ would be another independent set at least as large as I. Hence G is 1-expanding, and so by Theorem 6.2.1 G has a spanning, 1-expanding forest H, in which every vertex of C has degree 2. Every pendant vertex in these trees must lie in I, and so, by Proposition 10.2.5, each tree in H is a strong unique independence tree. It remains only to extend this forest to a spanning tree T of G by adding any necessary edges. Since all pendant vertices remain in I it is clear that T will also be a strong independence tree.   □

Exercises

10.2.1 △ Show that for any simple bipartite graph G the minimum number of stars the union of which is G equals the cardinality of a maximum matching in G.

10.2.2   Show that in a bipartite graph without isolated vertices the number of vertices in a maximum independent set is equal to the number of edges in a minimum edge covering.

10.2.3   Show that a tree T has two disjoint maximum independent vertex sets if and only if it has a perfect matching. (Slater (1978))

10.2.4 ▽ Show that for every $n \geq 1$ there exists a bipartite graph having exactly n distinct independent vertex sets. (Linek (1989))

10.2.5 ▽ An edge $e \in E(G)$ is critical if the cardinality of a minimum vertex covering in $G - e$ is strictly smaller than that in G. Show that any two critical edges of a bipartite graph are independent. (Dulmage, Mendelsohn (1958a))

10.2.6   Show that the n-cube has at least 2^{n-1} independent vertex coverings.

10.2.7   Show that a strong independence graph can be recognised in polynomial time.

10.2.8 ▽ Show that the set of vertices of a bipartite graph G with $\Delta(G) = k$ can be partitioned into k independent sets so that the cardinalities of any two subsets differ by at most 1. (Lih, Wu (1996))

10.3 Dulmage and Mendelsohn's canonical decomposition

In this section we shall present a decomposition of a bipartite graph which relies on the properties of its minimum vertex coverings. This method is due to Dulmage and Mendelsohn (1959), and later in this section we shall see that it has an attractive application in factorising the determinant of a non-singular matrix. For now, however, we shall turn to explaining the construction, for which we need two lemmas. If C is a vertex covering in a bipartite graph G with bipartition (V_1, V_2) it is convenient to denote by $C_1 = C \cap V_1$ and $C_2 = C \cap V_2$ the intersection of C with each colour class.

Lemma 10.3.1   *Let C and D be two minimum vertex coverings of a bipartite graph G. Then $C_1 \cup D_1 \cup (C_2 \cap D_2)$ and $(C_1 \cap D_1) \cup C_2 \cup D_2$ are also minimum vertex coverings of G.*

Proof. Let $|C| = |D| = k$ and $C' = C_1 \cup D_1 \cup (C_2 \cap D_2)$ and $D' = (C_1 \cap D_1) \cup C_2 \cup D_2$. Let e be an edge of G with ends $u \in V_1$ and $v \in V_2$. If $u \notin C_1 \cup D_1$ then $u \notin C_1$ and $u \notin D_1$. Thus, since C and D are coverings, v must be in $C_2 \cap D_2$, showing that C' is a vertex covering. That D' is also a vertex covering follows similarly. Then,

$$k \leq |C'| = |C_1| + |D_1| - |C_1 \cap D_1| + |C_2 \cap D_2|,$$
$$k \leq |D'| = |C_2| + |D_2| - |C_2 \cap D_2| + |C_1 \cap D_1|$$

and adding these inequalities we have

$$2k \leq |C'| + |D'| = |C_1| + |C_2| + |D_1| + |D_2| = |C| + |D| = 2k.$$

Thus we have equality throughout. So $|C'| = |D'| = k$ and the new coverings C' and D' are minimum as required. □

As we saw in the previous section there is an important interplay between maximum matchings and minimum vertex coverings. This interplay has a strong role in the canonical decomposition we will present. Indeed we shall begin by first splitting the graph G into two – those edges which lie in some maximum matching of G are called *admissible* and together they form a subset A of edges, and the remaining *inadmissible* edges form a subset N of edges.

Lemma 10.3.2   *An edge e is inadmissible if and only if there exists a minimum vertex covering C with both $C_1 \neq \emptyset$ and $C_2 \neq \emptyset$ such that $e \in E(G[C])$.*

Proof. If e is an admissible edge then there is a maximum matching M with $e \in M$. By König's Minimax Theorem, there is a vertex covering C which consists of precisely $|M|$ vertices. Then, since each edge of M is adjacent to exactly one vertex of C we must have that $e \notin E(G[C])$.

Conversely, suppose that e is an inadmissible edge with ends $u \in V_1$ and $v \in V_2$. Consider $G' = G - \{u, v\}$. Then a maximum matching M' in G' must contain at least two fewer edges than any maximum matching M in G, or e would be admissible. Applying König's Theorem once again, we have that G' has a minimum vertex covering C' with $|C'| = |M'|$. Then it is clear that $C = C' \cup \{u, v\}$ is a vertex covering of G. Furthermore $u \in C_1$, $v \in C_2$, $e \in E(G[C])$ and since $|C| = |M'| + 2 \leq |M|$, König's Theorem shows that C is a minimum vertex covering of G. $\square$

Consider $\mathcal{C}$ the set of all minimum vertex coverings in G. By applying Lemma 10.3.1, we have that the vertex subsets

$$L = \bigcap_{C \in \mathcal{C}} (C \cap V_1) \cup \bigcup_{C \in \mathcal{C}} (C \cap V_2) \quad \text{and} \quad R = \bigcup_{C \in \mathcal{C}} (C \cap V_1) \cup \bigcap_{C \in \mathcal{C}} (C \cap V_2)$$

are both minimum vertex coverings of G.

If $L \neq R$ then $L_1 \subset R_1$ and $R_2 \subset L_2$. Let us produce a sequence of intermediate minimum vertex coverings 'between' L and R. Consider L and choose a set $S_1 \subseteq R_1 \backslash L_1$ of smallest cardinality and an equally sized $T_1 \subseteq L_2 \backslash R_2$ so that $(L \backslash T_1) \cup S_1$ is a minimum vertex cover. Notice that we can certainly do this, because for the sets $S = R_1 \backslash L_1$ and $T = L_2 \backslash R_2$ we have that $(L \backslash T) \cup S = R$ which is a minimum vertex covering. In a similar fashion define $S_2 \subseteq R_1 \backslash (L_1 \cup S_1)$ and $T_2 \subseteq L_2 \backslash (R_2 \cup T_1)$ so that S_2 and T_2 are subsets of minimum cardinality with $(L \backslash (T_1 \cup T_2)) \cup S_1 \cup S_2$ a minimum vertex cover, and so on. Continuing in this way we eventually produce two sequences of sets $S_1, S_2, \ldots, S_k$ and $T_1, \ldots, T_k$ with the following properties:

(1) $S_1 \cup \ldots \cup S_k = R_1 \backslash L_1$ and $T_1 \cup \ldots \cup T_k = L_2 \backslash R_2$;

(2) $L^i = (L \cup S_1 \cup \ldots \cup S_i) \backslash (T_1 \cup \ldots \cup T_i)$ is itself a minimum vertex covering, for each $i = 1, \ldots, k$;

(3) for each $i = 1, \ldots, k$, there is no minimum vertex covering C such that $C_1 \subset L^i \cap V_1$ and $C_2 \supset L^i \cap V_2$.

Using the sequences we can define the following subgraphs of G:

$$G_0 = G[R], G_1 = G[L_1 \cup T_1], G_2 = G[(L_1 \cup S_1) \cup T_2], \ldots,$$
$$G_k = G[(L_1 \cup S_1 \cup \ldots S_{k-1}) \cup T_k],$$
$$H_L = G[L_1 \cup (V_2 \backslash L_2)], H_R = G[(V_1 \backslash R_1) \cup R_2]$$
and
$$H_i = G[S_i \cup T_i] \text{ for each } i = 1, \ldots, k.$$

These graphs are all we need to give the decomposition.

Theorem 10.3.3 (Dulmage and Mendelsohn) *Let G be a bipartite graph with bipartition (V_1, V_2). Then if L, R, $S_1, \ldots, S_k$ and $T_1, \ldots, T_k$ are defined as above we have that:*

(1) G is the edge disjoint union of the subgraphs $G_0, \ldots, G_k, H_1, \ldots, H_k, H_L$ and H_R where G_i, H_i, H_L and H_R are as defined above. Moreover, each edge of N lies on one of the subgraphs $G_0, \ldots, G_k$ and each edge of A lies in one of subgraphs $H_1, \ldots, H_k, H_L, H_R$.

(2) no-matter how the sequences $S_1, \ldots, S_k$ and $T_1, \ldots, T_k$ are constructed, the graphs in the decomposition will be the same up to reordering.

Proof. From Lemma 10.3.2 we have that any edge of N must lie in some subgraph induced by the vertices of a minimum vertex covering of G. Suppose that there is an edge of N which joins vertices $x \in S_1$ and $y \in T_1$. Then there must be a minimum vertex covering C with $x \in C_1$ and $y \in C_2$. But then since $(L \cup S_1) \backslash T_1$ is a minimum vertex covering, by Lemma 10.3.1 $((L_1 \cup S_1) \cap C_1) \cup ((L_2 \backslash T_1) \cup C_2)$ is a minimum vertex covering which contradicts property *(3)* of the definition of S_1 and T_1. In a similar way we can see that there can be no edge of N which has one end in S_i and the other in T_j with $i \geq j$. It is now easy to check that each edge of N lies in precisely one of the subgraphs $G_0, \ldots, G_k$ and, in a similar way, that together $H_1, \ldots, H_k, H_L$ and H_R provide a decomposition of the admissible edges A. This proves *(1)*.

By employing the minimality of the S_i and T_i as above, we see that each of the subgraphs $H_i = G[S_i \cup T_i]$ can have only two minimum vertex coverings S_i and T_i. Now, suppose that some other instance of the construction produces an essentially different sequence of subgraphs $H'_1, \ldots, H'_s$. Then some H'_i must be contained in the union of some of the H_i, $H_1 \cup H_2$ say. But then this contradicts the property that H'_i has only two minimum vertex coverings, since we can easily construct more, for instance $(S_1 \cap S'_i) \cup (T_2 \cap T'_i)$. Thus we have proved *(2)*. $\square$

We call the decomposition given in *(1)* of Theorem 10.3.3 the *canonical decomposition* of the bipartite graph G. Indeed when G has a perfect matching the subgraphs in this canonical decomposition have a much simpler interpretation.

Theorem 10.3.4 *Let G be a bipartite graph with bipartition (V_1, V_2) which has a perfect matching, and let $S_1, \ldots, S_k$ and $T_1, \ldots, T_k$ be defined as above. Then*

(1) the subgraphs $H_1, \ldots, H_k$ are the connected components of the induced subgraph $G[A]$;

(2) each H_i is an elementary bipartite graph;

(3) if $k \geq 2$, by permuting rows and columns $\mathbf{B}(G)$ can be put into the form

$$
\begin{pmatrix}
\mathbf{A}_1 & 0 & 0 & 0 \\
* & \mathbf{A}_2 & 0 & 0 \\
* & * & \ddots & 0 \\
* & * & * & \mathbf{A}_k
\end{pmatrix}
\tag{10.3.1}
$$

where the matrices $\mathbf{A}_1, \mathbf{A}_2, \ldots, \mathbf{A}_k$ are the square biadjacency matrices corresponding to the subgraphs $H_1, \ldots, H_k$.

Proof. Since G has a perfect matching, $|V_1| = |V_2|$ and both V_1 and V_2 are minimum vertex coverings. Thus H_R and H_L are empty and the edges in the subgraphs H_i form a decomposition of A alone. These subgraphs have disjoint vertex sets, and thus there can be no edge in A with ends in different subgraphs H_i and H_j. It follows from the definition of S_i and T_i that H_i has only two minimum vertex coverings S_i and T_i, and so must be connected. Thus the subgraphs H_i must be the connected components of $G[A]$, which proves *(1)*. Since each $|T_i| = |S_i|$ and S_i is a minimum vertex covering, König's Theorem implies that the subgraphs H_i each have a perfect matching. By definition each edge of A lies in a perfect matching of G. Thus since any perfect matching of G must induce a perfect matching in each H_i, each of these subgraphs must be elementary, which proves *(2)*.

We have already mentioned in the proof of Theorem 10.3.3 that there can be no edge of N which has one end in S_i and the other in T_j with $i \geq j$. Thus *(3)* follows by reordering the rows and columns of $\mathbf{B}(G)$ so that those rows corresponding to the vertices of S_1 come first, then those for S_2 and so on, and the columns corresponding to T_1 come first followed by T_2 and so on. □

If in the canonical decomposition of a graph G we have $H_1 = G$ then we say that the decomposition is *trivial*, otherwise it is *non-trivial*. Then, Theorem 10.3.4 implies that a graph with trivial canonical decomposition must be a connected, elementary bipartite graph.

Suppose that we have a non-singular, $n \times n$ matrix $\mathbf{A} = (a_{ij})$ whose non-zero entries are elements from $X = \{x_1, x_2, \ldots, x_l\}$, where each x_i is a variable defined over $\mathbb{R}$. We may associate with $\mathbf{A}$ a bipartite graph $G_\mathbf{A}$, by letting the colour classes of $G_\mathbf{A}$ be $V_1 = \{u_1, \ldots, u_n\}$ and $V_2 = \{v_1, \ldots, v_n\}$ and joining u_i to v_j exactly when $a_{ij} \neq 0$.

Lemma 10.3.5 *The graph $G_\mathbf{A}$ has a perfect matching.*

Proof. We have that $\det \mathbf{A} = \displaystyle\sum_{\pi \in S_n} \operatorname{sgn}(\pi) a_{1\pi(1)} a_{2\pi(2)} \cdots a_{n\pi(n)}$. Then any non-vanishing term in this sum corresponds directly to a perfect matching in $G_\mathbf{A}$ and so, since $\mathbf{A}$ is non-singular, there must be at least one perfect matching. $\qquad\square$

Hence we can apply Theorem 10.3.4 to $G_\mathbf{A}$. Let $H_1, \ldots, H_k$ be defined as in Theorem 10.3.4. If $k \geq 2$, i.e. the canonical decomposition were non-trivial, then, possibly by permuting rows and columns, we could consider the matrix $\mathbf{A}$ in the form shown in (10.3.1) where each matrix $\mathbf{A}_1, \mathbf{A}_2, \ldots, \mathbf{A}_k$ is a square matrix over X corresponding to the subgraphs $H_1, \ldots, H_k$. Furthermore, we would have a proper factorisation of $\det \mathbf{A}$ in $R[X]$, the ring of polynomials over X, as $\det \mathbf{A} = \pm \det \mathbf{A}_1 \cdot \det \mathbf{A}_2 \cdot \ldots \cdot \det \mathbf{A}_k$. What is perhaps surprising is that $\det \mathbf{A}$ can only be properly factorised if the canonical decomposition of $G_\mathbf{A}$ is non-trivial.

Theorem 10.3.6 (Yamada (1988)) *Let $\det \mathbf{A}$ be a non-singular matrix over X. Then $\det \mathbf{A}$ has a proper factorisation in $R[X]$ if and only if the canonical decomposition of $G_\mathbf{A}$ is non-trivial. Furthermore, the factors of $\det \mathbf{A}$ correspond to the components in the decomposition.*

Proof. Only the necessity remains to be proved. Let $\det \mathbf{A} = p_1(X)p_2(X)$ be a proper factorisation in $R[X]$. By Lemma 10.3.5, since $\mathbf{A}$ is non-singular, by possibly permuting rows and columns we can assume that the diagonal entries of $\mathbf{A}$ are all non-zero. Since $\det \mathbf{A}$ can be expanded along any row or column, as a function $\det \mathbf{A}$ must be linear with respect to each entry in each fixed row (or column). Thus one and only one of the functions p_1 and p_2 can be a non-constant polynomial function of the entries in any row or column. We say that that function is *affected* by that row (or column). Notice, however, that since the diagonal entries are all non-zero, if p_1 is affected by row i then it is also simultaneously affected by column i.

Consider now the directed graph $\overrightarrow{H}$ with vertex set $\{y_1, \ldots, y_n\}$ and a directed edge $\overrightarrow{y_i y_j}$ whenever $a_{ij} \neq 0$. Let $N_j = \{y_i : \text{column } i \text{ affects } p_j\}$, $j = 1, 2$. Then N_1 and N_2 form a partition of $V(\overrightarrow{H})$.

Suppose that $y_{i_0} y_{i_1} \ldots y_{i_{s-1}}$ is a directed cycle in $\overrightarrow{H}$ which ranges over vertices in both N_1 and N_2. Then there must be some r so that $y_{i_r} \in N_1$ and

$y_{i_{r+1}} \in N_2$. Let π be the permutation

$$\pi(j) = \begin{cases} i_{(k+1(\mathrm{mod}\, s))} & \text{if } j = i_k \text{ for some } 0 \le k \le s-1, \\ j & \text{otherwise.} \end{cases}$$

Then $a_{1\pi(1)} a_{2\pi(2)} \cdots a_{n\pi(n)}$ is a non-zero term in the sum defining $\det \mathbf{A}$ which contains the term $a_{i_r i_{r+1}}$. Observe that, since $y_{i_r} \in N_1$, p_2 cannot be affected by column i_r, and so also not by row i_r. Similarly, since $y_{r+1} \in N_2$, p_1 cannot be affected by column i_{r+1}. Thus $a_{i_r i_{r+1}}$ is a parameter in neither p_1 nor p_2; a clear contradiction. Thus in $\overrightarrow{H}$, if there is a directed path between $u \in N_1$ and $v \in N_2$ then there can be no directed path in the other direction, or we would have a cycle in $\overrightarrow{H}$.

Fix a vertex u and let Y be the set of vertices to which there is a directed path from u, together with u itself, and let Z be the set of vertices to which there is no directed path from u. Then Y and Z form a proper partition of $V(\overrightarrow{H})$. Furthermore, by definition there is no directed edge of $\overrightarrow{H}$ which starts at a vertex of Y and ends at a vertex of Z.

If we rearrange $\mathbf{A}$ so that the columns and rows corresponding to the vertices of Y come first, followed by the columns and rows corresponding to the vertices of Z, we will put $\mathbf{A}$ into the form

$$\begin{pmatrix} \mathbf{A_1} & 0 \\ * & \mathbf{A_2} \end{pmatrix},$$

since there is no directed edge in $\overrightarrow{H}$ which begins at a vertex in Y and ends at a vertex in Z.

Suppose now that the canonical decomposition of $G_{\mathbf{A}}$ is trivial. Then by Theorem 10.3.4, $G_{\mathbf{A}}$ must be a connected, elementary bipartite graph. Since $G_{\mathbf{A}}$ is connected there must be an edge e with ends u_{i_0} and v_{j_0} such that $y_{i_0} \in Z$ and $y_{j_0} \in Y$. What is more, since $G_{\mathbf{A}}$ has a perfect matching, we must have that $C = \{u_i : y_i \in Z\} \cup \{v_i : y_i \in Y\}$ is a minimum vertex covering. However, then e is an edge of $E(G_{\mathbf{A}}[C])$, which contradicts Lemma 10.3.2. Thus the canonical decomposition must be non-trivial. Moreover, it can easily be seen that by reapplying this factorisation process to each of p_1 and p_2, and so on, we obtain a finer and finer decomposition of $G_{\mathbf{A}}$, which will eventually correspond to the actual canonical decomposition. □

Exercise

10.3.1   Prove that a connected bipartite graph G with bipartition (V_1, V_2) is elementary if and only if it has only V_1 and V_2 as minimum vertex coverings.

Application

10.4 Decomposition of partially ordered sets into chains

One of the most famous results in the theory of partially ordered sets is due to Dilworth (1950). If we have a partially ordered set $\mathcal{P} = (P, \succeq)$, a *chain* in $\mathcal{P}$ is a non-empty subset $\mathcal{C} = \{p_1, p_2, \ldots, p_k\} \subseteq P$ such that

$$p_1 \succ p_2 \succ \ldots \succ p_k.$$

Two elements of P are called *comparable* if they appear together in some chain in $\mathcal{P}$; elements which are not comparable are called *incomparable*. A non-empty set of pairwise incomparable elements is called an *antichain*.

Since each single element in P is itself a chain, it is always possible to partition the elements of $\mathcal{P}$ into disjoint chains. Such a partition we call a *decomposition* of $\mathcal{P}$. Less clear is how to decompose $\mathcal{P}$ into as few chains as possible. A decomposition consisting of the smallest number of chains is called minimum.

If we have a decomposition $\mathcal{D}$ of $\mathcal{P}$ and an antichain $\mathcal{A}$ then $|\mathcal{A}| \leq |\mathcal{D}|$ since no two elements of $\mathcal{A}$ can lie in the same chain. Thus the maximum size of an antichain is at most the size of a minimum decomposition; Dilworth's Theorem asserts that in fact we have equality. Rather pleasingly, Dilworth's Theorem follows from König's Minimax Theorem and we shall devote this section to showing this implication using an argument due to Fulkerson (1956).

Let $\mathcal{P} = (P, \succeq)$ be a partially ordered set, with $P = \{a_1, a_2, \ldots, a_p\}$. We define a bipartite graph $G_{\mathcal{P}}$ with bipartition (V_1, V_2), where $V_1 = \{x_1, \ldots, x_p\}$, $V_2 = \{y_1, \ldots, y_p\}$ and an edge joining $x_i \in V_1$ to $y_j \in V_2$ whenever $a_i \succ a_j$.

Lemma 10.4.1 *Given any matching M in $G_{\mathcal{P}}$, there is a decomposition $\mathcal{D}$ of $\mathcal{P}$, with $|\mathcal{D}| + |M| = p$.*

Proof. Let $M = \{x_{i_1} y_{j_1}, x_{i_2} y_{j_2}, \ldots, x_{i_k} y_{j_k}\}$ be a matching in $G_{\mathcal{P}}$. Then we have that $a_{i_1} \succ a_{j_1}, a_{i_2} \succ a_{j_2}, \ldots, a_{i_k} \succ a_{j_k}$ and so by considering only distinct elements we can partition the set $\{a_{i_1}, a_{i_2}, \ldots, a_{i_k}, a_{j_1}, \ldots, a_{j_k}\}$ into chains, each containing two or more elements. This partial decomposition can be completed by adding all the remaining elements of $\mathcal{P}$ as one element chains. This gives a decomposition $\mathcal{D}$.

Now let l_j be the number of elements in the jth chain in $\mathcal{D}$. Then we have

$$p = \sum_{j=1}^{|\mathcal{D}|} l_j = |\mathcal{D}| + \sum_{j=1}^{|\mathcal{D}|} (l_j - 1) = |\mathcal{D}| + |M|,$$

since $l_j - 1$ counts zero for every one element chain, and the number of edges from M in each longer chain. □

Lemma 10.4.2 *Given any minimum vertex covering X of $G_\mathcal{P}$, there is an antichain $\mathcal{A}$ with $|X| + |\mathcal{A}| = p$.*

Proof. Let $X \cap V_1 = \{x_{i_1}, x_{i_2}, \ldots, x_{i_k}\}$ and $X \cap V_2 = \{y_{j_1}, y_{j_2}, \ldots, y_{j_l}\}$ and let $\mathcal{A} = P \backslash \{a_{i_1}, \ldots, a_{i_k}, a_{j_1}, \ldots, a_{j_l}\}$. Then, since X is a vertex covering the elements of $\mathcal{A}$ form an antichain.

Suppose that the indices $i_1, \ldots, i_k, j_1, \ldots, j_l$ are not distinct, and that without loss of generality $i_1 = j_1$. Then since X is a proper vertex covering there are a vertex $x_s \notin X \cap V_1$ with $x_s y_{j_1} \in E(G_\mathcal{P})$ and similarly a vertex $y_t \notin X \cap V_2$ with $x_{i_1} y \in E(G_\mathcal{P})$. Then, if $i_1 = j_1$, we have $a_s \succ a_{i_1}$ and $a_{i_1} \succ a_t$ and the transitivity of $\succ$ implies that $a_s \succ a_t$ and so $x_s y_t \in E(G_\mathcal{P})$, contradicting that X is a vertex covering. Thus the indices $i_1, \ldots, i_k, j_1, \ldots, j_l$ are distinct and $|X| + |\mathcal{A}| = p$. □

Theorem 10.4.3 (Dilworth's Theorem) *Given any partially ordered set $\mathcal{P}$, the maximum size of an antichain equals the size of a minimum decomposition.*

Proof. König's Minimax Theorem gives that the size of a minimum vertex covering of the graph $G_\mathcal{P}$ is equal to the size of maximum matching. Thus the result follows from Lemma 10.4.1 and Lemma 10.4.2. □

Exercises

10.4.1   Deduce König's Minimax Theorem from Dilworth's Theorem.

10.4.2   Show that in any partially ordered set, the size of a longest chain is equal to the size of a minimum antichain partition.

Chapter 11

Some combinatorial applications

11.1 Systems of distinct representatives

Let $\mathcal{F} = (S_1, \ldots, S_n)$ be a family of subsets of a finite set S. A sequence $F = (f_1, \ldots, f_n)$ of elements of S is called a *system of representatives* of $\mathcal{F}$ if $f_i \in S_i$, for $i = 1, 2, \ldots, n$. If the elements of F are distinct then F is called a *system of distinct representatives* (SDR) for $\mathcal{F}$.

Example 11.1.1 Let $S_1 = \{u_2, u_3, u_4\}$, $S_2 = \{u_1, u_2, u_3\}$ and $S_3 = \{u_3, u_4, u_5\}$. Then $F = (u_2, u_1, u_3)$ is an SDR for $\mathcal{F} = (S_1, S_2, S_3)$, since $u_2 \in S_1$, $u_1 \in S_2$ and $u_3 \in S_3$.

Many criteria for the existence of systems of representatives, under various restrictions, have been developed (see Mirsky (1971)). Bipartite graphs have proven to be a particularly useful tool in these investigations, since every collection of subsets $\mathcal{F}$ can be represented by the bipartite graph $G(\mathcal{F})$ with bipartition (V_1, V_2) where $V_1 = \{S_1, \ldots, S_n\}$, $V_2 = S$ and the vertices $S_i \in V_1$ and $u \in V_2$ are joined by an edge if and only if $u \in S_i$. We shall give a few examples of results on SDRs to demonstrate how this representation can be used, which employ a variety of different graph theoretic results. We begin with the principal result in this area, obtained by P. Hall (1935).

Theorem 11.1.2 (Hall's Theorem for SDRs) *A family* $\mathcal{F} = \{S_1, \ldots, S_n\}$ *of sets has an SDR if and only if the union of any k of the sets contains at least k elements, $k = 1, \ldots, n$, i.e.* $|\bigcup_{i \in I} S_i| \geq |I|$ *for all* $I \subseteq \{1, 2, \ldots, n\}$. $\qquad\square$

In fact we have already met this theorem earlier, but in different notation. It is clear that an SDR for $\mathcal{F}$ corresponds to a matching of V_1 into V_2 in the graph $G = G(\mathcal{F})$. Furthermore, for each $A \subseteq V_1$ in the graph $G(\mathcal{F})$ we have $|N_G(A)| = |\bigcup_{S_i \in A} S_i|$. Hence Theorem 11.1.2 is simply a reformulation of Theorem 6.1.1. Historically Hall proved his theorem in the language of systems of representatives, but this theorem has proved equally popular in graph theory. In fact Hall's Theorem follows in a very neat way (see below) from König's Minimax Theorem which was proved a few years earlier.

Proof of Theorem 11.1.2. We prove only the sufficiency (necessity is obvious). Let $C = V_1' \cup V_2'$ be a minimum vertex covering of $G = G(\mathcal{F})$, where $V_1' \subseteq V_1$ and $V_2' \subseteq V_2$. Clearly $N_G(V_1 \backslash V_1') \subseteq V_2'$, and by assumption $|N_G(V_1 \backslash V_1')| \geq |V_1 \backslash V_1'|$. Therefore, $|C| = |V_1'| + |V_2'| \geq |V_1'| + |N_G(V_1 \backslash V_1')| \geq |V_1'| + |V_1 \backslash V_1'| = n$.

On the other hand, $|C| \leq n$, and so $|C| = n$, and by König's Minimax Theorem, G has a matching with n edges, i.e. a matching of V_1 into V_2. Putting this into the set formulation, $\mathcal{F}$ has an SDR.   □

Theorem 11.1.2 establishes when a family $\mathcal{F}$ has an SDR, but when has $\mathcal{F}$ an SDR which contains all the elements of some given subset $S_0 \subseteq S$? It is customary to refer to the elements of S_0 as 'marginal elements'. The next result gives an answer to this problem.

Theorem 11.1.3 (Hoffman, Kuhn (1956)) *Let $\mathcal{F} = (S_1, \ldots, S_n)$ be a family of subsets of S, and let $S_0 \subseteq S$. Then $\mathcal{F}$ has an SDR which contains the elements of S_0 if and only if*

    (1) the union of any k sets of $\mathcal{F}$ contains at least k elements, $k = 1, 2, \ldots, n$,

    (2) for any subset $K \subseteq S_0$, the number of sets in $\mathcal{F}$ that intersect K is at least $|K|$.

Proof. Clearly, the required SDR exists if and only if in the graph $G(\mathcal{F})$ there is a matching of V_1 into V_2 which covers S_0. But by Theorem 5.1.6 such a matching exists if and only if there is a matching M_1 of V_1 to V_2 and a matching M_2 of S_0 to V_1. It remains only to notice that the conditions *(1)* and *(2)* are necessary and sufficient to ensure the existence of these two matchings, and the results follows.   □

Corollary 11.1.4 *Let $\mathcal{F} = (S_1, \ldots, S_n)$ be a family of subsets of S, and let $S_0 \subseteq S$ be an n element subset. Then $\mathcal{F}$ has an SDR which contains precisely the elements of S_0 if and only if for any subset $K \subseteq S_0$ the number of subsets in $\mathcal{F}$ that intersect K is at least $|K|$.*   □

Let $F_1, \ldots, F_k$ be SDRs of a family $\mathcal{F} = (S_1, \ldots, S_n)$, where $F_i = (f_{i1}, \ldots, f_{in})$, $1 \leq i \leq k$, and the elements $f_{1j}, \ldots, f_{kj}$ are distinct for each $j = 1, \ldots, n$. Then $F_1, \ldots, F_k$ are called *compatible* SDRs for $\mathcal{F}$.

In fact SDRs $F_1, \ldots, F_k$ $(k \geq 2)$ are compatible if F_{r+1} is an SDR for the family $(S_1 \backslash \{f_{11}, \ldots, f_{r1}\}, \ldots, S_n \backslash \{f_{1n}, \ldots, f_{rn}\})$, $1 \leq r \leq k - 1$. For example the family $\mathcal{F}$ which we gave in Example 11.1.1 has three compatible SDRs: $F_1 = (u_2, u_1, u_3)$, $F_2 = (u_3, u_2, u_4)$, and $F_3 = (u_4, u_3, u_5)$.

Theorem 11.1.5 (Asratian (1975)) *A family $\mathcal{F} = (S_1, \ldots, S_n)$ of subsets of S has k compatible SDRs if and only if*

$$\sum_{s \in S} \min\{k, |(\mathcal{F}(I), s)|\} \geq k|I|$$

for each subset $I \subseteq \{1, 2, \ldots, n\}$, where $(\mathcal{F}(I), s) = \{i \in I : s \in S_i\}$.

Proof. Once more we shall consider the graph $G = G(\mathcal{F})$. Then the condition of the theorem is equivalent to

$$\sum_{s \in V_2} \min\{k, |E_G(U, s)|\} \geq k|U| \tag{11.1.1}$$

for each $U \subseteq V_1$. It is not difficult to see that k compatible SDRs exist if and only if G has k disjoint matchings of V_1 into V_2.[†] If $|V_1| > |V_2|$ then G has no such matching. Hence we assume that $|V_1| \leq |V_2|$. Let G_1 be the bipartite graph with bipartition $(V_1 \cup \{z\}, V_2)$ which is obtained from G by adding a new vertex z, which is joined by k parallel edges to each vertex of V_2. Then (11.1.1) holds for all $U \subseteq V_1$ if and only if $\sum_{s \in V_2} \min\{k, |E_G(U^*, s)|\} \geq k|U^*|$ for all $U^* \subseteq V_1 \cup \{z\}$. (Note that if $z \in U^*$ then the left hand side of the last inequality is equal to $k|V_2|$.)

Define a function $f(x)$ so that $f(x) = k$ for each vertex $x \in V_1 \cup V_2$ and $f(z) = k(|V_2| - |V_1|)$. We shall show that G has k disjoint matchings of V_1 into V_2 if and only if G_1 has an f-factor.

Clearly, if G has the k required matchings, then G_1 has an f-factor. Conversely, if G_1 has an f-factor H_1, then G has $H = H_1 - z$ as a spanning subgraph, in which $d_H(x) = k$ for each $x \in V_1$ and $d_H(s) \leq k$ for each $s \in V_2$. Consider a proper edge k-colouring of H. Then the k matchings induced by the colouring provide k disjoint matchings of V_1 into V_2. $\qquad \square$

Clearly for $k = 1$ Theorem 11.1.5 is equivalent to Hall's Theorem. The theorem can be used to solve the following problem (see Broersma et al. (1993)).

In a certain year, each of k students at some school is required to spend probationary periods $p_1, \ldots, p_n$ at n different companies from

---

[†]The condition (11.1.1) for the existence of k disjoint matchings was also obtained later by Lebensold (1977). Folkman and Fulkerson (1969) found a 'two sided' condition for such matchings to exist.

a collection $\{c_1, \ldots, c_m\}$ of m companies. Each company can take only at most one student at a time. The question is whether and how the probationary requirements can be satisfied for all the students?

To solve this problem let S_i be the set of companies which are available to take a student in period p_i, for $i = 1, \ldots, n$. Clearly a probationary schedule for each student is an SDR of the family $\mathcal{F} = (S_1, \ldots, S_n)$. So the probationary requirements can be satisfied for all the students if and only if $\mathcal{F}$ has k compatible SDRs.

Now we shall consider two families of subsets of S, $\mathcal{F}_1 = (S_1, \ldots, S_n)$ and $\mathcal{F}_2 = (T_1, \ldots, T_n)$. A sequence $(f_1, \ldots, f_n)$ is called a *common system of distinct representatives* (common SDR) for $\mathcal{F}_1$ and $\mathcal{F}_2$ if there is a permutation π of $\{1, 2, \ldots, n\}$ such that $f_i \in S_i \cap T_{\pi(i)}$, for $i = 1, \ldots, n$.

The history of the appearance of the concept of a common SDR is quite interesting. It'first appeared in a paper of Miller (1910) on group theory. Let S be a finite group of nk elements, and A be a subgroup of S of order k. Recall that the *left (right) coset* containing $g \in S$ is the set $gA = \{ga : a \in A\}$ (respectively $Ag = \{ag : a \in A\}$) and that the families of distinct left and right cosets each form a partition of S. Miller showed that there always exist elements $g_1, g_2, \ldots, g_n$ so that we can write these partitions as $S = g_1 A \cup \ldots \cup g_n A$ and $S = A g_1 \cup \ldots \cup A g_n$. In other words the families of distinct left and right cosets have a common SDR. Van der Waerden (1927) pointed out that this result is still true for an arbitrary set S of nk elements and for every two partitions into k element subsets (see Theorem 11.1.6). Finally, (see König (1936) pp. 231–240) König showed that the result of Van der Waerden is a set reformulation of his result of 1916 that every k-regular bipartite graph has a perfect matching. The proof of this reformulation is given in one direction in the proof of Theorem 11.1.6 below, we leave the other direction as an exercise.

Theorem 11.1.6 *Let a set S of kn elements be partitioned into n sets in two different ways $S = S_1 \cup S_2 \cup \ldots \cup S_n = T_1 \cup T_2 \cup \ldots \cup T_n$ with each S_i and T_i containing precisely k elements. Then the families $\mathcal{F}_1 = (S_1, \ldots, S_n)$ and $\mathcal{F}_2 = (T_1, \ldots, T_n)$ have a common SDR.*

Proof. Consider the graph $G(\mathcal{F}_1, \mathcal{F}_2)$ with bipartition (V_1, V_2) where $V_1 = \{x_1, \ldots, x_n\}$, $V_2 = \{y_1, \ldots, y_n\}$. To form the edge set we join x_i to y_j by $|S_i \cap T_j|$ parallel edges which we label with the elements of $S_i \cap T_j$. Clearly, $G(\mathcal{F}_1, \mathcal{F}_2)$ is k-regular, and so has a perfect matching. This perfect matching corresponds to a common SDR for $\mathcal{F}_1$ and $\mathcal{F}_2$. $\square$

Now we shall give a criterion for the existence of a common SDR in the general case.

Theorem 11.1.7 (Ford, Fulkerson (1958)) *The families* $\mathcal{F}_1 = (S_1, \ldots, S_n)$ *and* $\mathcal{F}_2 = (T_1, \ldots, T_n)$ *of subsets of the set* S *possess a common SDR if and only if*

$$\left| \left(\bigcup_{i \in I} S_i \right) \cap \left(\bigcup_{j \in J} T_j \right) \right| \geq |I| + |J| - n \quad \text{for all } I, J \subseteq \{1, 2, \ldots, n\}. \quad (11.1.2)$$

Proof. We follow the proof of Perfect (1968). We define a simple bipartite graph G with bipartition (V_1, V_2), where $V_1 = \{a, b\} \cup S$ and $V_2 = X \cup Y$, $X = \{x_1, \ldots, x_n\}$ and $Y = \{y_1, \ldots y_n\}$, and with edge set given by $E(G) = \{ax_i, by_i : i = 1, \ldots, n\} \cup \{ux_i : u \in S_i, 1 \leq i \leq n\} \cup \{uy_j : u \in T_j, 1 \leq j \leq n\}$. Observe, now, that a common SDR of $\mathcal{F}_1$ and $\mathcal{F}_2$ exists if and only if there are n internally disjoint (a, b)-paths in G. We shall show that such paths exist if and only if condition (11.1.2) holds.

Suppose that I and J can be chosen so that (11.1.2) is false. Let $X_I = \{x_i : i \in I\}$ and $Y_J = \{y_j : j \in J\}$. Then the set $C = (X \backslash X_I) \cup (Y \backslash Y_J) \cup (N_G(X_I) \cap N_G(Y_J))$ is a vertex cut which disconnects a and b. Moreover,

$$|C| = |X \backslash X_I| + |Y \backslash Y_J| + |N_G(X_I) \cap N_G(Y_J)|$$

$$= (n - |I|) + (n - |J|) + \left| \left(\bigcup_{i \in I} S_i \right) \cap \left(\bigcup_{j \in J} T_j \right) \right|$$

$$< (n - |I|) + (n - |J|) + (|I| + |J| - n) = n,$$

and so, by Menger's Theorem, there cannot exist n internally disjoint (a, b)-paths, and so $\mathcal{F}_1$ and $\mathcal{F}_2$ cannot have a common SDR.

Conversely, let (11.1.2) be satisfied whenever $I, J \subseteq \{1, 2, \ldots, n\}$ and let $C = X_1 \cup Y_1 \cup R$ be a vertex cut disconnecting a and b, where $X_1 \subseteq X$, $Y_1 \subseteq Y$, and $R \subseteq S$. Let I and J denote the sets of indices of vertices in X_1 and Y_1, respectively. Clearly, $N_G(X \backslash X_1) \cap N_G(Y \backslash Y_1) \subseteq R$. Rephrasing this in terms of $\mathcal{F}_1$ and $\mathcal{F}_2$ gives that $\left(\bigcup_{i \in \bar{I}} S_i \right) \cap \left(\bigcup_{j \in \bar{J}} T_j \right) \subseteq R$, where $\bar{I} = \{1, 2, \ldots, n\} \backslash I$, and $\bar{J} = \{1, 2, \ldots, n\} \backslash J$. Then, from the last two observations we have

$$|C| = |I| + |J| + |R| \geq |I| + |J| + \left| \left(\bigcup_{i \in \bar{I}} S_i \right) \cap \left(\bigcup_{j \in \bar{J}} T_j \right) \right|$$

$$\geq |I| + |J| + (|\bar{I}| + |\bar{J}| - n) = n$$

and so the number of elements in any vertex cut disconnecting a and b is at least n. Hence, once again by Menger's Theorem, there are n internally disjoint (a, b)-paths, and $\mathcal{F}_1$ and $\mathcal{F}_2$ have a common SDR. $\qquad \square$

Finally, we note that Theorem 11.1.7 can be generalised for compatible SDRs, as follows. Let $\mathcal{F}_1 = (S_1, \ldots, S_n)$ and $\mathcal{F}_2 = (T_1, \ldots, T_n)$ be once again two families of sets, and $F_1, \ldots, F_k$ be compatible SDRs for the family $\mathcal{F}_1$. If there is a permutation π of $\{1, 2, \ldots, n\}$ such that $F_1, \ldots, F_k$ are also compatible SDRs for the family $(T_{\pi(1)}, \ldots, T_{\pi(n)})$ then the set $\{F_1, \ldots, F_k\}$ is called a *common set of k compatible SDRs* for $\mathcal{F}_1$ and $\mathcal{F}_2$.

Theorem 11.1.8 (Horák (1991)) *The families $\mathcal{F}_1 = (S_1, \ldots, S_n)$ and $\mathcal{F}_2 = (T_1, \ldots, T_n)$ have a common set of k compatible SDRs if and only if, for all $I, J \subseteq \{1, 2, \ldots, n\}$, we have*

$$\sum_{s \in S} \min\{k, |(s, \mathcal{F}_1(I))|, |(s, \mathcal{F}_2(J))|\} \geq k\,(|I| + |J| - n)\,. \qquad \square$$

In fact all of the results in this section can be generalised to matroids, indeed these results also have proofs which use matroid techniques. We draw the interested reader's attention to the books of Welsh (1976), Oxley (1992) and Mirsky (1971). In particular, Theorem 11.1.5 is generalised to matroids by Horák (1990) and Kochol (1994).

Exercises

11.1.1   Show that a family $\mathcal{F} = (S_1, \ldots, S_n)$ has k disjoint SDRs if and only if $|\bigcup_{i \in I} S_i| \geq k|I|$ for each $I \subseteq \{1, 2, \ldots, n\}$.

11.1.2   Let $\mathcal{F} = (S_1, \ldots, S_n)$ be a family of subsets of S. Show that $\mathcal{F}$ has k compatible SDR's if $|S_i| \geq k$ for $i = 1, \ldots, n$ and each element of S belongs to at most k subsets of $\mathcal{F}$.

11.1.3   Show that if a family of sets $\mathcal{F}$ possesses a set of k compatible SDRs $(k \geq 3)$, then this set is not unique.

11.1.4   Deduce Theorem 11.1.6 from Theorem 11.1.7.

11.1.5   Let $\mathcal{F}_1$ and $\mathcal{F}_2$ be two arbitrary families of sets and $G(\mathcal{F}_1, \mathcal{F}_2)$ be the graph defined in Theorem 11.1.6. What can be inferred from the fact that $G(\mathcal{F}_1, \mathcal{F}_2)$ has a perfect matching?

11.1.6   Let a set S be partitioned into n sets in two different ways: $S = S_1 \cup S_2 \cup \ldots \cup S_n = T_1 \cup T_2 \cup \ldots \cup T_n$. Prove that a common SDR for the families $\mathcal{F}_1 = (S_1, \ldots, S_n)$ and $\mathcal{F}_2 = (T_1, \ldots, T_n)$ exists if and only if the union of any k sets from $\mathcal{F}_1$ contains elements from at least k of the sets of $\mathcal{F}_2$. (P. Hall (1935))

11.1.7   Show that under the conditions of Theorem 11.1.6, S can be partitioned into k disjoint common SDRs.

11.2 Generation of subsets of a set

The solutions of many problems in computer science involve the systematic generation of some substructure from some large class of objects. Many of these problems can really be thought of as the systematic generation of subsets of some given set. Suppose, for instance that there is some conjecture about graphs which we would like to disprove by a computer search for a counterexample. We could regard the generation of candidate subgraphs for a counterexample as the generation of subsets of edges from the edge set of some large complete graph.

As we have seen, if we form a graph with vertex set $\{A : A \subseteq \{1,\ldots,n\}\}$, and an edge AB if and only if $A \subset B$ and $|B\backslash A| = 1$ or $B \subset A$ and $|A\backslash B| = 1$, then this graph is isomorphic to the n-dimensional cube Q_n. One way, then, to generate all subsets of the set $\{1,\ldots,n\}$ is to describe a Hamilton path on the n-cube. These Hamilton paths are often called *Gray codes* in deference to the simple and elegant recursive construction due to Gray (US Patent 2632058, March 1953) which we shall describe below.

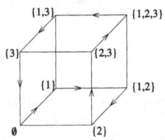

Figure 11.2.1 – The sequence of sets P_3

Let $P_0 = \emptyset$, and P_n denote a sequence of all the subsets of $\{1,\ldots,n\}$. Then recursively we define

$$P_n = P_{n-1}R_{n-1} \qquad \text{for } n \geq 1$$

where by R_n we mean the sequence of sets P_n taken in reverse order, but with the element $n + 1$ added to each set in the sequence. So $P_1 = \emptyset\{1\}$, $P_2 = \emptyset\{1\}\{1,2\}\{2\}$ and the sequence P_3 of subsets of $\{1,2,3\}$,

$$P_3 = \emptyset\{1\}\{1,2\}\{2\}\{2,3\}\{1,2,3\}\{1,3\}\{3\},$$

is shown in figure 11.2.1 above. It is a simple task to verify that this process does indeed generate a Hamilton cycle on the n-cube, and so every subset of a set of n elements.

Generating subsets using a Hamilton path in Q_n has several advantages over other possible methods. But the one that has proved most invaluable is that in this way consecutive subsets differ in only one element. Thus by following a Gray code, a computer can move from one test structure to the next extremely efficiently, by updating a minimum of the data it already holds. Of course, there are many possible Hamilton paths to choose from in Q_n, and we might do better to choose some other over the one described by Gray, to suit an application. In the remainder of this section we shall describe one such problem, and provide a construction of a Hamilton path which provides a much more satisfactory solution than Gray's.

Let us return to our computer search for a counterexample, and suppose that we have some data about the properties of a counterexample, should it exist; these data include an upper and lower bound on the density of edges. As so often, checking whether a given graph is a counterexample to our problem, or not, is rather expensive (in time), and this check becomes more expensive as the graph becomes denser. In addition we would be most interested in a counterexample with as few edges as possible. Of course, we would like to do the search as economically as possible. How, then, should we generate the graphs to check?

Using Gray's construction has the disadvantage that we have no control over the size of the subsets of edges which are generated. Some will lie within the interesting density range, and others will not. Also even after finding a counterexample the only way to know if it is a minimum one is to continue until we have tested all remaining smaller subgraphs within the density bounds, which may require stepping through many unnecessary subgraphs. Consider instead a Hamilton path of Q_n which exhausts every subset of size i before it ever reaches a subset of size $i + 2$, for every $i = 1, \ldots, n - 2$. Such a path is called a *monotone Hamilton path*. If we were to follow a monotone path to generate the subsets the sets of edges would arrive almost in order of increasing cardinality, and so, by following the appropriate subpath, we could check precisely those graphs with valid edge density by beginning with those with few edges and continuing later to those which are denser. Furthermore the first layer which contains a counterexample provides a counterexample with the minimum number of edges, there is no need to proceed further. We shall now exhibit a construction which provides a monotone Hamilton path in any hypercube.

Lemma 11.2.1   *Let $X_0 \subset X_1 \subset \ldots \subset X_n$ and $Y_0 \subset Y_1 \subset \ldots \subset Y_n$ be two sequences of subsets of $\{1, 2, \ldots, n\}$, with $|X_i| = |Y_i| = i$, for $i = 0, \ldots, n$. Then there is a permutation π of $\{1, 2, \ldots, n\}$ such that $\pi(X_i) = Y_i$ for $i = 1, \ldots, n$.*

Proof. For $i = 1, \ldots, n$ let $\{a_i\} = X_i \backslash X_{i-1}$ and $\{b_i\} = Y_i \backslash Y_{i-1}$ and define $\pi(a_i) = b_i$. $\square$

Theorem 11.2.2 (Savage, Winkler (1995)) *Every n-cube Q_n has a monotone Hamilton path.*

Proof. Consider Q_n as a lattice of subsets. We shall refer to the subset of vertices whose names are subsets with i elements by *level* i. We shall construct a monotone Hamilton path P of Q_n in the following way. First we find two sequences of vertices $x_0, \ldots, x_n$ and $y_0, \ldots, y_n$ with the properties that x_i and y_i are vertices of level i and that $x_i \subset x_{i+1}$ and $y_i \subset y_{i+1}$. We then define subpaths $P_0, P_1, \ldots, P_{n-1}$, where the path P_i uses only vertices from levels i and $i+1$ and begins at y_i and ends at x_{i+1} if i is odd, and begins at x_{i+1} and ends at y_i if i is even. Then $P = P_0 P_1 \ldots P_n$ (regarded as a sequence of vertices) is our monotone Hamilton path.

We shall proceed by induction on n. If $n = 1$ then we may let $x_0 = y_0 = \emptyset$ and $x_1 = y_1 = \{1\}$. Thus let us suppose that the hypothesis is true for an n-dimensional cube.

For $A \subseteq \{1, 2, \ldots, n\}$, let $A^+ = A \cup \{n+1\}$, and let Q_n^+ denote the subcube of Q_{n+1} whose vertex set consists of all subsets of $\{1, 2, \ldots, n+1\}$ which contain $n+1$. If P is a path in Q_n, let P^+ be the corresponding path in Q_n^+.

By induction Q_n contains a monotone Hamilton path $P = P_0 \ldots P_{n-1}$ with vertices $x_0, \ldots, x_n$ and $y_0, \ldots, y_n$ as specified above. By Lemma 11.2.1 there is a permutation π of $\{1, 2, \ldots, n+1\}$ so that $\pi(x_i) = y_i$, for $i = 0, \ldots, n$, and $\pi(n+1) = n+1$. Note that since P is a Hamilton path, the paths $P_0^+, \ldots, P_{n-1}^+$ are pairwise disjoint, and together contain all the vertices of Q_n^+. Thus the paths $\pi(P_1^+), \ldots, \pi(P_{n-1}^+)$ are also disjoint and contain all the vertices of Q_n^+.

The proof depends on the observations that when i is even the path $\pi(P_i^+)$ begins at $\pi(y_i^+)$ and ends at $\pi(x_{i+1}^+) = y_{i+1}^+$, and when i is odd, $\pi(P_i^+)$ starts at $\pi(x_{i+1}^+) = y_{i+1}^+$ and ends at $\pi(y_i^+)$. The vertices y_i and y_{i-1} are adjacent in Q_n, and therefore $\pi(y_i)$ and $\pi(y_{i-1})$ also are adjacent in Q_n^+. Thus, for odd $1 \le i < n$ the path $\pi(P_{i-1}^+)^{-1} \pi(P_i^+)^{-1}$ is a (y_i^+, y_{i+1}^+)-path in Q_n^+, and $P_{i-1} P_i$ is a (y_{i-1}, y_i)-path in Q_n. We can now combine these paths to form a Hamilton path in Q_{n+1}. The path differs slightly with the parity of n: for odd n we let

$$
\begin{aligned}
P^* =& P_0 P_1 \pi(P_0^+)^{-1} \pi(P_1^+)^{-1} P_2 P_3 \pi(P_2^+)^{-1} \pi(P_3^+)^{-1} \ldots \\
& P_{i-1} P_i \pi(P_{i-1}^+)^{-1} \pi(P_i^+)^{-1} \ldots \\
& P_{n-3} P_{n-2} \pi(P_{n-3}^+)^{-1} \pi(P_{n-2}^+)^{-1} P_{n-1} \pi(P_{n-1}^+)^{-1},
\end{aligned}
$$

and for even n,

$$
\begin{aligned}
P^* =& P_0 P_1 \pi(P_0^+)^{-1} \pi(P_1^+)^{-1} P_2 P_3 \pi(P_2^+)^{-1} \pi(P_3^+)^{-1} \ldots \\
& P_{i-1} P_i \pi(P_{i-1}^+)^{-1} \pi(P_i^+)^{-1} \ldots \\
& P_{n-2} P_{n-1} \pi(P_{n-2}^+)^{-1} \pi(P_{n-1}^+)^{-1}.
\end{aligned}
$$

It can be readily seen that both of these paths define monotone Hamilton paths of Q_{n+1} (for the appropriate parity). It remains only to check that they satisfy the other criteria of the construction. To see this we let

$$x_i^* = \begin{cases} x_i & \text{if } 0 \le i \le n, \\ \{1, \ldots, n+1\} & \text{if } i = n+1, \end{cases} \quad y_i^* = \begin{cases} y_0 & \text{if } i = 0 \\ \pi(y_{i-1}^+) & \text{if } 1 \le i \le n+1. \end{cases}$$

and

$$P_i^* = \begin{cases} P_0 & \text{if } i = 0, \\ P_i \pi(P_{i-1}^+)^{-1} & \text{if } i < n-1 \text{ is odd}, \\ \pi(P_{i-1}^+)^{-1} P_i & \text{if } i < n-1 \text{ is even}, \\ \pi(P_{n-1}^+)^{-1} P_i & \text{if } i = n. \end{cases}$$

Then $P^* = P_0^* P_1^* \ldots P_n^*$ and it is easy to check that the vertices $x_0^*, \ldots, x_{n+1}^*$ and $y_0^*, \ldots, y_{n+1}^*$ have the desired properties. $\square$

Exercises

11.2.1   Show that there exists a Hamilton path in the n-cube beginning at x and ending at y if and only if the distance between x and y is odd.

11.2.2   Let B be the bipartite graph induced by the middle two layers of Q_{2k+1}. Use Savage and Winkler's construction to show that B contains a path which contains half the vertices of B.[†]

11.3 Pebbling in hypergrids

$\mathbf{A}$s we have seen, results about bipartite graphs can have applications far from the field of graph theory. One graph for which this is especially true is the hypergrid, a generalisation of the n–dimensional cube. In this section we shall present a distributive game played on the hypergrid, which has a beautiful application to elementary number theory.

The n–dimensional grid $Q_{k_1, k_2, \ldots, k_n}$ is the product of disjoint paths $P_{k_1} \times P_{k_2} \times \ldots \times P_{k_n}$ (where P_k is the path on k vertices). Equivalently $Q_{k_1, k_2, \ldots, k_n}$ is the graph with the collection of vectors $(x_1, x_2, \ldots, x_n)$, where each x_i is an integer between 1 and k_i, $i = 1, \ldots, n$, as a vertex set, and in which an edge joins any two vertices whose vectors differ in precisely one position,

---

[†]Finding a Hamilton path in B would provide an answer to the 'Middle Layers Problem'. Showing that such a Hamilton path does not exist would disprove the conjecture of Lovász about the existence of a Hamilton path in a connected, vertex-transitive graph (see Lovász (1970c))

and in that position the values differ by 1. The game also requires n 'costs', positive integers p_i, $i = 1, \ldots, n$.

The game begins with a number of counters, *pebbles*, arranged over the vertices of a hypergrid and on these we carry out a series of moves. A *move in direction* i consists of removing p_i pebbles from a chosen vertex and placing one of them back onto an adjacent vertex whose vector differs only in the ith coordinate. The other pebbles are discarded. If some vertex has less than p_i pebbles then no immediate move in direction i can take place from it. The question we shall seek to answer is how many pebbles are sufficient to ensure that some sequence of moves will always take a pebble to the vertex $(0, \ldots, 0)$, regardless of the starting distribution. Intuition suggests that the worst case should be when all the pebbles are collected on the vertex furthest from $(0, \ldots, 0)$ and so $\prod_{i=1}^{n} p_i^{k_i - 1}$ pebbles ought to be sufficient. Intuition is indeed correct.

Proposition 11.3.1 (Chung (1989)) *For any distribution of* $\displaystyle\prod_{i=1}^{n} p_i^{k_i - 1}$ *pebbles over the vertices of* $Q_{k_1, k_2, \ldots, k_n}$ *there is a series of moves which takes a pebble to* $\mathbf{0} = (0, \ldots, 0)$. $\qquad\square$

We shall keep the proof until a little later. Instead, we turn to the number theory application. Let d be any divisor of a number m, and suppose that $a_1, a_2, \ldots, a_d$ are any d divisors of m, not necessarily distinct. It is obvious that we can always find some subset of $S = \{1, 2, \ldots, d\}$ so that $\sum_{i \in S} a_i$ is a multiple of d. More interesting is if we try to limit the size of this sum. Erdős and Lemke framed the question: can one always find such a multiple that is at most m? In particular when $m = d$: amongst any d divisors of d (not necessarily distinct) can one always find a collection that sum to d itself? This question was answered by Lemke and Kleitman (1989).

Theorem 11.3.2 *Given a positive integer d and any collection of d divisors of d, there is some some subcollection of the divisors which sum to d.* $\qquad\square$

Proof. We shall prove this result by using a reduction to the pebbling game. Let $d = p_1^{\alpha_1} p_2^{\alpha_2} \ldots p_n^{\alpha_n}$ be the prime factorisation of d. Then the playing field for the game is $G = Q_{\alpha_1 + 1, \alpha_2 + 1, \ldots, \alpha_n + 1}$ and p_i is the cost for a move in each direction i, $i = 1, \ldots, n$. Let the divisors be $a_1, \ldots, a_d$, then the divisors define a distribution of pebbles, by placing a pebble at $(b_1, b_2, \ldots, b_n)$ where $d/a_i = p_1^{b_1} \ldots p_n^{b_n}$, for each a_i, $i = 1, \ldots, d$.

Suppose now that some p_i divisors $x_1, x_2, \ldots, x_{p_i}$ correspond to p_i pebbles at vertex $(b_1, b_2, \ldots, b_n)$. Then it is easy to see that $\sum_{j=1}^{p_i} x_j = p_i \cdot x_1 = y$ and so we can replace $x_1, \ldots, x_{p_i}$ by the number y which is represented by a pebble on $(b_1, \ldots, b_i - 1, \ldots, b_n)$. Then, if by some sequence of moves we can place a pebble on $\mathbf{0}$, the implication is that there is some subset

$Q \subseteq \{1, 2, \ldots, d\}$ such that $\sum_{i \in Q} a_i = d$. But, by Proposition 11.3.1, if we begin with d pebbles, moving a pebble to $\mathbf{0}$ is always possible, and so the result follows. $\qquad\square$

It only remains now to give a proof of Chung's result.

Proof of Proposition 11.3.1. Let us first prove the result for the n–dimensional cube Q_n. We do this by induction on n, but we shall strengthen the induction hypothesis to

(1) $\displaystyle\prod_{i=1}^{n} p_i$ pebbles suffice to move one pebble to $\mathbf{0}$,

(2) if $k \geq 2$ and q vertices have at least 1 pebble then $1 + k \cdot \displaystyle\prod_{i=1}^{n} p_i - q$

pebbles suffice to move k pebbles to $\mathbf{0}$.

The hypotheses hold trivially for $n = 0$. Suppose now that they hold for any $n' < n$. Since $Q_n = Q_{n-1} \times K_2$ we can partition Q_n into two copies of Q_{n-1}, M_1 and M_2 say, so that $(0, \ldots, 0) \in V(M_1)$, $(1, 0, \ldots, 0) \in V(M_2)$ and there is a perfect matching M_0 between M_1 and M_2. Suppose that M_i contains m_i pebbles with q_i vertices having at least one pebble, for $i = 1, 2$. We first show (1). Suppose that $\prod_{i=1}^{n} p_i$ pebbles are distributed over Q_n. Then, by induction, if $m_1 \geq \prod_{i=2}^{n} p_i$ pebbles we can move a pebble to $\mathbf{0}$. Otherwise $m_1 < \prod_{i=2}^{n} p_i$ and we consider two cases.

Case A(i). $q_2 > m_1$

Since $m_2 = m - m_1 > \prod_{i=1}^{n} p_i - q_2$, by induction we can move p_1 pebbles to $(1, 0, \ldots, 0)$ and so 1 pebble to $\mathbf{0}$.

Case A(ii). $q_2 \leq m_1$

We may apply moves in direction 1 (along the edges of M_0) to each of the vertices in M_2 and move at least $(m_2 - q_2(p_1 - 1))/p_1$ pebbles to the vertices of M_1. Then in M_1, altogether we have

$$m_1 + \left\lfloor \frac{(m_2 - q_2(p_1 - 1))}{p_1} \right\rfloor \geq m_1 + \left\lfloor \frac{(m_2 - (p_1 - 1)m_1)}{p_1} \right\rfloor$$
$$= \frac{(m_2 + m_1)}{p_1} = \prod_{i=2}^{n} p_i$$

pebbles and by induction we can move a pebble to $\mathbf{0}$.

To prove (2) suppose that more than $k \cdot \prod_{i=1}^{n} p_i - q$ pebbles are distributed over Q_n. If $m_1 \geq k \cdot \prod_{i=2}^{n} p_i - q_1$ then by induction we can move k pebbles to $\mathbf{0}$. Suppose then that $m_1 \leq k \cdot \prod_{i=2}^{n} p_i - q_1$. The proof once again breaks into two cases.

Case B(i). $k \cdot \prod_{i=2}^{n} p_i - q_1 \geq m_1 \geq \prod_{i=2}^{n} p_i$

Since $m_1 \geq \prod_{i=2}^{n} p_i$, by case A we can move one pebble to **0**. We have

$$m_2 > k \cdot \prod_{i=1}^{n} p_i - q - \prod_{i=2}^{n} p_i - q_1 > (k-1)p_1 \cdot \prod_{i=2}^{n} p_i.$$

Thus, by induction, we have enough pebbles on M_2 to move $p_1(k-1)$ pebbles to $(1, 0, \ldots, 0)$ and so the remaining $k-1$ pebbles to **0**.

Case B(ii). $m_1 < \prod_{i=2}^{n} p_i$

Then we have

$$m_2 > k \cdot \prod_{i=1}^{n} p_i - q - m_1 = \left((k-1)p_1 \cdot \prod_{i=2}^{n} p_i - q_2 \right) + \left(\prod_{i=1}^{n} p_i - q_1 - m_1 \right).$$

This shows that taking $t = \prod_{i=2}^{n} p_i - \lceil (m_1 + q_1)/p_1 \rceil$ we can move t pebbles to M_1 along the edges of M_0, whilst leaving enough pebbles in M_2 to move $(k-1)p_1$ pebbles to $(1, 0, \ldots, 0)$ and so $k-1$ to **0**. It remains only to notice that we now have

$$\prod_{i=2}^{n} p_i + m_1 - \lceil (m_1 + q_1)/p_1 \rceil \geq \prod_{i=1}^{n} p_i$$

pebbles in M_1; enough to move the last pebble to **0**. This concludes the proof for Q_n, but now we must return to the general problem of playing the game on $Q_{k_1, k_2, \ldots, k_n}$. For this we must observe that $Q_{k_1, \ldots, k_n}$ can be regarded as a subgraph of $G = Q_{k_1 + k_2 + \ldots + k_n - n}$. If then we let the first $k_1 - 1$ directions of G have cost p_1 and the next $k_2 - 1$ have cost p_2, etc., and restrict our pebbling moves to this subgraph, then the n–cube case shows that $\prod_{i=1}^{n} p_i^{k_i - 1}$ pebbles are always sufficient to move a pebble to **0**, as required. $\qquad \square$

Exercise

11.3.1 ▽ Let $d = p_1^{d_1} \ldots p_k^{d_k}$ be the prime factorisation of d. Show that if $\sum_{i=1}^{k} (1/p_i) \leq 1$, then amongst any collection of d divisors of d there is some subcollection which sum to d which can also be ordered so that each divides its successor. (Denley (1997a))

11.4 Completing latin squares

Imagine a rectangular $r \times s$ array of cells containing n different **symbols**. If the symbols are arranged in the array so that no symbol appears more than once in any row or column then the array is called an $r \times s$ **latin rectangle**, and when $r = s = n$ the array is called an $n \times n$ **latin square**. Throughout this section we shall always assume that these symbols are the integers $1, 2, \ldots, n$. An example of a 6×6 latin square is shown in figure 11.4.1.

1	2	3	4	5	6
2	3	4	5	6	1
3	4	5	6	1	2
4	5	6	1	2	3
5	6	1	2	3	4
6	1	2	3	4	5

Figure 11.4.1

The study of problems concerning latin squares goes back at least to the time of Euler – in fact the term 'latin' square originated with Euler's use of latin letters for the symbols. However, not until much more recently have latin squares received serious attention as both combinatorial and algebraic objects (see Dénes and Keedwell (1974)).

We call an $n \times n$ array with some cells filled from the n symbols a **partial** $n \times n$ **latin square**, and if this partial structure can be embedded in an $n \times n$ latin square we say that it can be **completed**. In this section we shall concern ourselves with formulating sufficient conditions for when a partial latin square has a completion.

Theorem 11.4.1 (Ryser (1951)) *Let $n \geq \max\{r, s\}$. Then an $r \times s$ latin rectangle with entries from n symbols can be extended to an $n \times n$ latin square on those symbols if and only if each symbol occurs in the rectangle at least $r + s - n$ times.*

Proof. We shall give a proof due to Hilton and Johnson (1990). Suppose, without loss of generality, that the latin rectangle R lies, for convenience, in the top left hand corner, and that $s < n$.

We shall begin with necessity. Suppose that R has been completed to an $n \times n$ latin square. Then, each symbol can appear at most $n - s$ times in the rectangle formed by the first r rows and the last $n - s$ columns, and must appear precisely $n - r$ times in the last $n - r$ rows. Thus each symbol must appear at least $n - (n - r) - (n - s) = r + s - n$ times in R, as required.

To prove sufficiency, let us define a bipartite graph B with bipartition (V_1, V_2), where $V_1 = \{\rho_1, \ldots, \rho_r\}$, $V_2 = \{s_1, \ldots, s_n\}$ and there is an edge joining ρ_i to s_k precisely when symbol k does not appear in row i of R. Now since each row of R contains s symbols each vertex of V_1 must have degree $n - s$ in B. On the other hand, since by assumption each symbol occurs at least $r + s - n$ times in R, the degree of each vertex in V_2 is at most $n - s$. Thus applying König's Colouring Theorem B must have an $(n-s)$-colouring with colours $\{s + 1, \ldots, n\}$. Now, place symbol k in the (i, j)th cell if and only if the edge $\rho_i s_k$ in B is coloured j. It is easy to check that in this way we extend R to an $r \times n$ latin rectangle R'.

If r is also less than n then, since each symbol must occur exactly $r = r+n-n$ times in the $r \times n$ latin rectangle R', we may apply precisely the same construction once again to complete R' to an $n \times n$ latin square.   □

This theorem has several immediate corollaries.

Corollary 11.4.2 *Let P be a partial $n \times n$ latin square, all of whose entries lie within an $r \times s$ rectangle, with $r + s \leq n$. Then P can be completed to an $n \times n$ latin square.*

Proof. Let R be an $r \times s$ rectangle which contains all the entries of P. Then we can fill all the empty cells of R, since there can be at most $r+s-2 \leq n-2$ distinct symbols in the same row and column as any empty cell, leaving an available symbol. Once R has been filled, we may apply Theorem 11.4.1 to complete the latin square, since $r + s - n \leq 0$.   □

Corollary 11.4.3 *Any partial $n \times n$ latin square can be embedded into the top left hand corner of a $2n \times 2n$ latin square.*   □

Corollary 11.4.4 *Any $n \times n$ partial latin square which contains r symbols and $2n - r - 1$ unfilled cells can be embedded into the top left hand corner of a $(2n - 1) \times (2n - 1)$ latin square.*

Proof. The unused $2n - 1 - r$ symbols can certainly be used to fill the $2n - 1 - r$ unfilled cells. Then each of the symbols will be used at least once in the square and so the result follows by Theorem 11.4.1.   □

The theorem of Ryser and its corollaries provide the basic tools, but much of the impetus for the work on completing partial latin squares is due to a paper of Evans (1960). In this article Evans outlines the intrinsic interest of completing a variety of partial structures, including latin squares of course, but also groups and finite projective planes. He also made a number of conjectures, among them that every $n \times n$ partial latin square which has at most $n - 1$ filled entries can be completed. Evans' conjecture was finally proved by Häggkvist (1978) for $n \geq 1111$ and in its entirety by Smetaniuk (1981) and Andersen and Hilton (1983). We shall give a version of Smetaniuk's proof below, but first we need some new notation.

Given an $n \times n$ latin square A, we define the partial $(n+1) \times (n+1)$ latin square $T(A)$ by retaining the entries from A which lie above the back diagonal, placing the symbol $n+1$ in the cells which lie down the back diagonal, and leaving the remaining cells empty.

Theorem 11.4.5 (Smetaniuk (1981)) *If A is an $n \times n$ latin square, then $T(A)$ can be completed to an $(n+1) \times (n+1)$ latin square.*

Proof. Let G be a bipartite graph with bipartition (V_1, V_2) where $V_1 = \{r_1, \ldots, r_n\}$, $V_2 = \{s_1, \ldots, s_n, s_{n+1}\}$. Let an edge join $r_i \in V_1$ and $s_j \in V_2$ whenever symbol j is one of the last $i-1$ cells of row i in A, and let s_{n+1} be joined to each vertex of V_1. Furthermore, for each $s_j \in V_2 \backslash \{s_{n+1}\}$ let $D(s_j)$ be the set of columns in which symbol j does not appear in $T(A)$. We shall say that an edge $(n+1)$-colouring of G is *good* if it is proper and no vertex $s_j \in V_2$ is incident with an edge coloured with some colour in $D(s_j)$. We shall construct a good $(n+1)$-colouring of G in which $r_i s_{n+1}$ is coloured $n+2-i$, for each $i = 1, \ldots, n$, which will suffice.

Let the edge $r_i s_j \in E(G)$ be coloured k whenever the symbol j appears in the (i, k)th cell of A. Then observe that this is a good n-colouring of G, except that the edges incident with s_{n+1} receive no colour. Observe also that each vertex r_i has degree i in G and is incident with edges coloured $n - i + 2, \ldots, n$, for each $i = 2, \ldots, n$.

To produce our good colouring we shall recolour some of the edges of G. To begin, we colour $r_n s_{n+1}$ with colour 2. There is already an edge $r_n s_j \in E(G)$ which is coloured 2; this we recolour with $n+1$. The new colouring is still good, except for the uncoloured edges, and the edges incident with r_n are now coloured with the colours $2, 3, \ldots, n+1$. Now suppose that the edges $r_i s_{n+1}$ are coloured $n+2-i$ and that the edges incident with each r_i are coloured with the colours $n+2-i, \ldots, n+1$, for each $n \geq i > l$, whilst the edges incident with $r_{i'}$ ($i' \leq l$) are unchanged from the original colouring. We colour the edge $r_l s_{n+1}$ with colour $n+2-l$, although there is already an edge $r_l s_{j_1} \in E(G)$ coloured $n+2-l$; we recolour the edge $r_l s_{j_1}$ with colour $n+1$. If there is already an edge $s_{j_1} r_{i_1}$ coloured $n+1$, then we recolour this edge with $n+2-l$, and observe that by the order of the recolouring, since r_{i_1} was incident with an edge coloured $n+1$ it must also have been incident with an edge coloured $n+2-l$ which we can recolour $n+1$. By repeating this step, we produce a path beginning at s_{n+1} and ending at some other vertex of V_2, along which we have alternately recoloured the edges with colours $n+1$ and $n+2-l$. In this way we produce a new good colouring of G, in which the edges $r_i s_{n+1}$ are coloured $n+2-i$ and the edges incident with each r_i are coloured with the colours $n+2-i, \ldots, n+1$, for each $n \geq i \geq l$, but in which the colours of edges incident with vertices $r_{i'}$ ($i' < l$) remain unchanged. By repeating this procedure for each edge incident with s_{n+1}, we produce ψ, a good $(n+1)$-colouring of G, in which

$r_i s_{n+1}$ is coloured $n + 2 - i$, for each $i = 1, \ldots, n$. Now place symbol j in the empty (i, k)th entry of $T(A)$ whenever $\psi(r_i s_j) = k$. The properties of the colouring ψ ensure that in this way we fill the first n rows of $T(A)$ to give an $n \times (n + 1)$ latin rectangle. And, since each of the $n + 1$ symbols appears precisely n times in this latin rectangle, the last row of $T(A)$ can be completed by appealing to Theorem 11.4.1. $\qquad \square$

Theorem 11.4.5 contains the major step towards proving the Evans conjecture. To complete the proof requires only a simple lemma, and some observations.

Lemma 11.4.6   *Let A be a partial $n \times n$ latin square, with $n - 1$ filled cells, and let z be a specified entry. Then the rows and columns of A can be permuted so that z lies on the back diagonal, and all the other filled cells lie above it.*

Proof. The proof is by induction on n. The case $n = 2$ is trivial; we shall proceed to the induction step.

Suppose that the hypothesis holds for all smaller values of n. If all the filled cells lie in one row, then we simply permute A so that this row is row 1 and the entry z lies in the last column. The resulting square is as required.

Thus we may assume that some row, row i say, does not contain z, but does contain at least one filled cell. Since only $n - 1$ cells are filled in total, there is a column, column j, which contains no filled cells. We now permute the rows of A so that row i becomes the first row, and permute the columns so that column j becomes the last column. Then the lower left $(n-1) \times (n-1)$ square must contain at most $n-2$ filled cells, one of which is z. We need only apply the inductive hypothesis to this square to complete the proof. $\qquad \square$

Theorem 11.4.7   *A partial $n \times n$ latin square in which at most $n - 1$ cells are filled can always be completed.*

Proof. Let A be the partial latin square. We shall prove the theorem by induction on n. The cases $n = 1$ and 2 are trivial. Suppose that $n > 2$ and that the hypothesis holds for all smaller values of n. Without loss of generality, we may assume that A contains precisely $n - 1$ filled cells using symbols from $\{1, 2, \ldots, n\}$.

Suppose first that some symbol, which we shall call n, occurs only once in A. Then by Lemma 11.4.6 we may permute the rows and columns of A so that s lies on the back diagonal and all the other entries lie above that diagonal. Let A' be the $(n - 1) \times (n - 1)$ partial latin square obtained by deleting the last row and column from A and deleting the symbol n. Then A' is an $(n - 1) \times (n - 1)$ partial latin square containing $n - 2$ filled cells, and so can be completed by the induction hypothesis. If we now apply Theorem 11.4.5 we see that $T(A')$ is a completion of A.

Thus we may suppose that every symbol occurs in A at least twice; there are at most $\lfloor (n-1)/2 \rfloor$ symbols used in total. We can now employ the fact that the roles of row, column and symbol are completely interchangeable in the definition of a latin square: just as each symbol appears only once in each row and column, each row appears only once in association with each column and symbol, etc. Using this observation we can conclude that unless every row and column also contains at least two filled cells, we can complete the square with the argument above. Thus the filled cells lie in an $r \times r$ square with $r \leq \lfloor (n-1)/2 \rfloor$, and we may apply Corollary 11.4.2 to complete the proof. $\square$

The above result is, in some sense, best possible, since there are obvious configurations of n symbols in an $n \times n$ partial latin square which cannot be completed. However, if we insist on some additional structure for the filled cells and symbols, a variety of configurations can be shown to always have completions. In the remainder of this section we shall give several results of this type. In each, as well as the statement of the theorem, we also draw the reader's attention to the proof techniques which employ some results from Section 8.3.

Theorem 11.4.8 (Häggkvist (1983c)) *Let A be a partial $n \times n$ latin square whose occupied cells constitute the cells in the first q rows and the cells in the first q columns of A, and in which the upper left $q \times q$ square Q is latin. Then A can be completed.*

Proof. There is no loss of generality in assuming that the latin square Q uses the symbols $1, \ldots, q$, and that the rest of the symbols in A are $q+1, \ldots, n$. Let B be the bipartite graph with bipartition (V_1, V_2) where $V_1 = \{s_{q+1}, \ldots, s_n\}$ and $V_2 = \{r_{q+1}, \ldots, r_n\}$ correspond to the last $n-q$ rows and symbols, respectively, and let $s_i r_j$ be an edge of B precisely when the symbol i does not occur in row j of A. Then B is $(n-2q)$-regular, since symbols $1, \ldots, q$ do not appear in the last $n-q$ rows, but q symbols from $q+1, \ldots, n$ occur in each, and furthermore, each of these symbols occurs q times in the first q columns, but never in Q.

Let $L(s_i)$ be the set of unfilled columns which do not use the symbol i, for $i = q+1, \ldots, n$. Then each $|L(s_i)| = n - 2q$, since each such symbol is used q times in the first q rows and so q times in the last $n-q$ columns. Thus we may now apply Corollary 8.3.10 to obtain an edge L-colouring of B. By following this colouring we can fill symbols $q+1, \ldots, n$, by placing symbol i in the (j,k)th cell if and only if $s_i r_j$ was assigned colour k in the colouring. In this way we use each of the last $n-q$ symbols $n-q$ times in the lower right $(n-q) \times (n-q)$ square and n times in total.

It remains only to deal with the first q symbols, to complete the square. But to do this we need only notice that in each row and column of the lower right $(n - q) \times (n - q)$ square there are precisely q empty cells and so the corresponding bipartite graph is q-regular. Thus since any q-regular bipartite graph has a proper q-colouring we can complete the whole latin square, by using such a colouring in a similar way to that above. $\qquad\square$

Theorem 11.4.9 (Häggkvist (1987)) *Let $n \geq r$ and let R be a partial $r \times r$ latin square in which each row, column and symbol is used exactly $n - r$ times. Then R can be embedded into an $n \times n$ latin square.*

Proof. It suffices to embed R into an $r \times r$ latin square on n symbols, each of which is used at least $2r - n$ times. We can then apply Theorem 11.4.1 to extend this to an $n \times n$ latin square. Without loss of generality we shall assume that the symbols $\{1, 2, \ldots, r\}$ are used in R.

For the construction we form a bipartite graph B with bipartition $(V_1 \cup V_1', V_2 \cup V_2')$, where $V_1 = \{\rho_1, \ldots, \rho_r\}$, $V_2 = \{c_1, \ldots, c_r\}$, $V_1' = \{c_1', \ldots, c_r'\}$ and $|V_2'| = n - r$. The edge set of B is formed by placing an edge joining $r_i \in V_1$ to $c_j \in V_2$ if and only if the (i, j)th cell is an empty cell in R, by joining each $c_i \in V_2$ to $c_i' \in V_1'$ by $n - r$ parallel edges, and joining every vertex of V_1 to every vertex of V_2'.

To produce the latin square we shall edge colour this graph piece by piece, by employing various V_1-schemes. First observe that $B[V_1 \cup V_2]$ is a $(2r - n)$-regular bipartite graph. For each vertex $c_j \in V_2$, we define $L_1(c_j)$ to be the set of symbols, amongst those used in R, which are not used in column j. Then each $|L_1(c_j)| = 2r - n$, and so by Corollary 8.3.10 there is an edge L-colouring of $B[V_1 \cup V_2]$, ψ.

Now define $L_2(\rho) = \{1, \ldots, r\} \backslash \{\psi(\rho c) : c \in V_2\}$. Then observe that $|L_2(\rho)| = n - r$ and that amongst these 'lists' every symbol in $\{1, \ldots, r\}$ appears exactly $n - r$ times. Then $B[V_1 \cup V_2']$, in which each vertex of V_1 has degree $n - r$, has an edge L_2-colouring by Corollary 8.3.10. Observe now that the graph $B[V_1 \cup V_2 \cup V_2']$ is a bipartite graph which is r-regular for the vertices of V_1, and is r-coloured. To complete the colouring of B, for each $c \in V_2$ we colour the $n - r$ multiple edges with the colours $\{1, 2, \ldots, r\} \backslash L_1(c)$.

Let us now define two colouring schemes on the vertices of $V_1 \cup V_1'$. Let

$$L(x) = \begin{cases} \{1, 2, \ldots, r\} & \text{if } x \in V_1, \\ \{1, 2, \ldots, r\} \backslash L_1(c) & \text{if } x = c' \in V_1', \end{cases}$$

and

$$L'(x) = \begin{cases} \{1, 2, \ldots, n\} \backslash \{\text{symbols used in row } x \text{ in } R\} & \text{if } x \in V_1, \\ L(x) & \text{otherwise.} \end{cases}$$

Then it is clear that $L \prec L'$, but notice that the way that we have coloured B is exactly an edge L-colouring. Thus by Corollary 8.3.7, B also has an edge L'-colouring. If we now follow the restriction of this colouring to $B[V_1 \cup V_2]$

then we fill the empty cells of R to produce an $r \times r$ latin square with symbols from $\{1, 2, \ldots, n\}$. But moreover, it is easy to check that the construction ensures that every symbol will be used at least $r - (n - r) = 2r - n$ times. This completes the proof. □

Theorem 11.4.10 (Denley, Häggkvist (1995)) *Let A be a partial $3r \times 3r$ latin square with filled squares in the top left $2r \times 2r$ square T. Suppose that there is a pairing of the columns of T so that in each row there is a filled cell in at most one of each pair of columns. Then A can be completed if and only if there is some way to fill the cells of T.*

Proof. The necessity of the condition is trivial of course. Suppose then that the pairing exists and that all the cells of T have been filled. Observe that, possibly by permuting the rows, we may assume that column i is paired with column $i + r$, for each $i = 1, \ldots, r$. Our objective will be to redistribute the symbols in the fillable cells of T in such a way that every symbol appears at least $2r + 2r - 3r = r$ times; that A can be completed then follows from Theorem 11.4.1.

We define a bipartite graph H which has bipartition (V_1, V_2), where V_1 represents the set of rows of T and V_2 represents the set of $3r$ symbols. In H we join a vertex $\rho \in V_1$ to a vertex $s \in V_2$ with an edge $f_{\rho s}$ whenever the symbol s does not occur in row ρ in the original partial latin square A. Now we colour the edge $f_{\rho s}$ with colour c if when T was filled the symbol s was placed in the empty (ρ, c)th cell. We shall refer to those edges which receive no column in this way as the *uncoloured* edges. Finally we provide each vertex $s \in V_2$ with the list $U(s)$ of columns in which symbol s appeared in A, and the list $C(s)$ of the colours of the edges incident with the vertex s, and provide each vertex $\rho \in V_1$ with the list $F(\rho)$ of columns in which row ρ was filled in A.

Let s_0 be a symbol which is used less than r times in T. Then there must be an i, $1 \leq i \leq r$, so that symbol s_0 does not appear in columns i and $i + r$. Our aim will be to redistribute the symbols so that s_0 appears in one of these columns. Indeed we will ensure this for every symbol and every pair of columns, and thus ensure that every symbol appears r times in T.

Let B_0 be the subset of symbols which appear at most r times in T and appear in at most one of the columns i and $i + r$, and let H_0 be the subgraph of H induced by the uncoloured edges incident with the vertices of B_0. Then for each vertex $x \in B_0$, $d_{H_0}(x) \geq r$ and for each vertex $\rho \in V_1$, $d_{H_0}(\rho) \leq r$. Thus by Proposition 8.2.9 H_0 has a B_0-sequential colouring and the edges coloured 1 form a matching M of B_0 into V_1, which we shall regard as edges in H.

We shall construct a path P which begins at s_0 and ends somewhere outside B which we will use to redistribute the symbols. Let $f_{\rho_0 s_0}$ be the edge of

M incident with s_0. By assumption, either column i or column $i + r$ is not a member of $F(\rho)$, i.e. there is a column $c_0 \in \{i, i + r\} \backslash F(\rho)$. Then ρ_0 is incident with an edge $f_{\rho_0 s_1}$ which is coloured c_0 in H. If $s_1 \in B_0$ then we continue, otherwise we stop. If we continue, s_1 is incident with an edge of M joining it to a row ρ_1, and as before there must be some column $c_1 \in \{i, i + r\} \backslash F(\rho_1)$ and an edge coloured c_1 joining ρ_1 to s_2. We construct the path by repeating these steps, eventually stopping after some n steps at symbol s_n. P must be a path, since there is at most one edge coloured i or $i + r$ which is incident with each symbol other than the last, and the edges joining the symbols to the rows form a matching in M.

We recolour $f_{\rho_j s_j}$ with colour c_j, and uncolour $f_{\rho_j s_{j+1}}$, for $j = 0, \ldots, n - 1$ (or if you wish make the analogous changes to T). Then observe that after this change s_0 will appear in precisely one of columns i and $i + r$ and s_n will still appear either at least r times in T or in at least one of columns i and $i + r$. Furthermore, the only changes took place in columns i and $i + r$, and every other symbol appears amongst these columns as often in T as it did before.

By repeating this process for all offending symbols we eventually ensure that every symbol appears at least once in each pair of columns, and so at least r times in T in total. The result follows. $\qquad\qquad\square$

Theorem 11.4.10 provides a useful tool to ensure that a wide variety of configurations of filled cells which can be completed. Figure 11.4.2 shows but two important examples which we discuss in Theorems 11.4.11 and 11.4.12.

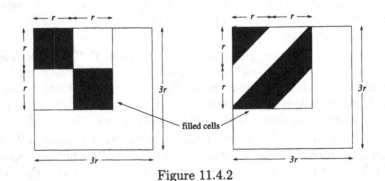

Figure 11.4.2

Theorem 11.4.11 *Let A be a $3r \times 3r$ partial latin square with filled cells which lie in two disjoint $r \times r$ squares. Then A can be completed.*

Proof. If the two $r \times r$ squares lie in the same set of r rows then the result follows immediately from Theorem 11.4.1.

If otherwise, then we can assume that the $r \times r$ squares lie down the diagonal of the top left $2r \times 2r$ square T, and observe that in each row a cell is filled in precisely one of columns i and $i + r$, for each $i = 1, \ldots, r$. Thus by Theorem 11.4.10 it is enough to simply fill the empty cells of T.

Let G be the bipartite graph with bipartition (R, C), where R and C are the sets of rows and columns of T respectively. In G we join a row ρ to a column c by an edge $e_{\rho c}$ precisely when the (ρ, c)th cell in T is empty. We also give each edge $e_{\rho c}$ a list $L_e(e_{\rho c})$ of the symbols which appear in neither row r nor column c. Then $|L_e(e)| \geq r$ for each $e \in E(G)$ and so since G is r-regular Theorem 8.3.2 ensures that G has an L_e-list colouring, which exactly corresponds to a valid completion of the empty cells of T. The result follows.   □

Theorem 11.4.12 *Let A be a $3r \times 3r$ partial latin square. Let the (i, j)th cell be filled only if $1 \leq i, j \leq 2r$ and either $i + j \leq r$ or $2r + 1 \leq i + j \leq 3r$. Then A can be completed.*

Proof. Observe that applying Theorem 8.3.2 as in the previous proof ensures that we can fill in the empty cells of the top left $2r \times 2r$ square. Thus, since this configuration has the property that in each row either the cell in the ith or that in the $(i + r)$th column is empty, for each $i = 1, \ldots, r$, the result follows from Theorem 11.4.10.   □

Exercises

11.4.1   Show that every partial $n \times n$ latin square is the union of four partial $n \times n$ latin squares which can each be completed.

11.4.2 ▽ Construct a partial 8×8 latin square with each row and column used exactly twice which cannot be completed.

11.4.3 ▽ Does there exist a partial 8×8 latin square where each row, column and symbol is used exactly twice which cannot be completed?

Chapter 12

Bipartite subgraphs of arbitrary graphs

12.1 Spanning bipartite subgraphs

Every graph has a spanning bipartite subgraph, and so it is natural to investigate the maximum number of edges such a bipartite subgraph can have. We have already seen that every connected graph G contains a spanning tree, i.e. a spanning connected bipartite subgraph. So a maximum bipartite subgraph will be connected. The number of edges in a maximum bipartite subgraph of G will be denoted by $b(G)$.

The problem of finding $b(G)$ is NP-hard even for cubic, triangle-free graphs (see Yannakakis (1978)). However, some good estimates for $b(G)$ can be obtained and our next result gives a sharp lower bound. The proof is by Poljak and Turzík (1982).

Theorem 12.1.1 (Edwards (1975)) *Let G be a connected simple graph with p vertices and q edges. Then $b(G) \geq q/2 + \lceil (p-1)/2 \rceil /2$.*

Proof. We proceed by induction on p, the cardinality of $V(G)$. The statement is trivial when $p = 1$. Suppose then that the statement holds whenever $p < n$ and that $|V(G)| = n$. The argument breaks into three cases.

Case A. G has a cut vertex v

Let $G_1, \ldots, G_k$ denote the connected components of $G - v$, and G_i' denote the subgraph of G induced by the vertex set $V(G_i) \cup \{v\}$ for each $i = 1, \ldots, k$. Then it is clear that $b(G) = \sum_{i=1}^{k} b(G_i)$. We may now apply the induction hypothesis to each summand to see that

$$b(G) \geq \sum_{i=1}^{k} |E(G_i)|/2 + \sum_{i=1}^{k} \lceil (|V(G_i)| - 1)/2 \rceil /2$$
$$\geq q/2 + \lceil (p-1)/2 \rceil /2$$

by observing that $\sum_{i=1}^{k} |E(G_i)| = q$, $\sum_{i=1}^{k}(|V(G_i)| - 1) = p - 1$, and $\lceil x \rceil + \lceil y \rceil \geq \lceil x+y \rceil$.

Case B. The graph has no cut vertex, but there is a vertex v with odd degree

Certainly, then, $b(G) \geq b(G - v) + (d_G(v) + 1)/2$. We now need only apply the induction hypothesis to obtain the required result for this case.

Case C. The graph has neither a cut vertex nor a vertex of odd degree.

Then there is some edge $uv \in V(G)$ so that $G - \{u, v\}$ is connected. This is certainly true if G is 3-connected. Suppose then that G is 2-connected and that $G - \{u, v\}$ is disconnected for every edge $uv \in E(G)$. Then $G - u$ has at least one cut vertex, and it is easy to see that there is a block (maximal 2-connected subgraph) B of $G - u$ which contains only one of these cut vertices, c say. Then, because of the 2-connectedness of G we must have that $N(u) \cap V(B - c) \neq \emptyset$. Let $v \in N(u) \cap V(B - c)$, then $G - \{u, v\}$ is connected as required. Now we may use a similar argument to that of case B. Clearly

$$b(G) \geq b(G - u) + d_G(v)/2$$

and since $d_{G-u}(v)$ is odd, $b(G-u) \geq b(G-\{u,v\}) + (d_{G-u}(v)+1)/2$. Hence, $b(G) \geq b(G - \{u, v\}) + (d_G(v) + d_G(u))/2$. The result now follows, simply by applying the induction hypothesis.   □

Indeed since there are polynomial algorithms to find cut vertices and blocks the proof also provides a polynomial algorithm to find a bipartite subgraph which satisfies the bound in the theorem. Weaker, but perhaps more appealing, is the following simple algorithm suggested by Erdős (1965), which builds the bipartition, (V_1, V_2), of a large spanning bipartite subgraph of G step by step from an arbitrary partition of $V(G)$.

Algorithm:

- **Let** $V_1 \cup V_2 = V(G)$ with $V_1 \cap V_2 = \emptyset$
- **Let** H be the bipartite subgraph of G induced by the bipartition (V_1, V_2)
- **While** there is a vertex $x \in V(G)$ with $d_H(x) < d_G(x)/2$
  - Move x to the other colour class and update H.
- **End while**

Clearly this process must terminate, since at each pass through the while loop the number of edges in the subgraph H increases. Furthermore, on termination this process gives a spanning bipartite graph H where $d_H(x) \geq d_G(x)/2$ for every vertex x of G.

Corollary 12.1.2 $b(G) \geq |E(G)|/2$. □

Of course, even a maximum bipartite subgraph of a general graph may have a cut vertex (consider for instance an odd cycle). What condition then guarantees that there is a spanning 2-connected bipartite subgraph? Our example shows very clearly that 2-connectivity is not sufficient, but our next result, due to Paulraja (1993), shows that 3-connectivity is sufficient.

Theorem 12.1.3 *Every 3-connected graph G has a 2-connected, spanning bipartite subgraph.*

Proof. Let G be a 3-connected graph. First let us notice that G contains a cycle of even length. If G is complete then this is evident. Otherwise let u and v be two non-adjacent vertices. Then by Menger's Theorem there are three internally disjoint (u, v)-paths. Clearly the lengths of two of these paths must be of equal parity, and so the cycle induced by them must be of even length. Thus G contains a 2-connected bipartite subgraph. We shall show that for any 2-connected, bipartite, proper subgraph H of G, and for any vertex $a \in V(G) \backslash V(H)$, there is a 2-connected, bipartite subgraph H' with $V(H) \cup \{a\} \subseteq V(H')$. Clearly this will suffice.

Add to G a new vertex b, and join it to all the vertices of H. The new graph G' is certainly still 3-connected, thus, by Menger's Theorem, there are three internally disjoint (a, b)-paths aP_1b_1b, aP_2b_2b and aP_3b_3b. It is easy to see that either two of the b_i's lie in the same colour class of H, and their corresponding paths are of the same parity, or two of the b_i's lie in different colour classes of H and their corresponding paths are of opposite parities. In either case, if we add a and the appropriate paths to H, we obtain a new 2-connected bipartite graph H' which contains a, as required. □

To close this section we shall consider one of many applications for spanning bipartite subgraphs. Combining what we have now discovered of their properties with a result from Section 8.3, we can give a result on list colourings, but this time for general graphs.

Theorem 12.1.4 (Borodin, Kostochka, Woodall (1996)) *Let G be a graph and for each edge $e \in E(G)$ $L_e(e)$ be a set of colours assigned to e. Then G has an L_e-list colouring if $|L_e(e)| = \lfloor 3\Delta(G)/2 \rfloor$ for each edge $e \in E(G)$.*

Proof. Consider a spanning bipartite subgraph H of G with $d_H(x) \geq d_G(x)/2$ for every $x \in V(G)$. Such a subgraph can be constructed by using the algorithm of Erdős we saw earlier in this section. Let $F = G - E(H)$. Then $d_F(x) \leq d_G(x)/2$ for each $x \in V(G)$. We can now construct an L_e-list coloring of F. This is possible since for every edge e with ends u and v we have $d_F(u) + d_F(v) \leq \Delta(G) \leq |L_e(e)|$.

For each edge e in H with ends x and y we have at least $l(e)$ colours, where

$$
\begin{aligned}
l(e) &= \lfloor 3\Delta(G)/2 \rfloor - d_F(x) - d_F(y) \\
&\geq \max\{d(x), d(y)\} + \lfloor \min\{d(x), d(y)\}/2 \rfloor - d_F(x) - d_F(y) \\
&= \max\{d(x) + \lfloor d(y)/2 \rfloor, d(y) + \lfloor d(x)/2 \rfloor\} - d_F(x) - d_F(y) \\
&\geq \max\{d_H(x), d_H(y)\},
\end{aligned}
$$

which can be still used for colouring of e. Therefore, by Theorem 8.3.4, the edges of H can be coloured from the remaining lists. The colourings of F and H together now give an L_e-list colouring of G. $\qquad\qquad\square$

In fact Theorem 12.1.4 is a generalisation of a well-known theorem of Shannon (1949) which says that for every graph G the chromatic index $\chi'(G)$ does not exceed $\lfloor 3\Delta(G)/2 \rfloor$.

Exercises

12.1.1   Construct a 2-connected graph G with minimum degree at least 3, which has no spanning, 2-connected bipartite subgraph.

12.1.2   Let G be a 3-connected, cubic, non-bipartite graph. Using Theorem 12.1.3 and Theorem 7.3.4 show that $G \times K_2$ is Hamiltonian.

12.1.3 ▽ Show that for any cubic, triangle-free graph G, $b(G) \geq 4|E(G)|/5$. (Hopkins, Staton (1982))

12.1.4 ▽ Let $I(n, r, s)$ be the graph with vertex set consisting of all r-sets chosen from $\{1, \ldots, n\}$, and an edge between two sets A and B if and only if $|A \cap B| = s$. Show that if H is a subgraph of $I(n, r, s)$ then

$$
\frac{b(H)}{|E(H)|} \geq \frac{b(I(n, r, s))}{|E(I(n, r, s))|}. \qquad\qquad \text{(Denley (1997b))}
$$

12.1.5   Show that every connected graph G contains a spanning tree with mutually non-adjacent (in G) pendant vertices unless G is a cycle, a complete graph, or a complete, balanced bipartite graph. (Böhme et al. (1997))

12.1.6 □ Show that there is a constant c so that $b(G) \geq m^2 + m/2 + c\sqrt{m}$ for every graph G with $2m^2$ edges, provided m is large enough. (Alon (1996))

12.1.7 □ Let G be a simple graph with $\Delta(G) = m$. Show that there is a partition of the vertices of G into two sets X and Y so that in the bipartite graph H induced by the edges $E_G(X, Y)$ we have

$$\frac{1}{2}d_G(v) - \frac{1}{2}\sqrt{m\log 3m} \leq d_H(v) \leq \frac{1}{2}d_G(v) + \frac{1}{2}\sqrt{m\log 3m}.$$
(Chetwynd, Häggkvist (1993))

12.1.8 ▽ Let G be a connected graph. Show that $V(G)$ can be partitioned into sets each of which induces a star except when every block of G is a complete graph on an odd number of vertices. (Saito, Watanabe (1993))

12.2 Covering the edges of a graph with bipartite subgraphs

In Chapter 10 we considered various types of covering problems set on bipartite graphs. Here we shall formulate some problems of covering a general graph G with bipartite subgraphs, by letting the covered set be $E(G)$ and the covering set be the set of bipartite subgraphs, each subgraph *covering* its edges in G. Accordingly, we call a collection of subgraphs $F_1, \ldots, F_n$ a *covering* of G if $\bigcup_{i=1}^{n} E(F_i) = E(G)$.

It is a natural question to ask how far from being bipartite a non-bipartite graph G is. One measure of this is $\beta(G)$, the cardinality of a minimum covering of G with bipartite subgraphs. The next result gives an exact formula for $\beta(G)$, but to formulate it we first need to make some new definitions.

A *proper vertex t-colouring* of a graph G is a mapping $\psi : V(G) \longrightarrow \{1, \ldots, t\}$ which has the property that whenever vertices x and y are adjacent $\psi(x) \neq \psi(y)$. The minimum t for which a vertex t-colouring exists is denoted by $\chi(G)$ and is called the *chromatic number* of G.

Theorem 12.2.1 (Harary, Hsu, Miller (1977)) *For any graph G, $\beta(G) = \lceil \log_2 \chi(G) \rceil$.*

Proof. We shall first prove an upper bound on $\beta(G)$ by induction to show that if $\chi(G) \leq 2^n$ then $\beta(G) \leq n$. This is obvious if $n = 1$. Suppose then that the induction hypothesis holds for $n = k$, and that $2^k < \chi(G) \leq 2^{k+1}$. Fix a vertex $\chi(G)$-colouring, ψ, of G. Let G_1 be the subgraph induced by those vertices x in G for which $\psi(x) \leq 2^k$ and let G_2 be the subgraph induced by the remaining vertices. Then, trivially, each of these subgraphs has chromatic number at most 2^k and so by induction we have

$$\beta(G) \leq 1 + \max_{i=1,2} \beta(G_i) \leq 1 + k.$$

To see the corresponding lower bound let $F_1, \ldots, F_t$ be a minimum covering of G with bipartite subgraphs. Possibly by adding isolated vertices, we may assume that each F_i has vertex set $V(G)$. We shall define a function $T : V(G) \longrightarrow B^t$, where as usual B^t is the set of all binary strings of length t constructed from the bipartitions of the covering subgraphs. Let F_i have bipartition (V_1^i, V_2^i) for each $i = 1, \ldots, t$. Then if $T_i(v)$ denotes the ith entry in the binary string for v

$$T_i(v) = \begin{cases} 1 & \text{if } v \in V_1^i, \\ 0 & \text{if } v \in V_2^i. \end{cases}$$

For each string $\sigma \in T(V(G))$, let $V_\sigma = T^{-1}(\sigma) = \{v \in V(G) : T(v) = \sigma\}$. Then whenever $u, v \in V_\sigma$, u and v are not adjacent in G since they are in the same colour class in every bipartite subgraph in the cover. Hence the V_σ's define a colouring of G and so $\chi(G) \leq |B^t| = 2^t$ as required. $\square$

Rather than covering with any bipartite subgraphs it is of interest to further restrict the covering subgraphs in various ways. First we shall present a result which bounds $\beta_c(G)$, the number of complete bipartite subgraphs needed to provide a covering of a graph G.

Theorem 12.2.2 (Tuza (1984)) *Let G be a simple graph on p vertices. Then $\beta_c(G) \leq p - \lfloor \log p \rfloor$.*

Proof. In fact we shall prove something rather stronger, namely that if G has at least $2^t + 2^{t-1}$ vertices then there is a covering of G consisting of at most $p - t$ complete bipartite subgraphs.

If the covering consists of the graphs $F_1, \ldots, F_k$, we shall write that F_i has bipartition (P_i, Q_i). It will be convenient in the proof not to work with a standard covering, but instead to seek a covering which has the added property that $P_1 \cup \ldots \cup P_k = V(G)$. We shall call this a *strong bipartite covering* of the edges of G and shall write $\beta_c^*(G)$ for the minimum number of complete bipartite graphs forming a strong bipartite covering of G. Observe that for any graph G we have $\beta_c^*(G) \leq |V(G)| - \alpha(G) + 1$ (where $\alpha(G)$ is the maximum cardinality of an independent set of vertices in G.)

Let us begin by proving the theorem for two special cases. Firstly when $G = K_p$, the complete graph on p vertices, it is easy to see that

$$\beta_c^*(K_p) = \lceil \log_2(p+1) \rceil,$$

which easily satisfies the required bound.

Our second special case is when G is triangle-free. For this case we shall appeal to our observation above and prove by induction on t that if $p \geq 2^t + 2^{t-1}$ then $\alpha(G) \geq t + 1$, which will suffice. The case when $t = 1$ is trivial. Therefore, consider when G is on $2^t + 2^{t-1}$ vertices and $t > 1$, and choose a vertex v. If $d_G(v) \geq t + 1$ then, since G is triangle-free, $N(v)$ forms an independent set of the required size. Suppose now that $d_G(v) \leq t$. Then there is a set of at least $2^t + 2^{t-1} - t \geq 2^{t-1} + 2^{t-2}$ vertices none of which is joined to v. Among these vertices, by induction, there must be an independent set of t vertices, which together with v gives an independent set of $t + 1$ vertices in G, as required.[†]

For the remaining cases we shall also proceed by induction on t. The case $t = 1$ is trivial. Suppose then that the hypothesis is true for every $1 \leq t \leq k$ and let $t = k + 1$. Let $T = \{x_1, x_2, x_3\}$ be a triangle in G such that the degree of x_1 is less than $p - 1$. Such a triangle must exist since G is neither complete nor triangle-free.

If $|N(x_1)| < 2^k + 2^{k-1}$ then the set $Y = V(G) \backslash (\{x_1\} \cup N(x_1))$ has cardinality at least $2^k + 2^{k-1}$ and, by induction, the subgraph induced by Y has a strong bipartite covering with at most $|Y| - k$ complete bipartite graphs. We can complete the covering of G by adding $|N(x_1)|$ stars to the collection; for $x \in N(x_1) \backslash \{x_2, x_3\}$ define $P_x = \{x\}$ and $Q_x = N(x)$, and for $x \in \{x_2, x_3\}$ define $P_x = N(x_1)$ and $Q_x = \{x\}$. Thus we have a covering with $|N(x_1)| + |Y| - k = p - (k + 1)$, as required.

Assume then that $|N(x_1)| \geq 2^k + 2^{k-1}$. There certainly exists a strong bipartite covering of $G[N(x_1)]$ by at most $|N(x_1)| - k$ complete bipartite graphs. Let these graphs be $F_1', \ldots, F_l'$ where F_i' has bipartition (P_i, Q_i'), $i = 1, \ldots, l$. Set $Q_i = Q_i' \cup \{x_1\}$, $i = 1, \ldots, l$. Then the complete bipartite subgraphs F_i with bipartition (P_i, Q_i), $i = 1, \ldots, l$, cover every edge in $G[\{x_1\} \cup N(x_1)]$, and $\bigcup P_i = N(x_1)$. Choose an arbitrary vertex $y_1 \in Y = V(G) \backslash (\{x_1\} \cup N(x_1))$. Then we can complete a covering of G by defining $P_y = \{y\}$ and $Q_y = N(y)$ for each $y \in Y$, and letting $P_{y_1} = \{x_1, y_1\}$ and $Q_{y_1} = N(x_1) \cap N(y_1)$. It is then clear that the F_i's together with the graphs induced by the P_y's form a strong bipartite covering of G consisting of at most $|Y| + |N(x_1)| - k = p - (k + 1)$ subgraphs as required. $\square$

Given a covering of G with bipartite subgraphs, $F_1, \ldots, F_t$, we can define the weight of this covering to be $\sum_{i=1}^k |V(F_i)|$. Rather than bounding the number of complete graphs, we may instead insist on minimising their weight in this sense.

Theorem 12.2.3 (Tuza (1984)) *For every simple graph G on p vertices there is a covering consisting of complete bipartite graphs of weight at most*

$$\left(\frac{3}{2} + o(1)\right) \frac{p^2}{\log p}.$$
$\square$

[†]In fact, a triangle-free graph with $\alpha(G) \leq t$ can have at most $O(t^2 / \log t)$ vertices (see Kim (1995)).

This result has an interesting application in complexity theory. Given a boolean function of the form $f(x_1, \ldots, x_p) = \bigvee_{k=1}^{t} (x_{i_k} \wedge x_{j_k})$ we may ask how many $\vee$ and $\wedge$ operations are required to calculate f. This is the *complexity* $C_{\{\vee, \wedge\}}(f)$ of f. We could also consider f in the form

$$f(x_1, \ldots, x_p) = ((y_{11} \vee \ldots \vee y_{1n_1}) \wedge (y_{21} \vee \ldots \vee y_{2n_2})) \vee \ldots$$
$$\vee ((y_{k1} \vee \ldots \vee y_{kn_k}) \wedge (y_{k+1n_{k+1}} \vee \ldots \vee y_{k+1n_{k+1}})) \qquad (12.2.1)$$

where each y_{ij} is one of the x_i. Written in this form $C_{\{\vee, \wedge\}}(f) = \sum_{l=1}^{k+1} n_l$ and it becomes clear that Tuza's result may be of use to bound $C_{\{\vee, \wedge\}}(f)$. Let G_f be the graph with vertex set $V(G_f) = \{x_1, \ldots, x_p\}$ and edge set $E(G_f) = \{x_{i_1} x_{j_1}, \ldots, x_{i_k} x_{j_k}\}$. Now any covering of G_f with complete bipartite graphs $K_{n_1, n_2}, \ldots, K_{n_k, n_{k+1}}$ corresponds precisely with a representation of f in the form (12.2.1). Hence, applying Theorem 12.2.3 bounds the minimal weight of such a covering and we see that

$$C_{\{\vee, \wedge\}}(f) \leq \left(\frac{3}{2} + o(1) \right) \frac{p^2}{\log p}.$$

Another natural restricted problem is to let the set of covering subgraphs be the set of forests. It is easy to see that any covering of G by forests can be transformed, by removing the common edges from one of every two forests which intersect, into a covering by forests in which the subgraphs are edge-disjoint. Such a covering is called an *edge-decomposition* of G into forests. Thus we define the *edge-arboricity* $\gamma_e(G)$ of a graph G to be the minimum number of forests in an edge-decomposition of G. Our next theorem is a famous result which provides an elegant formula for the edge-arboricity.

Theorem 12.2.4 (Nash-Williams (1961)) *Let G be a non-trivial, simple graph. Then*

$$\gamma_e(G) = \max_{H \subseteq G} \left\lceil \frac{|E(H)|}{|V(H)| - 1} \right\rceil.$$

Proof. We shall follow a short proof due to Chen et al. (1994). Observe that $\gamma_e(G)$ is certainly at least as big as this maximum, since for any subgraph H we must have $\gamma_e(H)(|V(H)| - 1) \geq |E(H)|$.

Suppose then that $\gamma_e(G)$ is strictly greater than the right hand side in the theorem, and that G is a graph with this property which minimises $|V(G)| + |E(G)|$. Then G must be connected with $\gamma_e(G) > 1$ and also must be critical with respect to γ_e; that is to say that $\gamma_e(G - e) < \gamma_e(G)$ for any edge $e \in E(G)$. The proof of the theorem now depends on the following lemma.

Lemma 12.2.5 *Let G be connected with $\gamma_e(G) > 1$ and let G be critical with respect to γ_e. Then for any $e \in E(G)$ any edge-decomposition of $G - e$ consisting of $\gamma_e(G) - 1$ forests is a decomposition into $\gamma_e(G) - 1$ spanning trees of $G - e$.*

Once we have this lemma we may note the equality

$$|E(G)| - 1 = |E(G - e)| = (|V(G)| - 1)(\gamma_e(G) - 1)$$

which leads to the following contradiction:

$$\gamma_e(G) > \lceil |E(G)|/(|V(G)| - 1) \rceil = \lceil (\gamma_e(G) - 1) + 1/(|V(G)| - 1) \rceil = \gamma_e(G).$$

It remains only to prove the lemma.

Proof of Lemma 12.2.5. Suppose the contrary and let $E_1, \ldots, E_{\gamma-1}$ be an edge-decomposition of $G - e$ into forests, where E_1 is not a spanning tree (for simplicity of notation we write γ for $\gamma_e(G)$). Since $E_1 + e$ must contain a cycle, both ends of e must lie in a connected component of E_1, T say. Then, since G is connected, and T is not a spanning tree of G, $K = G[T]$ must be a proper subgraph of G which contains the edge e. Moreover, by the criticality of G, K must have an edge-decomposition into $\gamma - 1$ forests $F_1, \ldots, F_{\gamma-1}$. Let $E_1, \ldots, E_{\gamma-1}$ be chosen so that $|V(K)|$ is maximal.

Consider the set S of all edge-decompositions into forests which have the form $E_1', \ldots, E_{\gamma-1}', \{e'\}$ for which a connected component of E_1' is a spanning tree of K and $e' \in E(K)$. The decomposition $E_1, \ldots, E_{\gamma-1}$ shows that S is non-empty. Thus let $\bar{E}_1, \ldots, \bar{E}_{\gamma-1}, \{\bar{e}\}$ be an element of S which maximises

$$\sum_{i=1}^{\gamma-1} |E(F_i) \cap E(\bar{E}_i)|. \tag{12.2.2}$$

Since $\bar{e} \in E(K)$ it must also be an edge of F_t for some t, and $\bar{E}_t \cup \{\bar{e}\}$ must contain a cycle C which also contains $\bar{e}$. If $t = 1$ then C itself is contained in K. If $t \neq 1$, and C is not contained in K, we can take an edge $f \in E(C)$ with one end in $V(K)$ and the other not. Since a connected component of $\bar{E}_1$ spans K, $\bar{E}_1 \cup \{f\}$ is acyclic and we may construct a new decomposition $\bar{E}_1 \cup \{f\}, \ldots, \bar{E}_t \cup \{\bar{e}\} \setminus \{f\}, \ldots, \bar{E}_{\gamma-1}$ which contradicts our maximising $|V(K)|$. On the other hand, if $C \subseteq K$ then, since F_t is acyclic, there exists an edge $f \in E(C) \setminus E(F_t) \subseteq E(K)$. Then the decomposition $\bar{E}_1, \ldots, \bar{E}_t \cup \{\bar{e}\} \setminus \{f\}, \ldots, \bar{E}_{\gamma-1}, \{f\}$ contradicts our maximising (12.2.2). $\square$

A nice application of Nash-Williams' formula is the following result which gives a sharp bound for the edge-arboricity in term of the number of edges.

Corollary 12.2.6 (Dean, Hutchinson, Scheinerman (1991)) *If G is a simple graph with q edges then $\gamma_e(G) \le \lceil \sqrt{q}/2 \rceil$.*

Proof. Theorem 12.2.4 implies the existence of a subgraph of G, G', with p' vertices, q' edges and $\gamma_e(G) = \gamma_e(G') = \lceil q'/(p'-1) \rceil$. If $q' \le (p'-1)^2/2$ then $(q'/(p'-1))^2 \le q'/2$ and $\gamma_e(G) \le \left\lceil \sqrt{q'/2} \right\rceil \le \left\lceil \sqrt{q/2} \right\rceil$. If $q' > (p'-1)^2/2$ we have $\gamma_e(G) = \gamma_e(G') \le \gamma_e(K_{p'}) = \lceil p'/2 \rceil$. However, by assumption $q' > (p'-1)^2/2$ and so $\sqrt{q'/2} > (p'-1)/2$. Thus

$$\gamma_e(G) \le \left\lceil \frac{p'}{2} \right\rceil \le \left\lceil \sqrt{\frac{q'}{2}} \right\rceil \le \left\lceil \sqrt{\frac{q}{2}} \right\rceil. \qquad \square$$

In fact there is an algorithm of Edmonds (1965) which constructs a minimal decomposition into forests, in polynomial time. However, suppose that we restrict the subgraphs in the decomposition further, from being any forest to being a fixed tree T. The problem now is to decide whether there exists an edge-decomposition consisting solely of copies of T. This we call a *T-decomposition*. For the case when T has only two edges, i.e. T is the path P_3 on three vertices, the answer is given by the following neat theorem.

Proposition 12.2.7 (Javorskiĭ (1978), Caro, Schönheim (1980)) *Let G be a simple graph. Then G has a P_3-decomposition if and only if every component of G has an even number of edges.*

Proof. The condition is clearly necessary, and so we shall turn to showing the sufficiency for which we shall induct on q, the number of edges in G. Without loss of generality we may also assume that G is connected. The result is trivial when $q = 0$ or 2. Assume then that the assertion is true for every even number $q' < q$.

Choose two incident edges of G, e_1 and e_2 where $e_1 = v_0 v_1$ and $e_2 = v_0 v_2$. If all the components of $G - \{e_1, e_2\}$ are even then we have our decomposition by induction. Otherwise let A_i, $i = 0, 1, 2$, be the component of $G - \{e_1, e_2\}$ containing v_i. Note that the A_i's may not be distinct.

Suppose that A_1 has an odd number of edges; then we can apply the induction hypothesis to the graphs $A_1 + e_1$ and $G - E(A_1) - e_1$. We proceed similarly if only A_2 has an odd number of edges. To complete the proof we need only notice that if A_0 contains an odd number of edges then so must either A_1 or A_2 and so this case reduces to the one above. The result follows. $\qquad \square$

Proposition 12.2.7 also implies a result about the structure of the so-called *line graph* of a graph G, $L(G)$. This graph has vertex set $V(L(G)) = E(G)$ and edge set $E(L(G)) = \{e_1 e_2 : e_1, e_2 \in E(G), \ e_1 \text{ and } e_2 \text{ share an end}\}$.

Corollary 12.2.8 *Let G be a simple graph. Then the line graph $L(G)$ has a perfect matching if and only if all components of G have an even number of edges.* □

In fact, the general problem of deciding the existence of a T-decomposition, when T has more than two edges, is NP-complete (see Dyer and Frieze (1985)). However, there are conjectures connected with special cases of this problem, not least the well-known conjecture of Ringel (1964) that K_{2m+1} has a T-decomposition for any tree T with m edges. In similar vein is the conjecture of Graham and Häggkvist that any $2m$-regular graph has a T-decomposition for any tree T with m edges. The following result gives a partial solution to this problem.

Theorem 12.2.9 (Häggkvist (1983a)) *Let T be a tree with m edges and G a $2m$-regular graph whose shortest cycle has length at least the diameter of T. Then G has a T-decomposition.*

Proof. We begin by deleting the pendant vertices from T to give the tree T'. Since the theorem is clearly true if $|V(T)| \leq 2$, we may assume that T' is non-trivial. We order the vertex set of T', $\{v_0, \ldots, v_n\}$, in such a way that $T'_i = T[\{v_0, \ldots, v_i\}]$ is connected, for $i = 1, \ldots, n$.

Claim. For any k, $0 \leq k \leq n$, G contains a spanning, $2k$-regular subgraph H_k which admits a T'_k-decomposition $H^1_k, H^2_k, \ldots, H^p_k$, where $p = |V(G)|$, and, if $v_{i,j}$ denotes the vertex corresponding to v_j in H^i_k, then $\{v_{1,j}, \ldots, v_{p,j}\} = V(G)$, for $j = 0, \ldots, p$.

To prove the claim we proceed by induction on k. The claim is certainly true for $k = 0$. Suppose, then, that it is true for some $k \geq 0$. We shall construct H_{k+1} from H_k.

Consider $G - E(H_k)$. This is a $2(m-k)$-regular graph and thus by a theorem of Petersen (see Exercise 7.3.11) contains a 2-factor F. We orient the edges of F to produce a directed graph $\vec{F}$ in which $d^+_{\vec{F}}(v) = d^-_{\vec{F}}(v) = 1$ for each $v \in V(\vec{F})$. Let v_s be the unique vertex joined to v_{k+1} in T'_{k+1}. Then, the vertices $v_{1,s}, \ldots, v_{k,s}$ have already been defined, since $s < k+1$. Let $v_{i,k+1}$ be the vertex in $N^+(v_{i,s})$ in $\vec{F}$, for each $i = 0, \ldots, p$. Then it is clear that $\{v_{1,k+1}, \ldots, v_{p,k+1}\} = V(G)$. Finally we put $H^i_{k+1} = H^i_k + v_{i,s}v_{i,k+1}$. Then H^i_{k+1} must be isomorphic to T'_{k+1}, or H^i_{k+1} would contain a cycle of length at most the diameter of T', contradicting our assumptions. Setting $H_{k+1} = \bigcup_i H^i_{k+1}$ proves the claim.

To complete the proof, it remains only to add the pendant vertices of T. This we do using results about edge L-colourings from Chapter 8.

Let a_i be the number of pendant vertices incident with v_i in T, $i = 0, \ldots, n$. Then, since every edge in T not in T' is incident with exactly one vertex of

T', we must have $\sum_{i=0}^{n} a_i = m - n$. We define a bipartite graph B with bipartition (V_1, V_2) where $V_1 = V(G) = \{x_1, \ldots, x_p\}$ and $V_2 = \{y_1, \ldots, y_p\}$ by joining x_s to y_i by a_j multiple edges if and only if $x_s = v_{i,j}$. Then B is an $(m - n)$-regular bipartite graph, and given any fixed indices s, j there is exactly one i, and similarly for each i, j there is exactly one s for which $x_s = v_{i,j}$.

Consider the $2(m-n)$-regular graph $R = G - E(H_n)$. By once again appealing to the theorem of Petersen, we can orient the edges of R so that $d_{\underset{R}{+}}^{-}(v) = m - n$ for each vertex $v \in V(\vec{R})$. We shall complete the T-decomposition by partitioning $\vec{R}$ into p edge-disjoint directed graphs $\vec{R}_1, \ldots, \vec{R}_p$ in such a way that $H_n^i \cup R_i$ is isomorphic to T. To do this we shall colour the edges of $\vec{R}$ with p colours in such a way that the edges of colour i, for $i = 1, \ldots, p$, induce a subgraph in which there are a_j arcs starting at $v_{i,j}$, but there are no arcs ending at $v_{i,j}$, for each $j = 0, \ldots, n$, and every other vertex is incident with at most one arc which ends there.

To find an appropriate colouring on R we shall find an edge L-colouring of B. At each vertex $x_s \in V_1$ we define the tight V_1-scheme $L(x_s)$ where $L(x_s) = \{j : \vec{x}_s \, x_j \in E(\vec{R})\}$. Then since $d_{\underset{R}{+}}^{-}(x_s) = m - n$ we have $|L(x_s)| = m - n = \Delta(B)$, and thus, by Corollary 8.3.10 there is an edge L-colouring of B. Fix a colour $1 \le i \le p$, and consider the vertices $v_{i,0} = x_{s_0}, \ldots, v_{i,n} = x_{s_n}$. Join the vertex x_{s_j} in colour i to those a_j vertices x_k, where k is a colour on an edge joining x_{s_j} to y_i. Then $d_{\underset{R_i}{+}}^{+}(x_s) = a_j$ and $d_{\underset{R_i}{-}}^{-}(x_s) = 0$ whenever $x_s = v_{i,j}$, or G would contain a cycle of length at most the diameter of T. Indeed $d_{\underset{R_i}{-}}^{-}(v) \le 1$ for every vertex $v \in V(\vec{R}_i)$, since the colouring of B is proper. Thus if we let $G_i = H_n^i \cup R_i$, then each G_i must be isomorphic to T.

It remains only to observe that every edge of R occurs in some R_i, since every colour in $L(x_s)$ is used in the colouring of B, and no edge can be assigned more than one colour, since each colour occurs at most once in each allowed list $L(x_s)$. Thus $G_1, \ldots G_p$ forms a T-decomposition of G. $\qquad\square$

Exercises

12.2.1 ▽ Show that $\beta_c(K_p) = p - 1$. (Graham, Pollack (1972))

12.2.2 ▽ Let $R(n, k)$ be the set of all k-regular simple bipartite graphs on $2n$ vertices, $n > k \ge 1$, and let $\beta(n, k) = \min_{G \in R(n,k)} \beta_{cd}(G)$ where $\beta_{cd}(G)$ is the minimum number of graphs in a decomposition of G consisting of complete bipartite graphs. Show that

(a) $n/k \leq \beta(n,k) \leq k + n/k,$

(b) $\beta(n,3) = \begin{cases} \lfloor n/3 \rfloor & \text{if } 3|n, \\ \lfloor n/3 \rfloor + 3 & \text{otherwise. (Gregory et al. (1991))} \end{cases}$

12.2.3   Let $m \leq n$ be positive integers and let G be the graph obtained by removing the edges of a complete graph K_m from a complete graph K_n. Show that $\beta_{cd}(G) = n - m$. (Jones et al. (1988))

12.2.4 □ Let G be a t-regular simple bipartite graph on $2n$ vertices. Show that the graph $K_{t^2!n,t^2!n}$ can be represented as a union of edge-disjoint copies of G. (Häggkvist (1989))

12.2.5 ▽ Let $m_1, \ldots, m_l$ be a sequence of positive integers, with $2m_i \leq n$, for $i = 1, \ldots, l$, and $\sum_{i=1}^{l} m_i = n(n-1)/2$. Show that K_n can be represented as the edge-disjoint union of the stars $K_{1,m_1}, \ldots, K_{1,m_l}$. (Tarsi (1986))

12.2.6   Show that a simple 3-regular graph G has a P_4-decomposition (P_4 is a path on four vertices) if and only if G has a perfect matching.

12.2.7   Show that the number of distinct induced subgraphs isomorphic to $K_{s,s}$ in any graph on p vertices does not exceed $\binom{\lfloor p/2 \rfloor}{s}\binom{\lceil p/2 \rceil}{s}$. Moreover equality is attained only for the graph $K_{\lfloor p/2 \rfloor, \lceil p/2 \rceil}$. (Bollobás, Nara, Tachibana (1986))

Applications

12.3 Optimal spanning trees and the Travelling Salesman Problem

Imagine that there are a number of towns which need to be connected by a network of roads. We know the cost of constructing a direct road between each pair of towns; our task is to construct the network of roads in a way which minimises the total construction cost. To translate this problem into a graph theoretic setting we shall consider the simple graph G with the set of towns forming the vertex set. We impose a weight function $w : E(G) \longrightarrow \mathbb{R}$ on the edges of the graph, so that the weight on an edge reflects the cost of building a road between the two towns it joins (as usual, we can make it impossible to join two towns by making the weight on that edge 'infinite'). The problem then amounts to finding a connected spanning subgraph of G so that the sum of the weights on the edges of this subgraph is minimum. It is clear that such a connected spanning subgraph must be a spanning tree, and we call such a spanning tree an *optimal spanning tree*. There are a number of algorithms to solve this problem. The best known are due to Prim (1957), which we describe below, and to Kruskal (1956), which we outline in Exercise 12.3.2.

Prim's Algorithm

•Choose a starting vertex u.

•**Let** $U = \{u\}$ and the tree T be the vertex u.

•**For all** $v \in V(G)\backslash\{u\}$ **let** $L(v) = w(uv)$.

•**While** $U \neq V(G)$

(1)•Choose a vertex w so that $L(w) = \min\{L(v) : v \in V(G)\backslash U\}$, and denote by u' the vertex of U for which $w(u'w) = L(w)$.

•Add the vertex w and the edge $u'w$ to the tree T, and w to the set U.

(2)•**For all** $v \in V(G)\backslash U$

•**If** $w(vw) < L(v)$ **then** $L(v) = w(vw)$.

•**End loop**

•**End while**

The algorithm works by building the tree T stage by stage. It begins with an arbitrary vertex u and then, as at each stage at step (1), looks for an edge of minimum weight which connects a vertex inside the tree to one outside. One such edge is then added extending the tree to span another vertex. This search is made easier by maintaining the labels $L(v)$ (loop (2)) so that each $L(v)$ is the minimum weight of an edge which could extend the tree from v. It remains only to prove that the algorithm does indeed produce a minimal weight spanning tree.

Theorem 12.3.1 *Prim's algorithm finds an optimal spanning tree in the weighted graph G.*

Proof. We shall proceed by induction on the cardinality of U, showing that T is always a subtree of a minimum weight spanning tree which spans the vertex set U.

It is clear that this hypothesis is true initially, since any spanning tree certainly must contain the vertex u. Let us assume then that it is also true when $V(T) = U$. Let w be the new vertex to be added, and consider the new graph $T + u'w$. Trivially this new graph is also a tree, and must span $U \cup \{w\}$, it remains only to show that it is still a subtree of a minimum weight spanning tree. By induction T is a subtree of such a spanning tree T_M. Suppose that $u'w \notin E(T_M)$. Then by Proposition 2.2.1 $T_M + u'w$ must contain a cycle C, and there is another edge xy of C joining U to $V(G)\backslash U$.

We can thus construct a new tree $T'_M = T_M + u'w - xy$ for which it is clear that $T + u'w$ is a subtree. Furthermore, the weight of this new tree differs from the old by $w(u'w) - w(xy)$ and so since w was chosen so that $u'w$ had minimal weight amongst all edges joining U to $V(G)\backslash U$, $w(xy) \geq w(u'w)$ and the new tree T'_M must also be a minimum weight spanning tree. Hence the induction hypothesis is satisfied. $\qquad\square$

In much the same vein we could consider not the problem of connecting the towns with a network, but instead making as short a tour of all the towns as possible, visiting each only once. In terms of graphs this means finding a Hamilton cycle with the smallest weight in a weighted complete graph. This is the famous Travelling Salesman Problem. The problem of finding such a tour in general is NP-hard, but as we shall see, Prim's algorithm provides a tour of length at most twice the optimum. The method is simply to number the vertices in the graph in the sequence in which they are added to the tree constructed by Prime's algorithm $x_1, \ldots, x_p$. Then the salesman's tour will be $x_1 x_2 \ldots x_p x_1$.

Theorem 12.3.2 (Christofides (1976)) *Let $w : E(K_p) \longrightarrow \mathbb{R}^+$ be a weight function on the edges of the complete graph. If w satisfies the triangle inequality then the salesman's tour constructed from Prim's tree is of weight at most twice the minimum weight of a Hamilton cycle.*

Proof. Let C be a minimum weight Hamilton cycle, in which $e \in E(C)$ is an edge of maximum weight. To prove the result we shall consider building our salesman's tour vertex by vertex; at each stage imagining that a new town is inserted into the old tour. We shall show a correspondence between the edges which must be inserted to join this new vertex, and the edges of $C - e$. The correspondence will be made in such a way that, at each stage, the edges needed to insert the new vertex into the present cycle will be of total weight at most twice a corresponding edge of $C - e$. To do this we envisage a graph which at each stage contains not only the present salesman's tour, but also a subset of the edges of $C-e$ – edges which connect those towns not yet in the tour to those which are, and which themselves are not yet accounted for by the correspondence.

Suppose that the salesman's tour so far is $x_1 x_2 \ldots x_n x_1$, and that we wish to insert the new town y next to x_i. Then the cost of doing this is $w(x_i y) + w(y x_{i+1}) - w(x_i x_{i+1})$. The vertex y is already connected to the tour by some path of edges of $C - e$ ending at x_j, say. Let $x_j z$ be the first edge of that path. Now, by construction we must have

$$w(x_i y) \leq w(x_j z) \qquad (12.3.1)$$

and by the triangle inequality we have

$$w(x_{i+1} y) \leq w(x_i x_{i+1}) + w(x_i y).$$

These two inequalities combine to give

$$w(x_{i+1}y) \leq w(x_ix_{i+1}) + w(x_jz). \qquad (12.3.2)$$

Thus adding (12.3.1) and (12.3.2) and rearranging we have

$$w(x_iy) + w(yx_{i+1}) - w(x_ix_{i+1}) \leq 2w(x_jz).$$

Hence inserting the new vertex y into the tour costs at most $2w(x_jz)$, where x_jz is an edge of $C - e$. Observe also that once the new vertex y has been inserted we may delete the edge x_jz from the graph, retaining the property that each vertex outside the present tour is connected to the tour by a path of edges from $C - e$. Repeating this procedure we see that the complete salesman's tour can be constructed so that each insertion has its corresponding edge of $C - e$.

Finally, summing the costs of all the insertions we see that the total weight of the tour produced by the algorithm is at most $2(w(C) - w(e))$. The theorem follows. $\qquad \qquad \square$

Exercises

12.3.1   Prove that the optimal spanning tree is unique if all the weights are distinct.

12.3.2   Kruskal's algorithm to construct an optimal spanning tree consists of beginning with an edge of minimum weight and then building the tree in such a way that at each stage the new edge added is the edge of minimal weight which can be added without creating a cycle. Show that this algorithm does indeed construct an optimal spanning tree. (Kruskal (1956))

12.3.3   Show that repeatedly deleting from the simple, connected, weighted graph G the edge of maximum weight whose removal does not disconnect G constructs an optimal spanning tree.

12.4 The optimal spanning tree and optimal path problems

In this section we shall consider the solutions to two further optimisation problems concerned with connecting paths in a weighted graph. Once again, let G be a connected simple graph with a weight function $w : E(G) \longrightarrow \mathbb{R}$ applied to the edges. For each pair of vertices x and y, let $P(x,y)$ denote the set of all (x,y)-paths of G.

The first of the two problems is to try to minimise the maximum weight edge in an (x,y)-path. In other words to find the path $P \in P(x,y)$ such that

$$\max_{e \in E(P)} w(e) = \min_{P' \in P(x,y)} \max_{e \in E(P')} w(e).$$

Such a path is called a *minimax* (x, y)-path of G and in fact, as the following theorem shows, we can find all minimax paths at once, by constructing an optimal spanning tree in G (see Prim's algorithm).

Theorem 12.4.1 (Bardadym (1990)) *Let G be a graph with edge weight function w and T an optimal spanning tree of G. Then for each pair of vertices $x, y \in V(G)$ the unique (x, y)-path in T is a minimax (x, y)-path of G.*

Proof. Let P be the (x, y)-path in T with $w_0 = \max_{e \in E(P)} w(e)$, let P^* be a minimax (x, y)-path of G with $w^* = \max_{e \in E(P^*)} w(e)$ and suppose that $w^* < w_0$. Delete from T an edge e_0 of P of weight w_0. Then $T - e$ contains two connected components, one containing x and the other y. However, now there is an edge e^* of P^* with one end in each of the two components of $T - e$, and so, since $w(e^*) \leq w^* < w(e_0)$, $T - e_0 + e^*$ must be a spanning tree of lesser weight than T, giving a contradiction. Hence P is a minimax (x, y)-path as required. □

The second problem is a generalisation of Lemma 3.1.4 which shows the existence of a distance preserving spanning tree. In a weighted graph we define the *length* of an (x, y)-path P to be

$$l(P) = \sum_{e \in E(P)} w(e).$$

We can then also define the length between any two vertices in the graph by $l_G(x, y) = \min_{P \in P(x,y)} l(P)$. Our problem is to find a path of this minimum length between the fixed vertex x and any other vertex. Happily, just as in the previous problem, we can find all such paths at once, by constructing a spanning tree in which the length between any two vertices in G is precisely the length of the unique path joining them in the tree.

Theorem 12.4.2 *Given a weighted graph G and a vertex x, there is a spanning tree $T = T(x)$ so that $l_G(x, y) = l_T(x, y)$ for each vertex $y \in V(G)$.*

Proof. In similar style to Prim's algorithm we shall build the tree vertex by vertex. We begin with the vertex x. At each stage we look at each of the vertices which is not yet in the tree T, but is a neighbour of a vertex which is. For each vertex $y \in N(V(T))$ we define

$$l_T'(y) = \min_{v \in N(y) \cap V(T)} (l_T(x, v) + w(vy)).$$

The tree is then extended by adding a new vertex y for which $l_T'(y)$ is minimal, joining y to the present tree by an edge vy which gives rise to this minimal value of $l_T'(y)$.

We shall show that this construction does indeed produce a tree which preserves all lengths from x by induction on the stages. Our hypothesis will be that at each stage of the construction the tree T is length preserving from x for the graph $G[V(T)]$. This hypothesis is clearly satisfied initially, thus consider when the new vertex y is added. If some (y, x)-path of length $l_G(x, y)$ contains an edge yw, for some $w \in V(T)$, then by the choice of y we must have that $l_G(x, y) = l'_T(y) = l_{T+y}(x, y)$ and the induction hypothesis remains satisfied. On the other hand, however, suppose that every (y, x)-path of length $l_G(x, y)$ goes from y to a vertex outside $V(T)$. Let $P' = y \ldots u \ldots x$ be one such path and u the last vertex on that path which is outside $V(T)$. Then we must have $l'_T(u) < l'_T(y)$ which contradicts the choice of y. The result follows.   □

Exercise

12.4.1 △ Construct a simple, connected, weighted graph G such that for each vertex $x \in V(G)$ every spanning tree $T = T(x)$ satisfying the condition of Theorem 12.4.2 is not an optimal spanning tree of G.

Appendix

Throughout the text we have mentioned a number of NP-complete decision problems concerning bipartite graphs. In this appendix we have collected these, together with several others.

Regular subgraph

INSTANCE: *A bipartite graph G and a positive integer $k \geq 3$.*

QUESTION: *Does G have a k-regular subgraph?*

This problem is NP-complete (see Plesník (1984)).

Embedding a tree in a hypercube (see p.39)

INSTANCE: *A tree T and an integer n.*

QUESTION: *Is T a subgraph of the hypercube Q_n?*

The NP-completeness of this problem was proved by Wagner and Corneil (1990). If we insist that the tree be an induced subgraph, then the problem can be decided in polynomial time (see Djoković (1973)).

Induced matching (see p.130)

INSTANCE: *A bipartite graph G and a positive integer k.*

QUESTION: *Is there an induced matching in G of cardinality k?*

This problem was proved to be NP-complete by Cameron (1989).

Restricted perfect matching

INSTANCE: *A balanced bipartite graph G, positive integers $n_1, \ldots, n_k$ and subsets $E_1, \ldots, E_k$ of $E(G)$.*

QUESTION: *Does there exist a perfect matching M such that $|M \cap E_i| \leq n_i$ for $i = 1, \ldots, k$?*

This problem is polynomially sovable for $k = 1$ and NP-complete in general (see Itai, Rodeh and Tanimoto (1978))

Matcher

INSTANCE: *A simple bipartite graph G with bipartition (V_1, V_2) where $|V_1| = |V_2| = 2m$.*

QUESTION: *Is there, for each $S \subset V_1$ with $|S| = m$, a set $T \subset V_2$, $|S| = |T|$ such that there is a matching of S into T?*

This problem is co-NP-complete (see Blum et al. (1981)).

Recognising expanders (see p.87)

INSTANCE: *A simple n by n bipartite graph G with maximum degree Δ.*

QUESTION: *Is G an $(n, 1/2, \Delta, 0)$-expander?*

Actually this is a reformulation of the previous problem, and so is co-NP-complete.

Minimum t-spanner

INSTANCE: *A graph G and two positive integers $m \geq 2$ and $t \geq 2$.*

QUESTION: *Does G contain a t-spanner with at most m edges, i.e. a spanning subgraph H with $|E(H)| \leq m$ where $d_H(u, v) \leq t \cdot d_G(u, v)$ for each pair of vertices $u, v \in V(G)$.*

The problem is NP-complete (see Peleg and Schäffer (1987)) and remains NP-complete if G is a bipartite graph and $t \geq 3$ (see Cai (1994)).

Hamilton cycle (see p.117)

INSTANCE: *A bipartite graph G.*

QUESTION: *Does G have a Hamilton cycle?*

Recognising hamiltonian graphs is an NP-complete problem for arbitrary bipartite graphs (see Krishnamoorthy (1975)), but in fact even for planar, cubic bipartite graphs it is an NP-complete problem (see Akiyama, Nishizeki and Saito (1980)).

Spanning eulerian subgraph (see p.116)

INSTANCE: *A connected bipartite graph G.*

QUESTION: *Is there a spanning eulerian subgraph of G?*

This problem is NP-complete even for planar, cubic bipartite graphs because a spanning eulerian subgraph in a such graph G is a Hamilton cycle of G.

Hamilton path

INSTANCE: *A bipartite graph G.*

QUESTION: *Does G have a Hamilton path?*

This problem is NP-complete (see Krishnamoorthy (1975)).

Embedded factors (see p.99)

INSTANCE: *A bipartite graph G, and two integral functions $g(x)$ and $f(x)$ on the vertex set $V(G)$ such that $g(x) \leq f(x) \leq d_G(x)$ for each vertex x.*

QUESTION: *Are there a maximum $(0, g)$-factor F_1 and a maximum $(0, f)$-factor F_2 so that $F_1 \subseteq F_2$?*

This problem is NP-complete (see Asratian (1981)).

V_1-interval t-colouring (see p.136)

INSTANCE: *A bipartite graph G and a positive integer t.*

QUESTION: *Has G an V_1-interval t-colouring?*

The NP-completeness of this problem was originally proved by Strusevich and Yeroshina in terms of scheduling (see the book of Tanajev, Sotskov and Strusevich (1989)).

Interval colourability (see p.136)

INSTANCE: *A bipartite graph G.*

QUESTION: *Has G an interval t-colouring for some t?*

The NP-completeness of this problem was proved by Sevast'janov (1990).

Edge L-colouring (see p.142)

INSTANCE: *A bipartite graph G and a tight V_1-scheme L.*

QUESTION: *Does G have an edge L-colouring?*

This problem is NP-complete even for bipartite graphs with maximum degree 3 (Even, Itai, Shamir (1976)).

V_1-sequential edge colouring (see p.137)

INSTANCE: *A bipartite graph G.*

QUESTION: *Has G a V_1-sequential colouring?*

The NP-completeness of this problem when $\Delta(G) = 3$ was proved by Asratian and Kamalian (1987).

List colouring (see p.138)

INSTANCE: *A bipartite graph G and lists L_e on the edges of G.*

QUESTION: *Has G an L_e-list colouring?*

This problem is NP-complete since an instance of **Edge L-colouring** can be reformulated as an instance of **List colouring** by putting $L_e(a) = L(x)$ for each vertex $x \in V_1$ and each edge a incident with x.

Colour-feasible sequence (see p.144)

INSTANCE: *A bipartite graph G and a sequence $N = (n_1, \ldots, n_t)$ of positive integers.*

QUESTION: *Is N a colour-feasible sequence for G?*

This problem is NP-complete even for bipartite graphs with maximum degree 3 (see Asratian and Kamalian (1987)).

Total colouring

INSTANCE: *A bipartite graph G and a positive integer t.*

QUESTION: *Is there a colouring of the vertices and edges of G with t colours so that no two adjacent vertices, no two incident edges, and no vertex and edge incident with it, are coloured with the same colour?*

This problem is NP-complete even for a k-regular bipartite graph when $t = k + 1$ (McDiarmid, Sánchez-Arroyo (1994)).

Harmonious colouring

INSTANCE: A tree T and an integer k.

QUESTION: Does there exist a harmonious colouring with k colours, i.e. a proper vertex colouring of T such that each pair of colours appears together on at most one edge?

This problem is NP-complete even for trees of radius 3 (see Edwards and McDiarmid (1995)).

Bandwidth

INSTANCE: A graph G and positive integer k.

QUESTION: Is there a one-to-one function $f : V \longrightarrow \{1, 2, \ldots, |V|\}$ such that, for every edge uv, $|f(u) - f(v)| \leq k$.

This problem is NP-complete even for trees with no vertex degree exceeding 3 (see Garey et al. (1978)). The problem corresponds to that of minimising the 'bandwidth' of a symmetric matrix by simultaneous row and column permutation.

Covering of V_2 by V_1 (see p.174)

INSTANCE: A bipartite graph with bipartition (V_1, V_2) and $k \geq 0$.

QUESTION: Is there a set of at most k vertices from V_1 whose neighbourhood is V_2?

This problem is NP-complete even if every vertex of V_1 has degree at most 3 (Karp (1972)).

Vertex dominating set (see p.176)

INSTANCE: A graph G and a positive integer k.

QUESTION: Has G a vertex dominating set with at most k vertices?

This problem is NP-complete even for chordal bipartite graphs (see Müller and Brandstädt (1987)), but can be solved in polynomial time for a convex bipartite graph (Damaschke, Müller, Kratsch (1990)).

Edge dominating set (see p.177)

INSTANCE: A graph G and a positive integer k.

QUESTION: Is there an edge dominating set of at most k edges in G?

This problem is NP-complete for bipartite graphs of maximum degree 3, but can be solved in polynomial time for trees (see Yannakakis and Gavril (1980))

Complete bipartite subgraph

INSTANCE: A graph G and a positive integer k.

QUESTION: Is there a complete bipartite subgraph of G which has k vertices?

This problem is NP-complete for general graphs (Yannakakis (1978)), but is solvable in polynomial time for bipartite graphs.

Balanced complete bipartite subgraph

INSTANCE: *A graph G and a positive integer n.*

QUESTION: *Is the complete bipartite graph $K_{n,n}$ a subgraph of G?*

This problem is NP-complete (see Garey and Johnson (1979))

Bipartite subgraph on k edges (see p.214)

INSTANCE: *A graph G and a positive integer k.*

QUESTION: *Is there a bipartite subgraph of G which has at least k edges?*

This problem is NP-complete even for triangle-free graphs with maximum degree 3 and also if we require the subgraph to be connected (see Yannakakis (1978)). There is a polynomial algorithm for planar graphs (see Hadlock (1975), Orlova and Dorfman (1972)).

Induced bipartite subgraph

INSTANCE: *A graph G and a positive integer k.*

QUESTION: *Is there a subset $V' \subseteq V(G)$ with $|V'| \geq k$ such that $G[V']$ is bipartite?*

This problem is NP-complete (see Yannakakis (1978) and Lewis (1978)).

Covering with complete bipartite subgraphs

INSTANCE: *A graph G, a positive integer k.*

QUESTION: *Do there exist subgraphs $F_1, \ldots, F_k$ of G, where each F_i is an induced complete bipartite subgraph and $\bigcup_i^k F_i = G$?*

The proof of the NP-completeness of this problem is due to Orlin (1976).

Partition into trees (see p.223)

INSTANCE: *A graph G, and a tree T with $|V(G)| = k|V(T)|$, for some $k \in \mathbb{N}$.*

QUESTION: *Is there a T-decomposition of G?*

This problem is NP-complete for any tree on four vertices or more (see Dyer and Frieze (1985)). In fact the analogous problem when the graphs in the decomposition are only required to have the same number of vertices is also NP-complete.

Steiner tree

INSTANCE: *A graph G, a subset of vertices $S \subseteq V(G)$, a positive integer k and weight function $w : E(G) \longrightarrow \mathbb{R}^+$.*

QUESTION: *Is there a tree T in G, such that $S \subseteq V(T)$ and $\sum_{e \in E(T)} w(e) \leq k$?*

This problem is NP-complete even for chordal bipartite graphs with the weight on each edge equal to 1 (Müller, Brandstädt (1987)).

References

AJTAI, M., KOMLÓS, J. AND SZEMERÉDI, E., Sorting in $c \log n$ steps, *Combinatorica* **3** (1983), 1–19. [91]

AKIYAMA, T., NISHIZEKI, T. AND SAITO, N., NP-completeness of the Hamilton cycle problem in bipartite graphs, *J. Inform. Proc.* **3** (1980), 73–76. [112,117,233]

ALAVI, Y., CHARTRAND, G., OELLERMAN, O.R. AND LESNIAK, L., Bipartite regularisation numbers, *Discrete Math.* **62** (1986), 113–118. [105]

ALEXANDROFF, A.D., K teorii smeshannykh ob'ëmov vypuklykh tel (Zur Theorie der gemischten Volumina von konvexen Körpern, *Mat. Sb.* **3 (45)** (1938), 227–251 (in Russian, German summary). [169]

ALON, N., Eigenvalues and expanders, *Combinatorica* **6** (1986), 83–96. [88]

ALON, N., Bipartite subgraphs, *Combinatorica* **16** (1996), 301–311. [217]

ALON, N. AND SPENCER, J.H., *The probabilistic method*, Wiley, New York, 1992. [viii]

ALON, N. AND TARSI, M., Colorings and orientations of graphs, *Combinatorica* **12** (1992), 125–134. [138]

ALT, H., BLUM, N., MEHLHORN, K. AND PAUL, M., Computing a maximum cardinality matching in a bipartite graph in time $O\left(n^{1.5}\sqrt{m/\log n}\right)$, *Inform. Proc. Let.* **37** (1991), 237–240. [60,62]

ALVAREZ, L.R., Undirected graphs realisable as graphs of modular lattices, *Canad. J. Math.* **17** (1965), 923–932. [38]

AMAR, D., Partition of a bipartite Hamiltonian graph into two cycles, *Discrete Math.* **58** (1986), 1–10. [112]

AMAR, D., FOURNIER, I. AND GERMA, A., Covering the vertices of a graph by cycles of prescribed lengths, *J. Graph Th.* **13** (1989), 323–330. [110]

ANDERSON, I., Perfect matchings of a graph, *J. Combin. Th. Ser. B* **10** (1971), 183–186. [80]

ANDERSEN, L.D. AND HILTON, A.J.W., Thank Evans!, *Proc. London Math. Soc.* **47** (1983), 507–522. [206]

ANDRÁSFAI, B., ERDŐS, P. AND SÓS, V.T., On the connection between chromatic number, maximal clique and minimal degree of a graph, *Discrete Math.* **8** (1974), 205–218. [9]

ASRATIAN, A.S., Compatible systems of distinct representatives, *Diskret. Analiz.* **27** (1975), 3–12 (in Russian). [194]

ASRATIAN, A.S., On the set of sequences which are color-feasible for one class of graphs, *Molod. Nauchn. Rabotnik* **2(28)** (1978), 65–75, (in Russian). [150]

ASRATIAN, A.S., *Investigation of some mathematical model of scheduling theory*, Doctoral dissertation, Moscow University (in Russian), 1980. [136,147]

ASRATIAN, A.S., Some NP-complete problems on embedded *c*-matchings of a bipartite multigraph, *Akad. Nauk Armyan. SSR Dokl.* **73** (1981), 259–264 (in Russian). [233]

ASRATIAN, A.S., Short solution of Kotzig's problem for bipartite graphs, *preprint* (1996). [151]

ASRATIAN, A.S. AND KAMALIAN, R.R., Interval edge colouring of multigraphs, *Appl. Math., Yerevan Univ.* **5** (1987), 25–34 (in Russian). [130,133,135,137,234]

ASRATIAN, A.S. AND KAMALIAN, R.R., Investigation on interval edge colourings of graphs, *J. Combin. Th. Ser. B* **61** (1994), 34–43. [132,133,135]

ASRATIAN, A.S. AND KHACHATRIAN, N.K., Stability of monotone graph properties, *manuscript* (in Russian) (1988). [109,112]

ASRATIAN, A.S. AND KHACHATRIAN, N.K., Some localization theorems on hamiltonian circuits, *J. Combin. Th. Ser. B* **49** (1990), 287–294. [111]

ASRATIAN, A.S. AND KHACHATRIAN, N.K., On stable properties of graphs, *Discrete Math.* **90** (1991), 143–152. [111]

ASRATIAN, A.S., KOSTOCHKA, A.V. AND MIRUMIAN, A.N., A criterion for the unique edge colorability of bipartite multigraphs, *Metody Discret. Analiz.* **45** (1987), 3–20 (in Russian). [158,159,168]

ASRATIAN, A.S. AND MIRUMIAN, A.N., Transformations of edge colourings of a bipartite multigraph and their applications, *Soviet Math. Dokl.* **43** (1991), 1–3. [151,156]

ASRATIAN, A.S. AND MIRUMIAN, A.N., Counterexamples to the Kotzig problem, *Diskret. Mat.* **4** (1992), 96–98 (in Russian). [156]

ASRATIAN, A.S. AND MIRUMIAN, A.N., On transformations of edge colourings of the complete bipartite graph $K_{n,n}$, *Akad. Nauk Respub. Armeniya Dokl.* **95** (1995), 10–12 (in Russian). [156]

ASRATIAN, A.S. AND SARKISIAN, G.V., Cyclic properties of some hamiltonian graphs, *Diskret. Mat.* **3** (1991), 91–104 (in Russian). [14]

BARDADYM, B.A., Minimax paths and minimum spanning trees, *Kibernetica* **No. 2** (1990), 122 (in Russian). [230]

BARNETTE, D., Problem 5, in *Recent Progress in Combinatorics*, Ed. W.T. Tutte, Academic Press, New York, 1969, p. 343. [112]

BASSALYGO, L.A., Asymptotically optimal switching circuits, *Prob. Infor. Trans.* **17** (1981), 206–211. [54]

BERGE, C., Two Theorems in Graph Theory, *Proc. Nat. Ac. Sciences, USA* **43** (1957), 842. [57]

BERGE, C., *Graphes et Hypergraphes*, Dunod, Paris, 1970. [102,120]

BERGE, C., Regularizable graphs, in *Advances in Graph Theory*, Ed. B. Bollobás, Ann. Disc. Math. 3, North-Holland, Amsterdam, 1978, pp. 11–19. [81]

BIGGS, N., *Algebraic Graph Theory*, Tracts in Mathematics, Vol. 67, C.U.P., Cambridge, 1974. [24]

BIRKHOFF, G., Tres observaciones sobre el algebra lineal, *Rev. Univ. Nac. Tucumán Sr. A.* **5** (1946), 147–151. [164]

BLUM, M., KARP, R.M., PAPADIMITROU, C.H., VORNBERGER, O. AND YANNAKAKIS, M., The complexity of testing whether a graph is a superconcentrator, *Inform. Proc. Let.* **13** (1981), 164–167. [84,87,232]

BÖHME, T., BROERSMA, H.J., GÖBEL ,F., KOSTOCHKA, A.V. AND STIEBITZ, M., A note on spanning trees with pairwise nonadjacent end-vertices, *Discrete Math.* **171** (1997), 219–222. [217]

BOLLOBÁS, B., *Random Graphs*, Academic Press, London, 1985. [viii]

BOLLOBÁS, B., NARA, CH. AND TACHIBANA, S., The maximal number of induced complete bipartite graphs, *Discrete Math.* **62** (1986), 271–275. [226]

BONDY, J.A. AND CHVÁTAL, V., A method in graph theory, *Discrete Math.* **15** (1976), 111–135. [108,109]

BONDY, J.A. AND MURTY, U.S.R., *Graph Theory with Applications*, North-Holland, Amsterdam, 1976. [3,71]

BOOTH, K.S. AND LUEKER, G.S., Testing for the consecutive 1's property, interval graphs and graph planarity using *PQ*-tree algorithms, *J. Comput. System Sci.* **13** (1976), 335–379. [66]

BORODIN, O.V., KOSTOCHKA, A.V. AND WOODALL, D.R., List edge and list total colourings of multgraphs, *preprint* (1996). [140,216]

BOSÁK, J., Hamiltonian lines in cubic graphs, in *Theory of Graphs*, Ed. P. Rosenstiehl, Gordon and Breach, New York, 1967, pp. 35–46. [111]

BRANDSTÄDT, A., SPINRAG, J. AND STEWART, L., Bipartite permutation graphs, *Discrete Appl. Math.* **3** (1987), 279–293. [15]

BRÈGMAN, L.M., Some properties of nonnegative matrices and their permanents, *Soviet Math. Dokl.* **14** (1973), 945–949. [173]

BROERSMA, H.J., FAUDREE, R.J., VAN DEN HEUVEL, J. AND VELDMAN, H.J., Decomposition of bipartite graphs under degree constraints, *Networks* **23** (1993), 159–164. [194]

BRUALDI, R.A. AND CSIMA, J., Extending subpermutation matrices in regular classes of matrices, *Discrete Math.* **62** (1986), 99–101. [82]

BRUALDI, R.A. AND GIBSON, P.M., Convex polyhedra of doubly stochastic matrices I — Applications of the permanent function, *J. Combin. Th. Ser. A* **22** (1977), 194–230. [168]

BUNEMAN P., A note on the metric properties of trees, *J. Combin. Th. Ser. B* **17** (1974), 48–50. [31]

BUSACKER, R.G. AND SAATY, T.L., *Finite Graphs and Networks: an Introduction with Applications*, McGraw-Hill, New York, 1965. [121]

CAI, L., NP-completeness of minimum spanner problems, *Discrete Appl. Math.* **48** (1994), 187–194. [233]

CAMERON, K., Induced matchings, *Discrete Appl. Math.* **24** (1989), 97–102. [232]

CARO, Y. AND SCHÖNHEIM, J., Decomposition of trees into isomorphic subtrees, *Ars Combin.* **9** (1980), 119–130. [223]

CAYLEY, A., A theorem on trees, *Quart. J. Math.* **23** (1889), 376–378. [123]

CAYLEY, A., On the theory of the analytical forms called trees, *Philos. Mag.* **13** (1857), 19–30. *Mathematical Papers, Cambridge* **3** (1891), 242–246. [vii]

CHEN, B., MATSUMOTO, M., WANG, J.F., ZHANG, Z.F. AND ZHANG, J.X., A short proof of Nash-Williams' theorem for the arboricity of a graph, *Graphs and Combinatorics* **10** (1994), 27–28. [221]

CHERIYAN, J., HAGERUP, T. AND MEHLHORN, K., Can a maximum flow be computed in $O(mn)$ time?, in *Automata, languages and programming (Coventry, 1990)*, Lecture Notes in Comput. Sci., Springer-Verlag, New York, 1990, pp. 235–248. [64]

CHETWYND, A. AND HÄGGKVIST, R., Some upper bounds on the total and list chromatic numbers of multigraphs, *J. Graph Th.* **16** (1992), 503–516. [138]

CHETWYND, A. AND HÄGGKVIST, R., *An improvement of Hind's upper bound on the total chromatic number*, Research Report, University of Umeå, No. 9, 1993. [218]

CHRISTOFIDES, N., *Worst-case analysis of a new heuristic for the Travelling Salesman Problem*, Technical Report of the Graduate School of Industrial Administration, Carnegie-Mellon Univ., Pittsburgh, Pa, 1976. [228]

CHUNG, F.R.K., Pebbling in hypercubes, *SIAM J. Disc. Math.* **2** (1989), 467-472. [202]

CHVÁTAL, V., On Hamilton's ideals, *J. Combin. Th. Ser. B* **12** (1972), 163-168. [110]

COLBOURN, C.J., *Graph generation*, Research Report CS-77-37, Dept. Computer Sci., University of Waterloo, Ontario, 1977. [120]

COULSON, C.A. AND ROUSHBROOKE, G.S., A note on the method of molecular orbitals, *Proc. Camb. Philos. Soc.* **36** (1940), 193-200. [17]

CVETKOVIĆ, D.M., DOOB, M. AND SACHS, H., *Spectra of Graphs, Theory and Applications*, Academic Press, New York, 1979. [19]

DAMASCHKE, P., MÜLLER, H. AND KRATSCH, D., Domination in convex bipartite graphs, *Inform. Proc. Let.* **36** (1990), 231-236. [177,235]

DEAN, A.M., HUTCHINSON, J.P. AND SCHEINERMAN, E.R., On the thickness and arboricity of a graph, *J. Combin. Th. Ser. B* **52** (1991), 146-151. [222]

DÉNES, J. AND KEEDWELL, A.D., *Latin Squares and their Applications*, English Universities Press, London, 1974. [205]

DENLEY, T.M.J., On a result of Lemke and Kleitman, *Combin. Probab. Comput.* **6** (1997a), 39-43. [204]

DENLEY, T.M.J., The odd girth of the generalised Kneser graph, *to appear in Europ. J. Comb* (1997b). [217]

DENLEY, T.M.J. AND HÄGGKVIST, R., Completing some partial latin squares, *preprint* (1995). [211]

DILWORTH, R.P., A decomposition theorem for partially ordered sets, *Ann. Math.* **51** (1950), 161-166. [190]

DJOKOVIĆ, D.Z., Distance preserving subgraphs of hypercubes, *J. Combin. Th. Ser. B* **14** (1973), 263-267. [39,40,232]

DOOB, M., Characterisations of regular magic graphs, *J. Combin. Th. Ser. B* **25** (1978), 94-104. [14,52]

DUGUID, A.M., *Structural Properties of Switching Networks*, Brown Univ. Progr. Rept. BTL-7, 1959. [161]

DULMAGE, A.L. AND MENDELSOHN, N.S., Coverings of bipartite graphs, *Canad. J. Math.* **10** (1958a), 517-534. [183]

DULMAGE, A.L. AND MENDELSOHN, N.S., Some generalizations of the problem of distinct representatives, *Canad. J. Math.* **10** (1958b), 230-241. [57]

DULMAGE, A.L. AND MENDELSOHN, N.S., A structure theory of bipartite graphs of finite exterior dimension, *Trans. Roy. Soc. Canada Ser. III* **53** (1959), 1–13. [184]

DYER, M.E. AND FRIEZE, A.M., On the complexity of partitioning graphs into connected subgraphs, *Disc. Appl. Math.* **10** (1985), 139–153. [224,236]

EDMONDS, J., Minimum partition of a matroid into independent subsets, *J. Res. Nat. Bur. Standards Sect.* **69B** (1965), 67–72. [223]

EDMONDS, J. AND JOHNSON, E.L., Matching, Euler tours, and the Chinese postman, *Math. Prog.* **5** (1973), 88–124. [116]

EDWARDS, C.S., An improved lower bound for the number of edges in a largest bipartite subgraph, in *Recent advances in graph theory (Proc. Second Czechoslovak Sympos., Prague 1974)*, Academia, Prague, 1974, pp. 167–181. [214]

EDWARDS, K. AND MCDIARMID C., The complexity of harmonious colouring for trees, *Discrete Appl. Math.* **57** (1995), 133–144. [235]

EGAWA, Y., URABE, M., FUKUDA, T. AND NAGOYA, S., A decomposition of complete bipartite graphs into edge-disjoint subgraphs with star components, *Discrete Math.* **58** (1986), 93–95. [179]

EGORYCHEV, G.P., Solution of van der Waerden's permanent conjecture, *Advances in Math.* **42** (1980), 299–305. [169,172]

ENOMOTO, H., MIYAMOTO, T. AND USHIO, K., C_k-factorizations of complete bipartite graphs, *Graphs and Comb.* **4** (1988), 111–113. [179]

ERDŐS, P., On some extremal problems in graph theory, *Israel J. Math.* **3** (1965), 113–116. [215]

ERDŐS, P., FAJTLOWITS, S. AND STATON, W., Degree sequences in triangle-free graphs, *Discrete Math.* **92** (1991), 85–88. [9]

ERDŐS, P., PACH, J., POLLACK, R. AND TUZA, Z., Radius, diameter and minimum degree, *J. Combin. Th. Ser. B* **47** (1989), 73–79. [25]

ERDŐS, P., RUBIN, A. AND TAYLOR, H., Choosability in graphs, *Congress. Numer.* **XXVI** (1979), 125–157. [138]

EULER, L., Solutio problematis ad geometriam situs pertinentis, *Comment. Acad. Petropolitanae* **8** (1736), 128–140. [117]

EVANS, T., Embedding incomplete latin squares, *Amer. Math. Monthly* **67** (1960), 958–961. [206]

EVEN, S., ITAI, A. AND SHAMIR, A., On the complexity of timetable and multicommodity flow problems, *SIAM J. Comput.* **5** (1976), 691–703. [141,234]

EVDOKIMOV, A.A., On the maximum length of a path in the unit n-dimensional cube, *Mat. Zametki* **6** (1969), 309–319 (in Russian). [39]

FALIKMAN, D.I., A proof of the van der Waerden conjecture on the permanent of a doubly stochastic matrix, *Mat. Zametki* **29** (1981), 931–938 (in Russian). [169,172]

FAUDREE, R.J., GYÁRFÁS, A., LESNIAK, L. AND SCHEINERMAN, E.R., Rainbow coloring the cube, *J. Graph Th.* **17** (1993), 607–612. [130]

FAVARON, O., MAGO, P. AND ORDAZ, O., On the bipartite independence number of a balanced bipartite graph, *Discrete Math.* **121** (1993), 55–63. [113]

FEIGENBAUM, J., HERSHBERGER, J. AND SCHÄFFER, A.A., A polynomial time algorithm for finding the prime factors of Cartesian product graphs, *Disc. Appl. Math.* **12** (1985), 123–138. [13]

FENCHEL, W., Inégalités quadratiques entre les volumes mixtes des corps convexes, *C.R. Acad. Sci. Paris* **203** (1936), 647–650. [169]

FIRSOV, V.V., On isometric embeddings of a graph in a Boolean cube, *Kibernetika* **1** (1965), 95–96 (in Russian). [44]

FOLKMAN, J. AND FULKERSON, D.R., Edge colourings in bipartite graphs, in *Combinatorial Mathematics and its Applications*, Eds. R.C. Bose and T.A. Dowling, Univ. N. Carolina Press, Chapel Hill, N.C., 1969, pp. 561–577. [144,145,149,194]

FOLDES, S., A characterisation of hypercubes, *Discrete Math.* **17** (1977), 155–159. [35,37]

FORD, L.R. AND FULKERSON, D.R., Network flow and systems of representatives, *Canad. J. Math.* **10** (1958), 78–85. [196]

FORD, L.R. AND FULKERSON, D.R., *Flows in Networks*, Princeton Univ. Press, Princeton, N.J., 1962. [viii]

FROBENIUS, G., Über Matrizen aus nicht negativen Elementen, *Sitzber. König. Preuss. Akad. Wiss.* **26** (1912), 456–477. [vii]

FROBENIUS, G., Über zerlegbare Determinanten, *Sitzber. König. Preuss. Akad. Wiss.* **XVIII** (1917), 274–277. [76]

FULKERSON, D.R., Note on Dilworth's Decomposition Theorem for partially ordered sets, *Proc. Amer. Math. Soc.* **7** (1956), 701–702. [190]

GABOW, H. AND KARIV, O., Algorithms for edge colouring bipartite graphs and multigraphs, *SIAM J. Comput.* **11** (1982), 117–129. [126]

GALE, D., A theorem on flows in networks, *Pacific J. Math.* **7** (1957), 1073–1082. [104]

GALE, D. AND SHAPLEY, L.S., College admissions and the stability of marriage, *Amer. Math. Monthly* **69** (1962), 9–15. [67]

GALLAI, T., Über extreme Punkt und Kantenmengen, *Ann. Univ. Sci. Budapest. Eötvös Sect. Math.* **2** (1959), 133–138. [181]

GALVIN, F., The list-chromatic index of a bipartite multigraph, *J. Combin. Th. Ser. B* **63** (1995), 153–158. [139]

GAREY, M.R., GRAHAM, R.L., JOHNSON, D.S. AND KNUTH, D.E., Complexity results for bandwidth minimization, *SIAM J. Appl. Math.* **34** (1978), 477–495. [235]

GAREY, M.R. AND JOHNSON, D.S., *Computers and Intractibility: A Guide to the theory of NP-completeness*, Freeman, New York, 1979. [6,236]

GELLER, D.P. AND HILTON, A.J.W., How to colour the lines of bigraphs, *Networks* **4** (1974), 281–282. [144]

GLOVER, F., Maximum matchings in convex bipartite graphs, *Naval Res. Logist. Quart.* **14** (1967), 313–316. [66]

GOLOMB, S.W., How to number a graph, in *Graph Theory and Computing*, Ed. R.C. Read, Academic Press, 1972, pp. 23–37. [14]

GOLUMBIC, M. AND GOSS, C., Perfect elimination and chordal bipartite graphs, *J. Graph Th.* **2** (1978), 155–163. [12,21,22]

GRAHAM, N. AND HARARY, F., Changing and unchanging the diameter of a hypercube, *Discrete Appl. Math.* **37–38** (1992), 265–274. [39]

GRAHAM, R.L. AND POLLACK, R., *On embeddings of graphs in squashed cubes*, Graph Theory and Applications, Lecture Notes no. 303, Springer-Verlag, Berlin, 1972. [41,44,225]

GREGORY, D.A, JONES, N.J., LUNDGREN, J.R. AND RULLMAN, N.J., Biclique coverings of regular bipartite graphs and minimum semiring ranks of regular matrices, *J. Combin. Th. Ser. B* **51** (1991), 73–89. [226]

GROSS, O., The Bottleneck Assignment Problem, *The RAND Corporation Paper P-1630*, 1959. [74]

GUPTA, R.P., A decomposition theorem for bipartite graphs (results), in *Theory of Graphs (International Symposium, Rome 1966)*, Ed. P. Rosenstiehl, Gordon and Breach, New York, 1967, pp. 135–136. [178]

HADLOCK, F.O., Finding a maximum cut of a planar graph in polynomial time, *SIAM J. Comput.* **4** (1975), 221–225. [236]

HÄGGKVIST, R., *in Unsolved problems, Proceedings of the Fifth Hungarian Colloquium on Combinatorics*, 1976. [111]

HÄGGKVIST, R., On F-hamiltonian graphs, in *Graph theory and related topics*, Eds. J.A. Bondy and U.S.R. Murty, Academic Press, London, 1977, pp. 219–231. [113]

HÄGGKVIST, R., A solution to the Evans conjecture for Latin squares of large size, *Colloqu. Math. Soc. Janos Bolyai* **18** (1978), 495–513. [59,105,206]

HÄGGKVIST, R., A remark on tree-decompositions, *manuscript* (1983a). [224]

HÄGGKVIST, R., Restricted edge colourings of bipartite graphs, *manuscript* (1983b). [142,143]

HÄGGKVIST, R., A family of completable Latin squares, *manuscript* (1983c). [209]

HÄGGKVIST, R., Embedding partial latin squares, *manuscript* (1987). [210]

HÄGGKVIST, R., Decompositions of complete bipartite graphs, *London Math. Soc. Lect. Notes Ser. C.U.P., Cambridge* **141** (1989), 115–147. [142,226]

HAKIMI, S.L., On the realizability of a set of integers as degrees of the vertices of a linear graph I, *J. Soc. Indust. Appl. Math.* **10** (1962), 496–506. [104]

HALL, M., Distinct representatives of subsets, *Bull. Amer. Math. Soc.* **54** (1948), 922–926. [81]

HALL, M., *Combinatorial Theory*, Wiley, New York, 1967. [89]

HALL, P., On representatives of subsets, *J. London Math. Soc.* **10** (1935), 26–30. [75,192,197]

HALMOS, P.R. AND VAUGHN, H.E., The marriage problem, *Amer. J. Math.* **72** (1950), 214–215. [75]

HAMACHER, H.W. AND QUEYRANNE, M., K-best solutions to combinatorial optimization problems, *Ann. Oper. Res.* **4** (1985), 123–143. [73]

HAMIDOUNE, Y.O. AND LAS VERGNAS, M., Local edge-connectivity in regular bipartite graphs, *J. Combin. Th. Ser. B* **44** (1988), 370–371. [49]

HAMMING, R.W., Error detecting and error correcting codes, *Bell Systems Tech. J.* **29** (1950), 147–160. [178]

HANSON, D. AND LOTEN, C., A lower bound for interval colouring bi-regular bipartite graphs, *Bull. Inst. Combin. Appl.* **18** (1996), 69–74. [137]

HARARY, F., HSU, D. AND MILLER, Z., The biparticity of a graph, *J. Graph Th.* **1** (1977), 131–133. [218]

HAVEL, V., Eine Bemerkung über die Existenz der endlichen Graphen, *Casopis Pest. Mat.* **80** (1955), 477–480. [104]

HEINRICH, K., HELL, P., KIRKPATRICK, D.G.AND LIU, G., A simple existence criterion for $(g < f)$-factors, *Discrete Math.* **85** (1990), 313–317. [97]

HELL, P. AND KIRKPATRICK, D.G., Algorithms for degree constrained graph factors of minimum deficiency, *J. Algorithms* **14** (1993), 115–138. [99]

HETYEI, G., Rectangular configurations which can be covered by 2×1 rectangles, *Pécsi Tan. Főisk. Közl.* **8** (1964), 351–367 (in Hungarian). [77,78]

HILTON, A.J.W. AND JOHNSON, P., A variation of Ryser's theorem and a necessary condition for the list-colouring problem, in *Graph colourings, Pitman Research Notes in Mathematics Series 218*, Eds. R. Nelson and R.J. Wilson, Longman Scientific and Technical, 1990, pp. 135–143. [205]

HOFFMAN, A.J., On the polynomial of a graph, *Amer. Math. Monthly* **70** (1963), 30–36. [18]

HOFFMAN, A.J. AND KUHN, H.W., Systems of distinct representatives and linear programming, *Amer. Math. Monthly* **63** (1956), 455–460. [193]

HOLTON, D.D., MANVEL, B. AND MCKAY, B.D., Hamiltonian cycles in cubic 3-connected bipartite planar graphs, *J. Combin. Th. Ser. B* **38** (1985), 279–297. [112]

HOLYER, I., The NP-completeness of edge-colouring, *SIAM J. Comput.* **10** (1981), 718–720. [125]

HOPCROFT, J.E. AND KARP, R.M., An $n^{2.5}$ algorithm for maximum matching in bipartite graphs, *SIAM J. Comput.* **2** (1973), 225–231. [viii,60,61]

HOPKINS, G. AND STATON, W., Extremal bipartite subgraphs of cubic triangle-free graphs, *J. Graph Th.* **6** (1982), 115–121. [217]

HOPKINS, G. AND STATON, W., Graphs with unique maximum independent sets, *Discrete Math.* **57** (1985), 245–251. [182,183]

HORÁK, P., Transversals and matroids, in *Topics in Combinatorics and Graph Theory*, Eds. R. Bodendiek and R. Henns, Physica-Verlag, Heidelberg, 1990, pp. 381–389. [197]

HORÁK, P., Common k-transversals of finite families, *Acta. Math. Univ. Comenianae* **LX** (1991), 253–256. [197]

HORTON, J.D., On two-factors of bipartite graphs, *Discrete Math.* **41** (1982), 35–41. [111,118]

ITAI, A., PAPADIMITROU, C.H. AND SZWAREFITER, J.L., Hamilton paths in grid graphs, *SIAM J. Comput.* **11** (1982), 676–686. [113]

ITAI, A., RODEH, M. AND TANIMOTO, S.L., Some matching problems for bipartite graphs, *J. Assoc. Comput. Mach.* **25** (1978), 517–525. [232]

JACKSON, B. AND LI, H., Hamilton cycles in 2-connected, regular bipartite graphs, *J. Combin. Th. Ser. B* **62** (1994), 236-258. [111]

JACOBS, K., Der Heiratsaatz, in *Selecta Mathematica 1*, Springer-Verlag, Heidelberg, 1969, pp. 103-141. [81]

JANSSEN, J.C.M., The Dinitz problem solved for rectangles, *Bull. Amer. Math. Soc.* **29** (1993), 243-249. [138]

JAVORSKIĬ, E.B., Representations of directed graphs and ϕ-transformations, in *Theoretical and Applied Questions of Differential Equations and Algebra*, Ed. A.N. Sharkovskii, "Naukova Dumka", Kiev (in Russian), 1978, pp. 247-250,272. [223]

JENSEN, T.R. AND TOFT, B., *Graph coloring problems*, Wiley, New York, 1995. [138]

JONES, K., LUNDGREN, J.R., PULLMAN, N.J. AND REES, R., A note on the covering number of $K_n - K_m$ and complete t-partite graphs, *Congress. Numer.* **66** (1988), 181-184. [226]

JORDAN, C., Sur les assemblages de lignes, *Journal für reine und angewändte Mathematik* **70** (1869), 185-190. [vii,30]

KAMALIAN, R.R., Interval colourings of complete bipartite graphs and trees, *Preprint of the Comp. Ctr. of the Acad. of Sci. of the Rep. Armenia, Yerevan* (in Russian), 1989. [137]

KANO, M. AND SAITO, A., $[a, b]$-factors of graphs, *Discrete Math.* **47** (1983), 113-116. [99]

KARP, R.M., Reducibility among combinatorial problems, in *Complexity of Computer Computatións*, Eds. R.E. Miller and J.W. Thatcher, Plenum Press, New York, 1972, pp. 85-103. [6,180,235]

KATERINIS, P., Two sufficient conditions for a 2-factor in a bipartite graph, *J. Graph Th.* **11** (1987), 1-6. [107]

KIM, J.H., The Ramsey number $R(3, t)$ has order of magnitude $t^2 / \log t$, *Random Structures Algorithms* **7** (1995), 173-207. [220]

KIRCHHOFF, G., Über die Auflösung der Gleichungen, auf welche man bei der Untersuchung der linearen Verteilung galvanischer Ströme geführt wird, *Ann. Phys. Chem.* **72** (1847), 497-508. [vii]

KNUTH, D.E., *The Art of Computer Programming, vol. 3, Sorting and Searching*, Addison-Wesley, Reading, Ma., 1973. [90]

KNUTH, D.E., *Mariages stables et leurs relations avec d'autres problèmes combinatoires*, Les Presses de l'Université de Montréal, Montréal, Quebec, Canada, 1976. [70]

KOCHOL, M., Compatible systems of representatives, *Discrete Math.* **132** (1994), 115–126. [197]

KÖNIG, D., Sur un problème de la théorie générale des ensembles et la théorie des graphes (Communication made April 7, 1914 in Paris to Congrés de Philosophie Mathématique), *Revue de Métaphysique et de Morale* **30** (1923), 443–449. [vii,76]

KÖNIG, D., Vonalrendszerek és determinások, *Math. Termész. Ért.* **33** (1915), 221–229 (in Hungarian). [vii]

KÖNIG, D., Über Graphen und ihre Anwendungen, *Math. Annalen* **77** (1916), 453–465. [vii,8,76,125,163]

KÖNIG, D., Graphok és matrixok, *Mathematikai és Fizikai Lapok* **38** (1931), 116–119 (in Hungarian). [180]

KÖNIG, D., Über trennende Knotenpunkte in Graphen (nebst Anwendungen auf Determinanten und Matrizen), *Acta Litterarum ac Scientiarum (Sectio Scientiarum Mathematicarum), Szeged* **6** (1933), 155–179. [180]

KÖNIG, D., *Theorie der endlichen und unendlichen Graphen*, Leipzig, 1936; english translation, Birkhäuser, Boston 1990. [vii,195]

KRISHNAMOORTHY, M.S., An NP-hard problem in bipartite graphs, *SIGACT News* **7** (1975), 1–26. [106,233]

KRUSKAL, J.B.JR, On the shortest spanning subtree of a graph and the travelling salesman problem, *Proc. Amer. Math. Soc.* **7** (1956), 48. [227,229]

KUHN, H.W., The Hungarian method for the assignment problem, *Naval Res. Logist. Quart.* **2** (1955), 83–97. [72]

KUZJURIN, N.N., Asymptotic investigation on the problem of covering, *Problemy Kybernetiky* **37** (1980), 19–56 (in Russian). [176]

KWAN, M-K, Graphic programming using odd or even points, *Chinese Math.* **1** (1960), 273-277. [115]

LAS VERGNAS, M., Transversales disjointes d'une famille s'ensembles, *preprint* (1970). [82]

LAWLER, E.L., *Combinatorial Optimization: Networks and Matroids*, Holt, Rinehart and Winston, New York, 1976. [viii]

LEBENSOLD, K., Disjoint matchings of graphs, *J. Combin. Th. Ser. B* **22** (1977), 207–210. [194]

LEMKE, P. AND KLEITMAN, D., An addition theorem on the integers modulo *n*, *J. Number Th.* **31** (1989), 335–345. [202]

LEWIS, J.M., On the complexity of the maximum subgraph problem, in *Proc. 10th Ann. ACC Symp. Theory Comput.*, A.C.M., New York, 1978, pp. 265–278. [236]

LIH, K-W. AND WU, P-L., On equitable coloring of bipartite graphs, *Discrete Math.* **151** (1996), 155–160. [184]

LINEK, V., Bipartite graphs can have any number of independent sets, *Discrete Math.* **76** (1989), 131–136. [183]

LITTLE, C.H., The parity of the number of 1-factors of a graph, *Discrete Math.* **2** (1972), 179–181. [173]

LITTLE, C.H., GRANT, D.A. AND HOLTON, D.D., On defect d-matching in graphs, *Discrete Math.* **13** (1975), 41–54. [77]

LONDON, D., Some notes on the van der Waerden conjecture, *Linear Algebra and Appl.* **4** (1971), 55–160. [171]

LONGE–HIGGINS, H.C., Resonance structures and MO in unsaturated hydrocarbons, *J. Chem. Phys.* **18** (1950), 265–274. [58]

LOVÁSZ, L., A generalization of Kőnig's theorem, *Acta Math. Acad. Sci. Hung.* **21** (1970a), 443–446. [82]

LOVÁSZ, L., Subgraphs with prescribed valences, *J. Comb. Th* **8** (1970b), 391–416. [97]

LOVÁSZ, L., Problem 11, in *Combinatorial Structures and their Applications*, Eds. R. Guy, H. Hanani, N. Sauer and J. Schönheim, Gordon and Breach, New York, 1970c, p. 497. [201]

LOVÁSZ, L., *Combinatorial problems and exercises*, North-Holland, Amsterdam, 1979. [15,19]

LOVÁSZ, L. AND PLUMMER, M.D., *Matching Theory*, North-Holland Mathematics studies 121, Ann. Disc. Math. 29, North–Holland, Amsterdam, 1986. [84]

MANTEL, W., Problem 28, *Wiskundige Opgaven* **10** (1907), 60–61. [11]

MARCUS, M. AND NEWMAN, M., On the minimum of the permanent of a doubly stochastic matrix, *Duke Math. J.* **26** (1959), 61–72. [171]

MARGULIS, G.A., Explicit constructions of concentrators, *Prob. Per. Infor.* **9** (1973), 71–80. [86]

MCCUAIG, W., A simple proof of Menger's theorem, *J. Graph Th.* **8** (1984), 427–429. [46]

MCDIARMID, C.J.H. AND SÁNCHEZ-ARROYO, A., Total colouring regular bipartite graphs is NP-hard, *Discrete Math.* **124** (1994), 155–162. [234]

MCKEE, T.A., Recharacterizing Eulerian: intimations of new duality, *Discrete Math.* **51** (1984), 237–242. [50]

MENGER, K., Zur allgemeinen Kurventheorie, *Fund. Math.* **10** (1927), 96–115. [45]

MILLER, G.A., On a method due to Gallois, *Quart. J. Pure Appl. Math.* **41** (1910), 382–384. [195]

MIRSKY, L., *Transversal Theory*, Mathematics in Science and Engineering 75, Academic Press, New York and London, 1971. [192,197]

MOON, J.W. AND MOSER, L., On hamiltonian bipartite graphs, *Israel J. Math.* **1** (1963), 163–165. [110]

MULDER, M., $(0, \lambda)$-graphs and n-cubes, *Discrete Math.* **28** (1979), 179–188. [35]

MULDER, M., *The interval function of a graph*, Mathematical Centre Tracts 132, Amsterdam, 1980. [36,37,38]

MÜLLER, H. AND BRANDSTÄDT, A., The NP-completeness of STEINER TREE and DOMINATING SET for chordal bipartite graphs, *Theoret. Comp. Sci.* **53** (1987), 257–265. [177,235,236]

MUNKRES, J., Algorithms for the assignment and transportation problems, *J. Soc. Indust. Appl. Math.* **5** (1957), 32–38. [72]

MURRAY, H.J.R., *A History of Chess*, Oxford University Press, London, 1913. [106]

MURTY, U.S.R., An algorithm for ranking all the assignments in order of increasing cost, *Oper. Res.* **16** (1968), 682–687. [73]

NASH–WILLIAMS, C.ST.J.A., Edge-disjoint spanning trees of finite graphs, *J. London Math. Soc.* **36** (1961), 445–450. [221]

VON NEUMANN, J., A certain zero-sum two-person game equivalent to the maximal assignment problem, in *Contributions to the Theory of Games II*, Ed. H.W. Kuhn, Ann. of Math. Stud. 28, Princeton Univ Press, Princeton, N.J., 1953. [164]

NORMAN, R.Z. AND RABIN, M.O., An algorithm for a minimum cover of a graph, *Proc. Amer. Math. Soc.* **10** (1959), 315–319. [178]

NUFFELEN, C. VAN, On the rank of the incidence matrix of a graph, *Cah. Ctr. Étud. Rech. Opér. (Bruxelles)* **15** (1973), 363–365. [19]

ORE, O., Graphs and matchings theorems, *Duke Math. J.* **22** (1955), 625–639. [79]

ORE, O., Studies in directed graphs I, *Ann. Math.* **63** (1956), 383–406. [101]

ORLIN, J., Contentment in graph theory: covering graphs with cliques, *manuscript* (1976). [236]

ORLOVA, G.I. AND DORFMAN, Y.G., Finding the maximum cut in a graph, *Engrg. Cybernetics* **10** (1972), 502–506. [236]

OXLEY, J.G., *Matroid Theory*, Oxford University Press, Oxford, 1992. [197]

PATERSON, M.S., Improved sorting networks with $O(\log n)$ depth, *Algorithmica* **5** (1990), 75–92. [91,96]

PAULRAJA, P., A characterization of hamiltonian prisms, *J. Graph Th.* **17** (1993), 161–171. [108,216]

PELEG, D. AND SCHÄFFER, A.A., Graph spanners, *J. Graph Th.* **13** (1987), 99–116. [233]

PELEG, D. AND ULLMAN, J.D., An optimal synchronizer for the hypercube, *SIAM J. Comput.* **18** (1989), 740–747. [39]

PERFECT, H., Applications of Menger's graph theorem, *J. Math. Analysis Appl.* **(22)** (1968), 96-111. [196]

PETERSEN, J., Die Theorie der regulären Graphen, *Acta Math.* **15** (1891), 193–220. [113]

PINSKER, M.S., On the complexity of a concentrator, in *Proc. 7th International Teletraffic Conference, Stockholm*, 1973. [52]

PIPPENGER, N., Superconcentrators, *SIAM J. Comput.* **6** (1977), 298–304. [52]

PLESNÍK, J., A note on the complexity of finding regular subgraphs, *Discrete Math.* **49** (1984), 161–167. [232]

PLESNÍK, J. AND ZNÁM, Š., On equality of edge-connectivity and minimum degree of a graph, *Archivum Math.* **25** (1989), 19–26. [50]

PLUMMER, M.D., On n-extendable graphs, *Discrete Math.* **31** (1980), 201–210. [81]

POLJAK, S. AND TURZÍK, D., A polynomial algorithm for constructing a large bipartite subgraph, with an application to a satisfiability problem, *Canad. J. Math.* **34** (1982), 519–524. [214]

PRIM, R.C., Shortest connection networks and some generalisations, *Bell Systems Tech. J.* **36** (1957), 1389–1401. [227]

RADCLIFFE, A.J. AND SCOTT, A.D., Every tree contains a large induced subgraph with all degrees odd, *Discrete Math.* **140** (1995), 275–279. [118]

RINGEL, G., Problem 25, in *Theory of Graphs and its Applications Int. Symp. Smolenice (June 1963)*, Czech Acad. Sci. Prague, 1964, p. 162. [224]

ROSA, A., On certain valuations of the vertices of a graph, in *Theory of Graphs*, Ed. P. Rosenstiel, Gordon and Breach, New York, 1967, pp. 349–355. [14]

RYSER, H.J., A combinatorial theorem with an application to Latin squares, *Proc. Amer. Math. Soc.* **2** (1951), 550–552. [205]

RYSER, H.J., *Combinatorial Mathematics*, Carsus Math. Mon. no.14. Math. Assoc. of America, Providence R.I., 1963. [105]

RYSER, H.J., Combinatorial properties of matrices of zeros and ones, *Canad. J. Math.* **9** (1957), 371–377. [104]

SABIDUSSI, G., Graph multiplication, *Math. Zeitsch.* **72** (1960), 446–457. [3,13]

SACHS, H., Über Teiler, Faktoren und charakteristische Polynome von Graphen. Teil 1, *Wiss. Z. Th.* **12** (1966), 7–12. [18]

SACHS, H., Über Teiler, Faktoren und charakteristische Polynome von Graphen. Teil 2, *Wiss. Z. Th.* **13** (1967), 405–412. [19]

SAITO, A. AND WATANABE, M., Partitioning graphs into induced stars, *Ars Combin.* **36** (1993), 3–6. [218]

SAPOZHENKO, A.A., On the complexity of disjunctive normal forms, obtained with the help of the gradient algorithm, *Diskret. Analiz.* **21** (1972), 62–71 (in Russian). [175]

SAVAGE, C.D. AND WINKLER, P.M., Monotone Gray codes and the middle levels problem, *J. Combin. Th. Ser. A* **70** (1995), 230–248. [200]

SCHRIJVER, A.P., Bounds on permanents, and the number of 1-factors and 1-factorizations of a bipartite graph, in *Surveys in Combinatorics*, Ed. E.K. Lloyd, London Math. Soc. Lecture Note Ser. 82, C.U.P., Cambridge, 1983, pp. 107–134. [169]

SEBÖ, A., Finding the *t*-join structure of graphs, *Math. Prog.* **36** (1986), 123–134. [116]

SEBÖ, A., A quick proof of Seymour's Theorem on *t*-joins, *Discrete Math.* **64** (1987), 101–103. [115]

SEVAST'JANOV, S.V., Interval colourability of the edges of a bipartite graph, *Metody Diskret. Analiz.* **50** (1990), 61–72 (in Russian). [131,137,234]

SEYMOUR, P., On odd cuts and plane multicommodity flows, *Proc. London Math. Soc.* **42** (1981), 178–192. [115]

SHANNON, C.E., A theorem on coloring the lines of a network, *J. Math. Physics* **28** (1949), 148–151. [217]

SKUPIEN, Z, On counting maximum path-factors of a tree, in *Algebra und Graph (Proc. Siebenlehn 1985 Conf.)*, Bergakademie Freiberg, Section Math., 1986, pp. 91–94. [100]

SLATER, P.J., A constructive characterization of trees with at least k disjoint maximum matchings, *J. Combin. Th. Ser. B* **25** (1978), 326–338. [183]

SLEPIAN, D., Two theorems on a particular crossbar switching network, *manuscript* (1952). [161]

SMETANIUK, B., A new construction for latin squares I. Proof of the Evans conjecture, *Ars Combin.* **11** (1981), 155–172. [206,207]

SMOLENSKIĬ, E.A., On a method of linear writing of graphs, *Zhurnal Vychisl. Mat. i Matem. Fiz.* **2** (1962), 371–372 (in Russian). [28]

STEGER, A. AND YU, M.L., On induced matchings, *Discrete Math.* **120** (1993), 291–295. [130]

STEINITZ, E., *Über die Construction der Configurationen* n_3, Doctoral Dissertation, Breslau, 1894. [77]

SUŠKOV, J.A., $(1, q)$-matchings, *Vestnik Leningrad Univ. no. 19, Mat. Meh. Astronom. vyp.* **4** (1975), 50–55 (in Russian). [85]

SYLVESTER, J.J., On recent discoveries in mechanical conversion of motion, *Proc. Roy. Inst. Great Britain* **7** (1873), 179–198. [vii,30]

SYSLO, M.M., On tree and unicyclic realizations of degree sequences, *Demonstratio Math.* **15** (1982), 1071–1076. [120]

TALANOV, V.A. AND ŠEVČENKO, V.N., A certain problem on a dynamic transport network, *Izvestia Vyss. Ucebn. Zaved. Radiofizika* **15** (1972), 1113–1114 (in Russian). [150]

TANAJEV, V.S., SOTSKOV, YU., N. AND STRUSEVICH, V.A., *Scheduling Theory*, Nauka, Moscow (in Russian), 1989. [233]

TANNER, R.M., Explicit concentrators from generalised N-gons, *SIAM. J. Alg. Disc. Meth.* **5** (1984), 287–293. [87]

TARAKANOV, V.E., On the properties of the substitution operation in classes of $(0, 1)$-matrices, *Mat. Zametki* **53** (1993), 131–141 (in Russian). [105]

TARSI, M., On the decomposition of complete bipartite graphs into edge-disjoint subgraphs with star components, *Discrete Math.* **58** (1986), 93–95. [226]

THOMASON, A.G., Hamiltonian cycles and uniquely edge colourable graphs, *Ann. Disc. Math.* **3** (1978), 259–268. [157]

THOMASSEN, C., On the number of hamiltonian cycles in bipartite graphs, *Combin. Probab. Comput.* **5** (1996), 437–442. [112]

TOMPA, M., Time-space tradeoffs for computing functions, using connectivity properties of their circuits, *J. Comput. System Sci.* **20** (1980), 118–132. [55]

TRINAJSTIĆ, N., *Chemical Graph Theory*, CRC Press, Boca Raton, Fl., 1983. [118]

TUCKER, A.C., A structure theorem for the consecutive 1's property, *J. Combin. Th. Ser. B* **12** (1972), 153–162. [14]

TUTTE, W.T., The factorization of linear graphs, *J. London Math. Soc* **22** (1947), 107–111. [80]

TUTTE, W.T., On the 2-factors of bicubic graphs, *Discrete Math.* **1** (1971), 203–208. [118]

TUZA, Z., Coverings of graphs by complete bipartite graphs; complexity of 0–1 matrices, *Combinatorica* **4** (1984), 111–116. [219,220]

VALIANT, L.G., Graph-theoretic properties in computational complexity, *J. Comput. System Sci* **13** (1976), 278–285. [52,55]

VIZING, V.G., The cartesian product of graphs, *Vychisl. Sistemy* **9** (1963), 30–43 (in Russian). [13]

VIZING, V.G., On the estimate of the chromatic class of a p-graph, *Diskret. Analiz* **3** (1964), 25–30 (in Russian). [125]

VOLKMANN, L., Bemerkungen zum p-fachen Kantenzusammenhang von Graphen, *An. Univ. Bucuresti Mat.* **37** (1988), 75–79. [52]

WAERDEN, B.L. VAN DER, Ein Satz über Klasseneinteilungen von endlichen Mengen, *Abh. mth. Sem. Hamburg Univ.* **5** (1927), 185–188. [195]

WAGNER, A. AND CORNEIL, D.G., Embedding trees in a hypercube is NP-complete, *SIAM J. Comput.* **19** (1990), 570–590. [232]

WELSH, D., *Matroid Theory*, Academic Press, New York, 1976. [197]

DE WERRA, D., Some combinatorial problems arising in scheduling, *CORS J.* **8** (1970), 165–175. [128]

DE WERRA, D., Balanced Schedules, *INFOR Canad. J. Oper. Res. and Inform. Proc.* **9** (1971a), 230–237. [129]

DE WERRA, D., Investigation on an edge colouring problem, *Discrete Math.* **1** (1971b), 167–179. [145,147,149,150]

DE WERRA, D., Equitable colorations of graphs, *Rev. Française Informat. Recherche Opérationnelle* **5** (1971c), 3–8. [129]

DE WERRA, D. AND SOLOT, P.H., Compact cylindrical chromatic scheduling, *SIAM J. Disc. Math.* **4** (1991), 527–534. [138]

WHITNEY, H., Non-separable and planar graphs, *Trans. Amer. Math. Soc.* **34** (1932), 339–362. [46]

WILF, H.S., *Algorithms and Complexity*, Prentice-Hall, London, 1986. [5]

WINKLER, P.M., Proof of the squashed cube conjecture, *Combinatorica* **3** (1983), 135–139. [41]

WINKLER, P.M., Factoring a graph in polynomial time, *Europ. J. Comb.* **8** (1986), 209–212. [13]

WOODALL, D.R., The binding number of a graph and its Anderson number, *J. Combin. Th. Ser. B* **15** (1973), 225-255. [107]

YAMADA, T., A note on sign-solvability of linear systems of equations, *Linear and Multilinear Algebra* **22** (1988), 313–323. [188]

YAMAMOTO, S., IKEDA, H., SHIGE-EDA, S. AND USHIO, K., On claw-decomposition of complete graphs and complete bigraphs, *Hiroshima Math. J.* **5** (1975), 33–42. [179]

YANNAKAKIS, M., Node- and edge-deletion, NP-complete problems, in *Proc. 10th Ann. A.C.M. Symp. Theory Comp.*, A.C.M., New York, 1978, pp. 253–264. [214,235,236]

YANNAKAKIS, M. AND GAVRIL, F., Edge dominating sets in graphs, *SIAM J. Appl. Math.* **38** (1980), 364–372. [177,235]

YU, Q., A note on n-extendable graphs, *J. Graph Th.* **16** (1992), 349–353. [82]

YUSHMANOV, S.V., Representation of a tree with p hanging vertices by $2p - 3$ elements of its distance matrix, *Matem. Zametki* **35** (1984), 877–887 (in Russian). [30]

ZARETSKIĬ, K.A., Constructing a tree using the collection of distances between pendant vertices, *Uspekhi Mat. Nauk* **20** (1965), 94–96 (in Russian). [28]

ZVEROVICH, V.E., Domination-perfect graphs, *Matem. Zametki* **48** (1990), 66–69 (in Russian). [179]

ZVEROVICH, I.E. AND ZVEROVICH, V.E., Contributions to the theory of graphical sequences, *Discrete Math.* **105** (1992), 293–303. [105]

Index